536·461:535·338·1:542·51

HANDBOOK OF
FLAME SPECTROSCOPY

HANDBOOK OF
FLAME SPECTROSCOPY

M. L. Parsons, B. W. Smith, and G. E. Bentley
Department of Chemistry
Arizona State University
Tempe, Arizona

PLENUM PRESS • NEW YORK AND LONDON

Library of Congress Cataloging in Publication Data

Parsons, Michael Loewen, 1940-
  Handbook of flame spectroscopy.

  1. Flame spectroscopy. I. Smith, Benjamin William, 1951-   joint author.
II. Bentley, Glenn Edward, joint author. III. Title.
QD96.F5P37   543'.085   75-17865
ISBN 0-306-30856-X

© 1975 Plenum Press, New York
A Division of Plenum Publishing Corporation
227 West 17th Street, New York, N.Y. 10011

United Kingdom edition published by Plenum Press, London
A Division of Plenum Publishing Company, Ltd.
Davis House (4th Floor), 8 Scrubs Lane, Harlesden, London, NW10 6SE, England

Printed in the United States of America

Acknowledgment

The authors would like to express their appreciation to the Arizona State University Computing Center for both computer time and help in preparing the computer-generated portions of the book.

PREFACE

     Analytical flame spectroscopy is a rich and growing disci-
pline, rooted in the broad fields of physics and chemistry. Its
applications abound not only in these large areas, but also thrive
in the geosciences, materials science, and clinical and biochemical
analysis. As an inevitable corollary of the field's growth, the
scientist seeking to develop a fluent expertise has been forced to
assimilate and master a rapidly increasing quantity of information.
Our guiding hope in creating the present work has therefore been to
provide the investigator with a single reference source for nearly
all the material ever likely to be required in the daily conduct
of basic or applied research.

     Flame spectroscopy is not a new analytical field. It has seen
at least three major eras, in each of which much new information
was developed - the Bunsen-Kirchhoff years, the Beckman D.U. years,
and finally the atomic absorption years. In the Bunsen-Kirchhoff
era, several new elements were discovered. During the Beckman years
- nearly all the early flame emission data were taken on modified
Beckman D.U. spectrometers - trace metal analysis for the alkaline
metals and for many alkaline earth elements reached a new high (low?)
- the parts per million level. More recently, trace metal analysis
has in general achieved a new maturity with the advent of atomic
absorption analysis, which was co-discovered by C. Th. J. Alkemade
and Alan Walsh in 1955. Many elements are now analyzed in the
parts per million range, and at least several to the parts per
billion level. The next era is undoubtedly just around the corner.

     Many researchers, we fear, do not appreciate the similarities
among the flame techniques, nor do they appreciate fully the dif-
ferences between flames and the other, often much more energetic,
excitation sources such as the d.c. arc or the spark. The information in this volume - which has been rather painstakingly compiled,
calculated, tabulated, and critically evaluated - applies nearly
equally to <u>all</u> three flame spectrometric techniques.

But because one must stop somewhere within reason, the book is in certain senses woefully incomplete. Further, the literature itself suffers gaping holes. As an example, transition probability data are only sporadically provided, and their improvement must become a higher priority among the research community active in the field. Undoubtedly, too, we have accidentally overlooked some data sources, and therefore hope for the forebearance of our audience, and their aid in locating these data for future editions of the work.

In sum then, the book, however incomplete, represents our best efforts to bring together for the first time most of the important and useful information in the field of flame spectroscopy. I sincerely trust we have achieved a small measure of success.

CONTENTS

LIST OF TABLES

CHAPTER I

BASIC INTRODUCTION

A. THEORY AND PRACTICE

It is felt that the user of this book will be a practicing flame spectroscopist, either a bench chemist in an industrial laboratory or a student (or professor) doing research. As such, it is felt that this person would already have a basic knowledge of flame spectroscopic theory and practice. If this is not the case, it would be highly recommended that the reader select one of the many introductory texts on the subject. (See Chapter I, B.) This chapter is included, however, to complement such texts and to remind the reader of the important fundamental concepts and experimental parameters in these techniques.

The basic concept behind this book was to compile as much of the useful information required by the practicing flame spectroscopist as was humanly possible and put it into one document. This information was obtained from many sources due to its diverse nature and the many extremely specific publications. Literally thousands of references have been consulted in preparing the manuscript.

Nomenclature

Fortunately, in 1974, the International Union of Pure and Applied Chemistry (IUPAC) approved Part III of their "Nomenclature, Symbols, Units and Their Usage in Spectrochemical Analysis." Section III concerns itself with "Analytical Flame Spectroscopy and Associated Procedures." Atomic spectroscopists should be forever grateful to Professor C. Th. J. Alkemade and his committee for their seven-year struggle to reach the necessary compromises on the recommended terminology, symbols, and definitions (1). These recommendations are utilized in Table I.A.1 in this chapter, Table III.A.1 and in part 9 of Section B in Chapter III.

Table I.A.1

International Union of Pure and Applied
Chemistry (IUPAC) Recommended
Classification of Methods and Instruments[1]

	ABSORPTION	EMISSION	FLUORESCENCE
METHODS:			
General[2]	Absorption Spectroscopy	Emission Spectroscopy	Fluorescence Spectroscopy
When atomic lines are used add[3]	Atomic (AAS)	(AES)	(AFS)
When a flame[4] cell is used add	Flame (FAS)	(FES)	(FFS)
When both a flame and atomic lines are observed add	Flame Atomic (FAAS)	(FAES)	(FAFS)
Instruments:			
General[5]	Absorption Spectrometer	Emission Spectrometer	Fluorescence Spectrometer

1. C. Th. J. Alkemade, Ref. 1 at end of Section A.

2. "Spectrometry" may be used if detection is performed with a photoelectric detector.

3. If molecular species are observed, the term "molecular" is used, i.e. MAS, etc.

4. If a non-flame cell is to be specified, the appropriate adjective should be substituted for flame, i.e. Furnace absorption spectroscopy, etc.

5. "Spectrometer" may be used if detection is performed with a photoelectric device.

Basic Atomic Theory

Many people are probably not aware of how much atomic theory
they already know. A rather extensive presentation is generally
given in college freshman courses. Of course you remember learn-
ing about the four basic quantum numbers, the Pauli exclusion
principle, and the like. A review of an old freshman text might
surprise you. As in all scientific fields, there are also extensive treatises in atomic theory available, e.g., the classic
texts of Kuhn (2) or Mitchell and Zemansky (3).

In a discussion of atomic spectroscopy, or to be more specific,
the study of electronic excitation phenomena, it is pertinent to
realize that any given element has a rather large number of possible excited energy levels and therefore a large number of
potential transitions. And whereas, the phenomena involved in
absorption, emission, and fluorescence can be (and usually is)
demonstrated by example of only one excited state as in Figure I.A.1,
there are generally many levels available to any element, depending, of course, on the energy available for excitation. This is
also true of the ground level (or whatever the lower level). It
is a fact that the common analytical flames have the capability
of excitation of over 1000 atomic transitions for about 70 different elements (See Chapter II).

Figure I.A.1
Excitation Phenomena

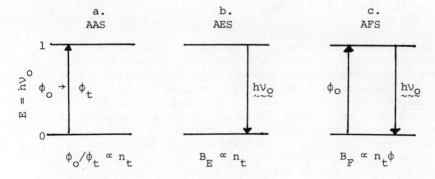

Symbols are defined in Chapter III.A.

In AAS Figure I.A.1-a , radiation of energy equal to $h\nu_0$ is focused through the flame (or non-flame) cell containing the atoms of the appropriate element, and the decrease in the intensity of this radiation is measured. This decrease is related to the number of absorbing atomic species just as in molecular ultra-violet-visible absorption in solution, and a relationship complete-ly analogous to Beer's Law can be derived. The fundamental relationships are given in Chapter III.A. It should be pointed out that in FAAS the flame cell acts only as a sample cell and that its only purpose is to dissociate the molecules and render the element of analytical interest in the atomic state. Molecules of any kind, as well as ions, do not have the same electronic energy levels, and their transitions are therefore not observed at the wavelength corresponding to the energy, $h\nu_0$, for the atoms.

In FES (Figure I.A.1-b, the flame acts not only as a sample cell, i.e., the source of free atoms, but also provides the energy for electronic excitation of the atomic species. This energy is generally in the form of kinetic energy. Once an atom's electrons are in the excited state, a certain number will relax to the ground state and emit radiation corresponding to the dif-ference in energy levels, $h\nu_0$. The intensity of this emitted radiation is proportional to the total number of analytical species, and this gives the analytical usefulness to the technique (See Chapter III.A).

In AFS (Figure I.A.1-c), aspects of both AAS and AES are in-corporated. Excitation is accomplished as in AAS with an ex-ternal source of radiational energy, and the intensity of the radiation emitted by the excited atomic species is the source of the analytical measurement. Here, as in AAS, the flame is simply the sample cell needed to provide the free atomic species (See Chapter III.A).

As stated, over 1000 transitions have been observed by the flame techniques; however, almost all of the analytically useful transitions involve the ground state and the first excited state. Any transition involving the ground state is said to be a reso-nance transition. These transitions involve some of the most periodical phenomena associated with elements and are detailed in Table I.A.2. This information should be digested in conjunction with the information in Table III.B.3 which gives the percent of atoms in the ground state levels (or multiplets) as a function of the temperature of the common analytical flames.

The sensitivity of all three techniques is directly propor-tional to the number of free atoms in the lower energy level from which the transition originates. For example, if elent 'Q' has a singlet ground level with 100% of its free atoms in the ground state

at the temperature of the flame being used, and element 'F' has
a triplet ground state with only 50% of its _free_ atoms in the
particular level from which the analytically useful transition
originiates, element 'Q' has a built-in sensitivity potentially
twice that of element 'F'. Of course, the number of _free_ atoms is
a complicated function of the flame and the experimental conditions and the matrix of the sample being analyzed.

Explanation of Table I.A.2

     This table was included to demonstrate the periodicity of the
electronic structure of the elements as well as show the under-
lying simplicity of the resonance atomic spectra which are generally
observed in analytical flame spectrometry.

     Elements are presented in their normal periodic groupings.
The term symbol for the ground state is given. (It is generally
the same for the entire group; the only exceptions are in the transition metals where there are often more than one very low lying
energy level or multiplet. The exceptions in term symbol are listed
in parenthesis below the ground state.) The multiplicity of the
lower and upper levels is also given for the group. It should be
pointed out that the upper level information is for the first excited state only and that other transitions to other multiplets are
indeed possible.

     The Table also gives the electronic configuration for the
ground level, the electronic transition and representative wavelengths for the designated transition. In some cases, it was possible to give all of the possible wavelengths but in many only a few
of the more important transitions are detailed.

Basic Experimental Setup

     The simplest practical experimental configuration for each
method is presented in Figure I.A.2. AAS systems typically consist of a hollow cathode lamp (HCL) used for the excitation source
(See Chapter I.D), simple quartz lenses or front surfaced mirrors
to focus the excitation radiation through the flame (or non-flame)
cell and into the monochromator. Often a reference beam used to
compensate for excitation source fluctuations is added. On better
instruments, a background correction system is added to compensate
for flame background or molecular absorption or scatter from the
sample matrix. The flame cells (Chapter I.C) generally used are
the air-acetylene, nitrous oxide-acetylene, or hydrogen (or pro-
pane)-air using chamber-type aspirator-burners with 5 to 10 cm slots.
Flameless cells are generally of the furnace or filament types
(Chapter I.C), although the Delves Cup seems to be gaining popularity. Monochromators are usually one-quarter to one-half meter

Table I.A.2

General Information on Periodicity
of Resonance Transitions

Group Ia	Term Symbol $^2S_{\frac{1}{2}}$		Singlet - Doublet
Element	Ground State	Transition	Wavelength (Å)
Li	(He) 2s	2s - 2p	6707.8
Na	(Ar) 3s	3s - 3p	5889.95
			5895.92
K	(Kr) 4s	4s - 4p	7664.91
			7698.98
Rb	(Xe) 5s	5s - 5p	8521.10
			8943.50

Group Ib			
Cu	(Ar) $3d^{10}4s$	4s - 4p	3247.54
			3273.96
Ag	(Kr) $4d^{10}5s$	5s - 5p	3280.68
			3382.89
Au	(Xe) $5d^{10}6s$	6s - 6p	2427.95
			2675.95

Group IIa	Term Symbol ^1S_o		Singlet - Singlet
Be	$1s^2 2s^2$	$2s^2 - 2s2p$	2348.61
Mg	(Ne) $3s^2$	$3s^2 - 3s3p$	2852.13
Ca	(Ar) $4s^2$	$4s^2 - 4s4p$	4226.73
Sr	(Kr) $5s^2$	$5s^2 - 5s5p$	4607.33
Ba	(Xe) $6s^2$	$5s^2 - 5s5p$	5535.48

Group IIb			
Zn	(Ar) $3d^{10}4s^2$	$4s^2 - 4s4p$	2138.56
Cd	(Kr) $4d^{10}5s^2$	$5s^2 - 5s5p$	2288.02

Group IIb (cont'd)	Term Symbol $^1S_0$		Singlet - Singlet
Element	Ground State	Transition	Wavelength ($\mathring{A}$)
Hg	(Xe) $4f^{14}5d^{10}6s^2$	$6s^2 - 6s6p$	(Vacuum UV)
Group III		Term Symbol $^2P_{\frac{1}{2}}^{0}$	Doublet - Singlet
B	$1s^2 2s^2 2p$	$2p - 3s$	2497.72 2496.77
Al	(Ne) $3s^2 3p$	$3p - 4s$	3961.52 3944.01
Ga	(Ar) $3d^{10}45^2 4p$	$4p - 5s$	4032.98 4172.06
In	(Kr) $4d^{10}5s^2 5p$	$5p - 6s$	4101.76 4511.31
Tl	(Xe) $4f^{14}5d^{10}6s^2 6p$	$6p - 7s$	3775.72 5350.46
			Doublet - Doublet
B	Same as above	$2s^2 2p - 2s2p^2$	2089.57 2088.84
Al	" " "	$3p - 3d$	3082.15 3092.71 3092.84
Ga	" " "	$4p - 4d$	2874.24 2943.64 2944.18
In	" " "	$5p - 5d$	3039.36 3256.09 3258.56
Tl	" " "	$6p - 6d$	2767.87 3519.24 3529.43

Group IV		Term Symbol $^3P_0$	Triplet - Triplet
Element	Ground State	Transition	Wavelength (Å)
C	$1s^2 2s^2 2p^2$	$2s^2 2p^2 - 2s 2p^3$	(Vacuum UV)
Si	(Ne) $3s^2 3p^2$	$3p^2 - 3p4s$	2506.90
			2514.32
			2516.11
			2519.20
			2524.11
			2528.51
Ge	(Ar) $3d^{10} 4s^2 4p^2$	$4p^2 - 4p5s$	2592.54
			2651.18
			2651.58
			2691.34
			2709.63
			2754.59
Sn	(Kr) $4d^{10} 5s^2 5p^2$	$5p^2 - 5p6s$	2706.51
			2839.99
			2863.33
			3009.14
			3034.12
			3175.05
Pb	(Xe) $4f^{14} 5d^{10} 6s^2 6p^2$	$6p^2 - 6p7s$	2833.06
			3639.58
			3683.48
			4057.83

Group V		Term Symbol $^4S^0_{3/2}$	Singlet - Triplet
N	$1s^2 2s^2 2p^3$	$2p^3 - 2p^2 3s$	(Vacuum UV)
P	(Ne) $3s^2 3p^3$	$3p^3 - 3p^2 4s$	1779.99
			1782.87
			1787.68
As	(Ar) $3d^{10} 4s^2 4p^3$	$4p^3 - 4p^2 5s$	1936.93
			1971.97
Sb	(Kr) $4d^{10} 5s^2 5p^3$	$5p^3 - 5p^2 6s$	2068.33
			2175.81
			2311.47
Bi	(Xe) $4f^{14} 5d^{10} 6s^2 6p^3$	$6p^3 - 6p^2 7s$	3067.72

Group VI		Term Symbol $^3P_2$	Triplet - Singlet
Element	Ground State	Transition	Wavelength (Å)
O	$1s^2 2s^2 2p^4$	$2p^4 - 2p^3 3s$	(Vacuum UV)
S	(Ne) $3s^2 3p^4$	$3p^4 - 3p^3 4s$	1807.34
			1820.36
			1826.26
Se	(Ar) $3d^{10} 4s^2 4p^4$	$4p^4 - 4p^3 5s$	1960.26
			2039.85
			2062.79
Te	(Kr) $4d^{10} 5s^2 5p^4$	$5p^4 - 5p^3 6s$	2142.75
			2383.25
			2385.76

Group VII	No Information
Group VIII	No Information

Group IIIb		Term Symbol $^2D_{3/2}$	Doublet - Doublet
Sc	(Ar) $3d4s^2$	$4s^2 - 4s4p$	3996.61
			4020.40
			4023.69
			4047.79
Y	(Kr) $4d5s^2$	$5s^2 - 5s5p$	4039.83
			4128.31
			4142.85
			4235.94
La	(Xe) $5d6s^2$	$6s^2 - 6s6p$	5158.69
			5455.15
			5501.34
			5839.79

Group IVb		Term Symbol $^3F_2$	Triplet - Triplet
Ti	(Ar) $3d^2 4s^2$	$4s^2 - 4s4p$	2631.54
			2632.42
			2644.26
			2646.64
Zr	(Kr) $4d^2 5s^2$	$5s^2 - 5s5p$	2985.39
			3519.60

Group IVb (cont'd)	Term Symbol $^3F_2$		Triplet - Triplet
Element	Ground State	Transition	Wavelength (Å)
Hf	(Xe) $5d^2 6s^2$	$6s^2 - (6s6p)$	3072.88 / 2866.37

Group Vb		Term Symbol $^4F_{3/2}$	Quartet - Quartet
V	(Ar) $3d^3 4s^2$	$4s^2 - 4s4p$	3185.40 / 3183.41 / 3183.98
Nb*	(Kr) $4d^4 5s$ $(^6D_{\frac{1}{2}})$	$5s - 5p$ (quint. - quint.)	4079.73 / 4058.94
Ta	(Xe) $5d^3 6s^2$	$6s^2 - 6s6p$	2714.67 / 2775.88

*This represents the first anomolous behavior in the ground state of the atoms. The $^6D_{\frac{1}{2}}$ term is near the ground state for V&Ta and the $^4F_{3/2}$ term is near the ground state for Nb.

Group VIb		Term Symbol $^7S_3$	Singlet - Triplet
Cr	(Ar) $3d^5 4s$	$3d^5 4s - 3d^4 4s4p$	3578.69 / 3593.49 / 3605.33
Mo	(Kr) $4d^5 5s$	$4d^5 5s - 4d^4 5s5p$	3132.59 / 3170.33 / 3193.97
W	(Xe) $5d^4 6s^2$ $(^5D_0$ quint.$)$	$6s^2 - 6s6p$	2551.35

Group VIIb		Term Symbol $^6S_{5/2}$	Singlet - Triplet
Mn	(Ar) $3d^5 4s^2$	$4s^2 - 4s4p$	2794.82 / 2798.27 / 2801.06
Tc	(Kr) $4d^5 5s^2$	?	?
Re	(Xe) $5d^5 6s^2$	$6s^2 - 6s6p$	3451.88 / 3460.46 / 3464.73

Group VIIIb (Fe)	Term Symbol $^5D_4$	Quint. - Quint.

Element	Ground State	Transition	Wavelength (Å)
Fe	(Ar) $3d^6 4s^2$	$4s^2 - 4s4p$	2483.27
			2522.85
Ru	(Kr) $4d^7 5s$ $[^5F_5]$	$5s - 5p$	3728.03
			3498.94
Os	(Xe) $5d^6 6s^2$	$6s^2 - 6s6p$	2909.06
			3058.66

Group VIIIb (Co)	Term Symbol $^4F_{9/2}$	Quartet - Quartet

Co	(Ar) $3d^7 4s^2$	$4s^2 - 4s4p$	2424.93
			2384.86
Rh	(Kr) $4d^8 5s$	$5s - 5p$	3434.89
Ir	(Xe) $5d^7 6s^2$	$6s^2 - 6s6p$	2088.82

Group VIIIb (Ni)	Term Symbol $^3F_4$	Triplet - Triplet

Ni	(Ar) $3d^8 4s^2$	$4s^2 - 4s4p$	2320.03
Pd	(Kr) $4d^{10}$ $[^1S_0]$	$4d^{10} - 4d^9 5p$	2447.91
Pt	(Xe) $5d^9 6s$ $[^3D_3]$	$6s - 6p$	2659.45

Figure I.A.2

Basic Experimental Setup

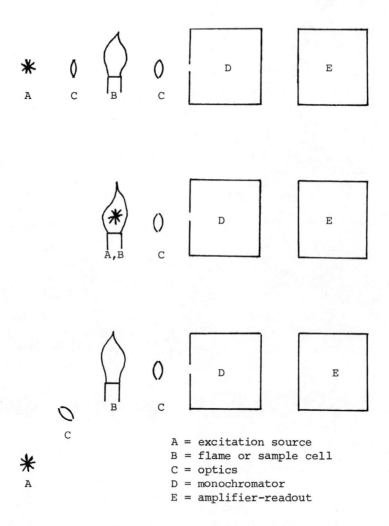

A = excitation source
B = flame or sample cell
C = optics
D = monochromator
E = amplifier-readout

optics with resolution down to about 0.2Å. Photomultiplier tubes
are invariably used with synchronous amplification systems. Signals
are processed by means of strip-chart recorders or more sophisticated electronic signal averaging systems with digital readout.

AES systems are the simplest of the techniques, experimentally,
in that the excitation source and sample cell are one and the same.
Further, the optics required to direct the radiation into the monochromator are generally simpler. Most of the commercial instruments are designed specifically for the alkali metals and/or the
alkline earths. These elements are most easily excited in the
cooler flames such as the air-hydrogen or air-propane (Chapter I.C).
In some systems, the much hotter oxy-hydrogen or oxy-acetylene
total-consumption aspirator burners have been used; however, more
recently, the chamber-type aspirator-burners using the very reducing nitrous oxide-acetylene flame has become more popular.

Because of the simplicity of the AES system, electrometer
readouts which directly measure the current produced by the photomultiplier tubes are often used. However, the synchronous readout systems used with most commercial AAS instrumentation is usually adaptable to AES measurements. Resolution is more important in
AES so the higher resolution monochromators are sometimes favored
even though they have less light gathering ability. As usual, there
is a resulting trade-off with sensitivity and a compromise is necessary.

In AFS, the analytical signal is directly proportional to the
intensity of the excitation source (Chapter III.A); therefore,
much attention has been paid to this aspect of experimentation. The
more intense excitation sources such as electrodless discharge
lamps (EDL) and laser sources (Chapter I.D) have been extensively
researched. HCL's used in the pulse-mode and high intensity continuum sources have also been studied. The optics and monochromator requirements are similar to AES, and the readout devices are
often similar to AAS. Because of the problem of collisional de-
activation of the excited state, inert gases such as argon have
been utilized to sheath the flame cell. Once again the flame acts
only as an atom cell, and most of the common flames have been used.

Experimental Procedure

These atomic techniques find their analytical application by
use of standards comparison in most cases. The fundamental relationships which give rise to the possibility of abolute analysis
are presented in Chapter III.A; however, only in very favorable
cases can this be done with any accuracy. In practice, direct
comparison with standards is preferred. The closer the standard
composition approximates the sample matrix, the more accurate the
analyses can be accomplished. Of course, this matching is not
always possible so one of the many compensating techniques must often
be used, such as standard additions, standard dilutions, and oc-
casionally internal standarization. In AAS, a background correction is often required to compensate for molecular absorption,
particulate scattering, or other continuum phenomena. In AES and
AFS these effects must be corrected by scanning techniques.

Ionization suppression is often used for analytes which are
easily ionized (Chapter III.B.4). This is accomplished by the
addition of an easily ionizable element such as Cs or K. In some
cases, an element such as La is added to "release" chemical effects.
It appears that the elements which are most prone to chemical effects are those which form stable compounds with oxides or halides
(Chapter III.B.5&6).

Aside from determining the matrix of the sample in which the
analyte is determined, other aspects of the sample should also be
considered. For example, the viscosity of the sample is critical
in a naturally aspirating burner system and variations from sample
to sample can cause serious errors. The author's group has ex-
perienced signal readings which have varied from zero to full scale
with samples of the same analyte concentration but with varying
viscosity. This type of problem can only be overcome by force-
feeding the sample into the burner system.

Another problem which must be faced in trace analysis is that
of surface adsorption both in the sample containers and in the
burner system. Some elements are much harder to keep in solution
than others, lead is a notorious example. This problem exists in
both standard and sample solutions and is often alleviated by con-
trolling the pH of both. This has often been mistaken as an acid
interference.

Standards

Standard solutions can be made from the compounds and according to the procedures described in Table I.A.3. It is of utmost
importance to have standards and samples with the same solution
matrix; therefore, acid concentrations should be kept the same in

both as well as the total ion concentrations. If the composition
of the sample solution is known, it should be duplicated in the
standard solutions; if not, comparison of results using two different techniques should always be investigated, i.e., compare the
results of the analytical curve technique with those of the standard addition technique. Samples should always be checked to see
if background corrections are required.

Explanation of Table I.A.3

The compounds chosen for inclusion in Table I.A.3 followed a
rather stringent set of criteria. These include stability, purity,
ease of preparation, availability, high molecular weight, and
toxicity. Of great importance is the ability to take a compound
off the shelf and know that it can be dried, weighed and dissolved
with comparative ease. The list provided by and large meets these
goals insofar as is possible. It should be appreciated that this
list does not include all possible compounds which meet these requirements nor are all the compounds listed all that easy to dissolve. In fact, some elements do not have compounds which meet
the requirements. Many halide compounds are quite hydroscopic
and this is of course disastrous for a standard solution.

So far as solvents are concerned, it is often that any acid
(or base) would do the job, in other instances, the solvent listed
is critical and no substitutions are permitted. In some cases, a
considerable time is required for dissolution.

In this table the significant figures in all columns repre-
sent the accuracy with which the atomic weights of the elements are
known.

Table I.A.3

Standard Solutions - Compounds and Procedures[1,2]

Elements Compd.	Formula Weight (Grams)	Weight in Grams for One Liter of Solution		Solvent	Comments
		1000 ppm (µg/g)	0.1 Molar		
Aluminum					
Al-Metal	26.982	1.0000	2.6982	Hot dil. HCl~2M	b
Antimony					
$KSbOC_4H_4O_6 \cdot \frac{1}{2}H_2O$	333.92	2.7427	33.392	Water	
$KSbOC_4H_4O_6$	324.92	2.6687	32.492	Water	f
Sb-Metal	121.75	1.0000	12.175	Hot Aqua Regia	
Arsenic					
As_2O_3	197.84	1.3203	9.892	1:1 NH_3	a,c,n
Barium					
$BaCO_3$	197.35	1.4369	19.735	Dil. HCL	h
$BaCl_2$	208.25	1.5163	20.825	Water	g
Beryllium					
Be-Metal	9.0122	1.0000	0.9012	HCL	c
$BeSO_4 \cdot 4H_2O$	177.135	19.6550	17.7135	Water & Acid	i
Bismuth					
Bi_2O_3	465.96	1.1148	23.297	HNO_3	
Bi-Metal	208.980	1.00000	20.8980	HNO_3	
Boron					
H_3BO_3	61.84	5.720	6.184	Water	o
Bromine					
KBr	119.01	1.4894	11.901	Water	b
Cadmium					
CdO	128.40	1.1423	12.840	HNO_3	
Cd-Metal	112.40	1.0000	11.240	Dil. HCl	
Calcium					
$CaCO_3$	100.09	2.4972	10.009	Dil. HCl	h
Cerium					
$(NH_4)_2Ce(NO_3)_6$	548.23	3.9126	54.823	Water	
Cesium					
Cs_2SO_4	361.87	1.3614	18.094	Water	

Standard Solutions - Compounds and Procedures[1,2]

Elements	Compd.	Formula Weight (Grams)	Weight in Grams for One Liter of Solution		Solvent	Com- ments
			1000 ppm (μg/g)	0.1 Molar		
Chlorine						
NaCl		58.442	1.6485	5.8442	Water	a
Chromium						
$K_2Cr_2O_7$		294.19	2.8290	14.710	Water	a
Cr-Metal		51.996	1.0000	5.1996	HCl	
Cobolt						
Co-Metal		58.933	1.0000	5.8933	HNO_3	b
Copper						
Cu-Metal		63.546	1.0000	6.3546	Dil. HNO_3	b
CuO		79.545	1.2517	7.9545	Hot HCl	b
$CuSO_4.5H_2O$		249.678	3.92909	24.9678	Water	
Dysprosium						
Dy_2O_3		373.00	1.1477	18.650	Hot HCl	e
Erbium						
Er_2O_3		382.56	1.1435	19.128	Hot HCl	e
Europium						
Eu_2O_3		351.92	1.1579	17.596	Hot HCl	e
Fluorine						
NaF		41.988	2.2101	4.1988	Water	j
Gadolinium						
Gd_2O_3		362.50	1.1526	18.125	Hot HCl	e
Gallium						
Ga-Metal		69.72	1.000	6.972	Hot HNO_3	k
Germanium						
GeO_2		104.60	1.4410	10.460	Hot 1M NaOH or with 50g oxalic acid+water	
Gold						
Au-Metal		196.97	1.0000	19.697	Hot Aqua Regia	b
Hafnium						
Hf-Metal		178.49	1.0000	17.849	HF, Fusion	l
Holmium						
Ho_2O_3		377.86	1.1455	18.893	Hot HCl	e

Standard Solutions - Compounds and Procedures[1,2]

Elements	Compd.	Formula Weight (Grams)	Weight in Grams for One Liter of Solution		Solvent	Comments
			1000 ppm (μg/g)	0.1 Molar		
Indium						
In_2O_3		277.64	1.2090	13.882	Hot HCl	
In-Metal		114.82	1.0000	11.482	Dil. HCl	
Iodine						
KIO_3		214.00	1.6863	21.400	Water	a
Iridium						
Na_3IrCl_6		473.8	2.466	47.38	Water	
Iron						
Fe-Metal		55.847	1.0000	5.5847	Hot HCl	b
Lanthanum						
La_2O_3		325.82	1.1728	16.291	Hot HCl	e
Lead						
$Pb(NO_3)_2$		331.20	1.5985	33.120	HCl	b,p
Lithium						
Li_2CO_3		73.890	5.3243	3.6940	Dil. HCl	b,h
Lutetium						
Lu_2O_3		397.94	1.1372	19.987	Hot HCl	e
Magnesium						
MgO		40.311	1.6581	4.0311	HCl	
Mg-Metal		24.312	1.0000	2.4312	Dil.HCl	
Manganese						
$MnSO_4 \cdot H_2O$		169.01	3.0764	16.901	Water	q
Mercury						
$HgCl_2$		271.50	1.3535	27.150	Water	c
Hg-Metal		200.59	1.0000	20.059	5M HNO_3	
Molybdenum						
MoO_3		143.94	1.5003	14.394	1M NaOH or 2M NH_3	
Neodymium						
Nd_2O_3		336.48	1.1664	16.824	HCl	e
Nickel						
Ni-Metal		58.71	1.000	5.871	Hot HNO_3	b

Standard Solutions - Compounds and Procedures[1,2]

Elements	Compd.	Formula Weight (Grams)	Weight in Grams for One Liter of Solution		Solvent	Comments
			1000 ppm ($\mu g/g$)	0.1 Molar		
Niobium						
Nb_2O_5		265.81	1.4305	13.290	HF,Fusion	r,s
Nb-Metal		92.906	1.0000	9.2906	$HF+H_2SO_4$	s
Osmium						
Os-Metal		190.20	1.0000	19.020	Hot H_2SO_4	d
Palladium						
Pd-Metal		106.40	1.0000	10.640	Hot HNO_3	
Phosphorus						
KH_2PO_4		136.09	4.3937	13.609	Water	
$(NH_4)_2HPO_4$		209.997	6.77983	20.9997	Water	
Platinum						
K_2PtCl_4		415.12	2.1278	41.511	Water	
Pt-Metal		195.05	1.0000	19.505	Hot Aqua Regia	
Potassium						
KCl		74.555	1.9067	7.4555	Water	b
$KHC_8H_4O_4$		204.22	5.2228	20.422	Water	a,n
$K_2Cr_2O_7$		294.19	3.7618	36.922	Water	a,n
Praseodymium						
Pr_6O_{11}		1021.43	1.20816	17.0247	HCl	e
Rhenium						
Re-Metal		186.2	1.000	18.62	HNO_3	
KReO4		289.3	1.554	28.93	Water	
Rhodium						
Rh-Metal		102.91	1.0000	10.291	Hot H_2SO_4	
Rubidium						
Rb_2SO_4		267.00	1.5628	13.357	Water	
Ruthenium						
RuO_4		165.07	1.6332	16.507	Water	
Samarium						
Sm_2O_3		348.70	2.3193	17.435	Hot HCl	e
Scandium						
Sc_2O_3		137.91	1.5339	6.8955	Hot HCl	

Standard Solutions - Compounds and Procedures[1,2]

Elements	Compd.	Formula Weight (Grams)	Weight in Grams for One Liter of Solution		Solvent	Comments
			1000 ppm (μg/g)	0.1 Molar		
Selenium						
Se-Metal		78.96	1.000	7.896	Hot HNO_3	
SeO_2		110.9	1.405	11.09	Water	
Silicon						
Si-Metal		28.086	1.0000	2.8086	NaOH, conc.	
SiO_2		60.085	2.1393	6.0085	HF	
Silver						
$AgNO_3$		169.875	1.57481	16.9878	Water	b,t
Ag-Metal		107.870	1.0000	10.7870	HNO_3	
Sodium						
NaCl		58.442	2.5428	5.8442	Water	a
$Na_2C_2O_4$		134.000	2.91432	6.70000	Water	a,n
Strontium						
$SrCO_3$		147.63	1.6849	14.763	Dil. HCl	b,h
Sulfur						
K_2SO_4		174.27	5.4351	17.427	Water	
$(NH_4)_2SO_4$		114.10	3.5585	11.410	Water	
Tantalum						
Ta_2O_5		441.893	1.22130	22.0992	HF, Fusion	r,s
Ta-Metal		180.948	1.0000	18.0948	$HF+H_2SO_4$	s
Tellurium						
TeO_2		159.60	1.2507	15.960	HCl	
Terbium						
Tb_2O_3		365.85	1.1512	18.292	Hot HCl	e
Thallium						
Tl_2CO_3		468.75	1.1468	23.437	Water	b,c
$TlNO_3$		266.37	1.3034	26.637	Water	
Thorium						
$Th(NO_3)_2 \cdot 4H_2O$		552.118	2.37943	55.2118	HNO_3	
Thulium						
Tm_2O_3		385.87	1.1421	19.293	Hot HCl	e

Standard Solutions - Compounds and Procedures[1,2]

Elements	Compd.	Formula Weight (Grams)	Weight in Grams for One Liter of Solution		Solvent	Comments
			1000 ppm ($\mu g/g$)	0.1 Molar		
Tin						
Sn-Metal		118.69	1.0000	11.869	HCl	
SnO		134.69	1.1348	13.469	HCl	
Titanium						
Ti-Metal		47.90	1.000	4.790	1:1 H_2SO_4	b
Tungsten						
$Na_2WO_4 \cdot 2H_2O$		329.86	1.7942	32.986	Water	
Na_2WO_4		293.83	1.5982	29.383	Water	f
Uranium						
UO_2		270.03	1.1344	27.003	HNO_3	
U_3O_8		842.09	1.1792	28.068	HNO_3	a,n
$UO_2(NO_3)_2 \cdot 6H_2O$		502.13	2.1095	50.213	Water	
Vanadium						
V_2O_5		181.88	1.7852	9.0942	Hot HCl	
NH_4VO_3		116.98	2.2963	11.698	Dil. HNO_3	
Ytterbium						
Yb_2O_3		394.08	1.1386	19.704	Hot HCl	e
Yttrium						
Y_2O_3		225.81	1.2700	11.291	Hot HCl	e
Zinc						
ZnO		81.37	1.245	8.137	HCl	b
Zn-Metal		65.37	1.000	6.537	HCl	
Zirconium						
Zr-Metal		91.22	1.000	9.122	HF, Fusion	l
$ZrOCl_2 \cdot 8H_2O$		322.2	3.533	32.22	HCl	

Footnotes:

a. Primary standard.
b. These compounds conform very well to the criteria and approach primary standard quality.
c. Highly toxic.
d. Very Highly toxic.
e. The rare earth oxides, because they absorb CO_2 and water vapor.

from the air, should be freshly ignited prior to weighing.

f.   Loses water at 110°C. Water is only slowly regained, but rapid weighing and desiccator storage are required.

g. $-2H_2O$ at 250°C, rapid weighing and desiccator storage are required.

h.   It is recommended to add a quantity of water then dilute acid and swirl until the CO_2 has ceased to bubble out, then dilute.

i·   Dissolve in water then add 5ml of acid and dilute.

j.   Sodium fluoride solutions will etch glass and should be freshly prepared.

k.   mp=29.6°C. The metal may be warmed and weighed as the liquid.

l.   Zincronium and hafnium compounds were not investigated in the laboratory. The following methods have been recommended for dissolution of zirconium and hafnium.

1,000 g of the powdered metal is placed in a platinum dish with 5-10 ml of water and 1-2 ml of HF (1:5) and covered with a platinum lid or a paraffined watch glass. Once dissolved, the fluorine may be removed by adding 1-2 ml of sulfuric acid (cold) and evaporating to dense fumes or to dryness if required.

A fusion method may also be used. A 5-10 fold excess of $K_2S_2O_7$ is placed in a platinum or quartz crucible along with the sample. After melting to a homogeneous molten mass in a muffle furnace or burner the fusion produce is dissolved in 2N sulfuric acid.

A third method avoids the use of platinum ware. The sample of the metal is finely ground and placed in a small heat-resistant beaker. Two to four grams of ammonium sulfate and 3-6 ml of sulfuric acid are then added. A homogeneous melt is obtained on a hot plate and dissolved in 2 N sulfuric acid.

m.   No suitable compound was found by the author of reference (1).

n.   These compounds are sold as primary standards by the National Bureau of Standards, Office of Standard Reference Materials, Washington, D. C. 20234.

o.   Boric acid may be weighed accurately directly from the bottle. It will loose 1 H_2O molecule at 100°C and is difficult to dry to a constant weight.

p.   Several references have suggested the addition of acid to help stabilize the solution.

q. $MnSO_4 \cdot H_2O$ may be dried at 100°C without losing the water of hydration.

r.   Niobium and tantalum pentoxides are slowly soluble at 40% HF. The addition of H_2SO_4 accelerates the solution process. They may also be dissolved by a fusion technique. $K_2S_2O_7$ is an often used flux. The pentoxides are fully decomposed at $650-800^\circ$C in the presence of an 8-10 fold amount of potassium pyrosulfate. A quartz or porcelain crucible is suitable and the resulting melt may be dissolved in sulfuric acid.

s.   Dissolve in 20 ml hot HF in a platinum dish; add 40 ml H_2SO_4 and evaporate to SO_3 fumes, dilute with 8M H_2SO_4.

t.   When kept dry, silver nitrate crystals are not affected by light. Solutions of silver nitrate should be stored in brown bottles.

u.   Sodium tungstate loses both water molecules at $110^\circ C$. The water is not rapidly regained but the compound should be kept in a desiccator after drying and should be weighed quickly once it is removed.

References:

1.   B. W. Smith & M. L. Parsons, J. Chem. Ed., 50, 679 (1973).

2.   J. A. Dean & T. C. Rains, in "Flame Emission and Atomic Absorption Spectrometry," Vol. 2, Marcel Dekker, New York, 1971, pp. 327-341.

Experimental Parameter Settings

In the flame techniques there are several independent parameters which can be adjusted for maximum response; these can be
optimized independently from all other parameters. They are listed
in Table I.A.4a. Independent parameters are obviously the best
kind in that when the values of all other parameters are varied,
there is no effect on the optimum value of the independent variable.

Unfortunately, there are even more dependent, or interdependent, variables as well. These are listed in Table I.A.4b. In
optimizing interdependent variables, it must be remembered that
each time the value for one variable is changed it affects the
optimum value for each of the variables which is interdependent on
it. This means that all of the interdependent variables must be
simultaneously optimized. This necessitates the use of statistical
design experiments such as the Simplex (4), factorial design (5) or
more sophisticated techniques (5,6). Using these techniques offers
the quickest most efficient optimization of the parameters, with
the least amount of data.

If sensitivity and low limits of detection are not a require-
ment for a particular determination, the parameters can probably
be optimized in a random way by the operator with no overall ill
effect.

References

1. C. Th. J. Alkemade, "Nomenclature, Symbols, Units and Their
Usage in Spectrochemical Analysis III. Analytical Flame Spectroscopy
and Associated Procedures," IUPAC, 1972. Plus personal communication,
1973.

2. H. G. Kuhn, "Atomic Spectra," Academic Press, New York,
1962.

3. A. C. G. Mitcheel & M. W. Zemansky, "Resonance Radiation
and Excited Atoms," Cambridge Univ. Press, New York, 1961.

4. S. L. Morgan & S. N. Deming, Anal. Chem., 46, 1170 (1974).

5. W.G. Cochran & G. M. Cox, "Experimental Designs," Wiley,
New York, 1957.

6. W. Mendenhall, "The Design and Analysis of Experiments,"
Wadsworth Publishing, Belmont, California, 1968.

Table I.A.4

Critical Parameters for Optimization
in Flame Spectrometric Techniques

a.   Independent Parameters

Parameter	AAS	FAAS[1]	AES	AFS
Excitation Source Power	Yes	Yes	N.A.[2]	Yes
Photomultipler[3] Voltage	Yes	Yes	Yes	Yes
Readout Gain[4]	Yes	Yes	Yes	Yes
Filtering Circuit[5] Setting	Yes	Yes	Yes	Yes

b.   Dependent (Interdependent) Parameters

	AAS	FAAS	AES	AFS
Oxidant Gas Flow Rate	Yes	N.A.	Yes	Yes
Fuel to Oxidant Ratio	Yes	N.A.	Yes	Yes
Sheath Gas[6] Flow Rate	Yes	Yes	Yes	Yes
Solution Flow Rate[7]	Yes	N.A.	Yes	Yes
Sample Size	N.A.	Yes	N.A.	N.A.
Height of Measurement	Yes	Yes	Yes	Yes
Monochromator Slit Width	Yes	Yes	Yes	Yes
Burner Variables[8]	Yes	N.A.	Yes	Yes
Furnace or Filament Variables[9]	N.A.	Yes	N.A.	N.A.

Footnotes:

1. FAAS stands for furnace or filament atomic absorption spectrometry (flameless techniques).

2. N.A. stands for not applicable.

3. The photomultiplier voltage will not affect the signal-to-noise ratio unless too great a voltage is used. It merely specifies the intensity of the observed signal.

4. The readout gain will not affect the signal-to-noise ratio unless too large a gain is used. It also merely specifies the intensity of the observed signal.

5. The filtering circuit settings dictate the frequency response of the readout device. The greater the filtering the longer the time required for measurement. This setting is the ultimate control on the noise observed, and should always be specified.

6. In many burner systems, there is no sheath gas; however, if there is one, it will affect the flame composition and temperature.

7. In many systems, the solution flow rate is dictated by the oxidant gas flow. However, in some systems it is determined by an independent source and should be treated separately in those cases.

8. Some burner systems have parameters such as orifice adjustments, heated chamber systems, etc. If so, these are valid interdependent parameters.

9. With the filament or furnace systems, the time for each heating cycle (drying, ashing, and atomization, each is an interdependent variable for each sample matrix as well as the power (temperature) for each.

Reference:

M. L. Parsons & J. D. Winefordner, Appl. Spectrosc., 21, 368 (1967).

B. SCIENTIFIC LITERATURE OF FLAME SPECTROSCOPY

It is the intent of this section to provide the flame spectroscopist with a general feel for the literature in the field.
Several really good references have been written on one or another
of the flame areas and a couple of outstanding references on most
aspects of the flame techniques. Table I.B.1 presents a tabulation
of reference works.

In the realm of abstracts, there are four which are most pertinent: Chemical Abstracts, Analytical Abstracts, Atomic Absorption
Abstracts, and Spectrochemical Abstracts.

The most common journals which frequently publish information
about the flame techniques are listed in Table I.B.2 along with an
indication of the frequency of articles cited in recent review articles. Review articles appear more or less regularly in the most
useful journals such as Anal. Chem. and J. Appl. Spectrosc. Recently,
several review journals have appeared, notably Appl. Spectrosc.
Reviews & C. R. C. Reviews.

In terms of bibliographic material, R. Mavrodineanu has produced
an exhaustive work covering 1800-1966 in "Bibliography on Flame Spectroscopy," NBS Miscellaneous Publication #281, which is available
from the Superintendent of Documents, U. S. Government Printing Office, Washington, D. C. 20402. He has extended this work to cover
through 1968 in the excellent reference "Analytical Flame Spectroscopy" (See Table I.B.2). Other sources of bibliographic material
are those of Sabina Slavin, which appear semiannually in Atomic
Absorption Newsletter, and J. D. Winefordner and T. J. Vickers,
whose reviews appear every other year in Analytical Chemistry.

Finally, a classified list or related references is given in
Table I.B.3.

Table I.B.1

Selected Reference Works in Atomic Spectroscopy

F. Burriel-Martí & J. Ramíerz-Muñoz, "Flame Photometry,"
Elsevier, New York, 1957.

J. A. Dean, "Flame Photometry," McGraw-Hill, New York, 1960.

R. Herrmann & C. Th. J. Alkemade, "Chemical Analysis by Flame
Photometry," Trans. P. T. Gilbert, Interscience, New York, 1963.

R. Mavrodineanu & H. Boiteux, "Flame Spectroscopy," Wiley,
New York, 1965.

E. Pungor, "Flame Photometric Theory," Van Nostrand, London,
1967.

W. T. Elwell & J. A. F. Gidley, "Atomic Absorption Spectrometry," Pergamon, New York, 1966.

J. Ramierz-Muñoz, "Atomic Absorption Spectrometry, Elsevier,
New York, 1968.

W. Slavin, "Atomic Absorption Spectroscopy," Interscience,
New York, 1968.

I. Rubeska & B. Moldan,, "Atomic Absorption Spectrophotometry,"
CRC, Cleveland, 1969.

J. A. Dean & T. C. Rains, "Flame Emission and Atomic Absorption Spectrometry, Vol. I, 1969; Vol. II, 1971, Vol. III, in prep.,
Dekker, New York, 1969.

R. Mavrodineanu, Ed., "Analytical Flame Spectroscopy,"
Springer-Verlag, London, 1970.

B. V. L'vov, "Atomic Absorption Spectrochemical Analysis,"
J. H. Dixon Trans., Hilger, London, 1970.

J. D. Winefordner, Ed., "Spectrochemical Methods of Analysis,"
Wiley, New York, 1971.

R. J. Reynolds & K. Aldous, "Atomic Absorption Spectroscopy,
A. Practical Guide," Griffin, London, 1970.

W. J. Price, "Analytical Atomic Absorption Spectrometry,"
Heyden & Sons, London, 1972.

J. D. Winefordner, S. G. Schulman, & T. C. O'Haver, "Lumi-
nescence Spectrometry in Analytical Chemistry," Wiley, New York, 1972.

Table I.B.2

Common Basic Journals and Frequency of Citation
in Bibliography and Review Articles

Journal[1]	Ref. A		Ref. B		Ref. C	
	#	%	#	%	#	%
Anal. Chem.	70	3.42	265	12.34	136	10.58
Anal. Chim. Acta	38	1.86	203	9.45	114	8.86
Appl. Spectrosc.	20	0.98	97	4.52	41	3.19
Spectrochim. Acta	15	0.73	142	6.61	34	2.64
Analyst	14	0.68	89	4.14	38	2.95
At. Absorpt. Newslett.	60	2.93	NA[2]	NA	94	7.31
J. Anal. Chem. (USSR)	17	0.83	78	3.63	31	2.41
Talanta	9	0.44	92	4.28	28	2.18
Nature	44	2.15	4	0.19	16	1.24
Comb. Flame	37	1.81	94	4.38	0	0
Ind. Lab. (USSR)	23	1.12	26	1.21	14	1.09
Opt. Spectrosc. (USSR)	15	0.73	17	0.79	4	0.31
Z. Anal. Chem.	18	0.88	43	2.00	36	2.80
Anal. Letters	NA	NA	41	1.91	20	1.56
Spectrosc. Letters	NA	NA	44	2.04	16	1.24
Appl. Optics	9	0.44	86	4.00	0	0
J. Opt. Soc. Am.	13	0.64	17	0.79	0	0
Bunseki Kagakin	11	0.54	52	2.42	31	2.41
Chem. Listy	11	0.54	11	0.51	9	0.70
Clin. Chem.	13	0.64	21	0.98	25	1.94
J. Chem. Phys.	42	2.05	25	1.16	2	0.16
J. Appl. Spectrosc. (USSR)	13	0.64	30	1.39	46	3.58
J. Assoc. Off. Anal. Chem.	25	1.22	19	0.88	41	3.19
Proc. Roy. Soc.	21	1.03	23	1.07	2	0.16
Trans. Faraday Soc.	12	0.59	28	1.30	1	0.08
J. Quant. Spectrosc. Rad. Transfer	6	0.29	36	1.68	1	0.08
Rev. Sci. Instruments	7	0.34	49	2.28	4	0.31
J. Chem. Ed.	11	0.54	5	0.23	4	0.31
Compt. Rend.	34	1.66	11	0.51	0	0
Methods Phys. Anal.	2	.10	35	1.63	4	0.31
Environ. Sci. Technol.	1	.05	4	0.19	27	2.10
Chem. Phys. Letters	NA	NA	10	0.47	3	0.23
J. Phys.	NA	NA	28	1.30	2	0.16
Clin. Chem. Acta	NA	NA	10	0.47	18	1.40
Proc. Soc. Anal. Chem.	NA	NA	10	0.47	9	0.70
J. Chromatog.	3	.15	7	0.33	2	0.16
Anal. Biochem.	5	.24	5	0.23	4	0.31
Can. Spectrosc.	5	.24	7	0.33	4	0.31
Chim. Anal.	8	.39	7	0.33	5	0.39

Journal[1]	Ref. A		Ref. B		Ref. C	
	#	%	#	%	#	%
Mikrochim. Acta	4	.20	8	0.37	6	0.47
Microchem. J.	3	.15	8	0.37	3	0.23
Science	2	.10	1	0.47	3	0.23
Totals	641	31.31	1779	82.82	878	68.27

Footnotes:

[1]Journals are abbreviated according to Chem. Abstracts.

[2]NA means not applicable.

Reference A [2047 Total Citations], R. Mavrodineau, "Analytical Flame Spectroscopy," Springer-Verlag, 1970, pp. 651-741.

Reference B [2148 Total Citations], J. D. Winefordner & T. J. Vickers, "Flame Spectrometry," Anal. Chem., 42, 206R (1970); plus 44, 150R (1972); plus 46, 192R (1974).

Reference C [1286 Total Citations], Sabina Slavin, "Atomic Absorption Bibliography," At. Abs. Newsletter, 11, 74 (1972); plus 12, 9 (1973); plus 13, 11 (1974); plus 13, 84 (1974).

Table I.B.3

Miscellaneous Supporting Library Material

I.   Data Compilations

     A. Journals
         1. J. Res. NBS
         2. J. Phys. Chem. Ref. Data

     B. Books
         1. NSRDS - NBS Series
         2. NBS Monographs, Special Publications, etc.
         3. MIT Wavelength Tables

II. New Instrumentation & Chemicals

     A. Journals
         1. American Lab.
         2. Industrial Research
         3. Research & Development
         4. Buyer's Guide in Anal. Chem.

     B. Books
         1. Handbook of Commercial Scientific
               Instrumentation, Atomic Absorption,
               Dekker, N. Y., 1974
         2. "Chem. Sources"

III. Reports of Meetings, Conference, etc.

         1. Annual Reports on Anal. Spectrosc.
         2. Recent Developments in Applied Spectrosc.

(See also references following the various tables.)

C. FLAMES & SAMPLE CELLS

Flames

Historically, the flame has provided the researcher with the
most stable, versatile source of atoms with which to study electronic transitions. Numerous flame compositions and burner types
have been studied; however, the most common flames in use today
are the air-acetylene, nitrous oxide-acetylene, and the air-hydrogen. The most common method of sample introduction is by pneumatic
aspiration although ultrasonic nebulizers are making some inroads.
The laminar-flow chamber-type burner has virtually taken the
place of the more noisy (both audio and electrical) total-consumption burner.

Flame temperatures and general references for temperature
measurement techniques are listed in Table I.C.1. Composition of
the major flame gas combustion products are presented in Table I.C.2.
Emission spectra of the most common analytical flames are shown in
Chapter III Part B.8 along with information with respect to the
most useful analytical lines for the elements. These figures de-
monstrate the molecules which are spectrally important in these
flames.

Although the analytical techniques based on atomic spectra
are exponentially related to the temperature of the sample cell,
it has been amply demonstrated that the chemical composition of
the flame cell is equally, if not more important, than the temperature in determining the number of free atoms in the cell. The
flames using acetylene are generally more reducing than the ones
with hydrogen as a fuel. The nitrous oxide-acetylene flame is
particularly useful in atomizing elements with strong oxide bonds.
A good indication of the reducing nature of flames and the compound
forming capabilities of the elements is obtained by studying
Tables III.B.5 & 6. These tables deal with the energy required
to break molecular bonds of the elements and with the free atom
fraction of the elements in the various flames, respectively.

Non-Flame Cells

1. Atomic Absorption. The use of non-flame cells has gained
the most popularity in atomic absorption spectroscopy. The
furnace, the filament, and the metal cup (used in conjunction with
a flame) are the most commonly used non-flame cells.

The furnace techniques were probably the first non-flame
cells, and analytical methods were first developed by Woodriff in
the United States, L'vov in Russia, and Massmann in Germany. The
major commercial versions of the furnace utilize a very small

Table I.C.1

Flame Temperatures and References
on Temperature Measurements

Flame	Typical Experimental Temperatures	Calculated Stoichiometric	Typical*
Ac/A	2360-2600 °K	2523 °K	2450 °K
Ac/N	2830-3070	3148	2950
H/O/Ar	2100	----	2100
H/A	2100-2300	2373	2300
H/N	2500	2960	2500
H/O	2500-2900	3100	2800
Ac/O	2900-3300	3320	3100
NG/A	1900-2150	2228	2000

*This value represents the approximate temperature most often reported for a flame in a typical analytical situation.

The following sources contain detailed discussions of the techniques available for the measurement of flame temperature:

1. A. G. Gaydon and H. G. Wolfhard, Flames, Their Structure Radiation and Temperature, Chapman and Hall Ltd., London, 1970.

2. R. M. Fristrom and A. A. Westenberg, Flame Structure, Chap. 8, McGraw-Hill, 1965.

3. R. H. Tourin, Spectroscopic Gas Temperature Measurement, Elsevier Publishing Co., Amsterdam, 1966.

4. A. G. Gaydon and H. G. Wolfhard, "The Spectrum-Line Reversal Method of Measuring Flame Temperature," Proc. Phys. Soc. (London), 65A, 19-24, 1952.

5. R. F. Browner and J. D. Winefordner, "Measurement of Flame Temperatures by a Two-Line Atomic Absorption Method," Anal. Chem., 44, 247, 1972.

6. N. Omenetto, P. Benetti and G. Rossi, "Flame Temperature Measurements by Means of Atomic Fluorescence Spectrometry," Spectrochimica Acta, 27B, 453, 1972.

7. C. M. Herzfeld, Temperature, Its Measurement and Control in Science and Industry, Vol. III, Part 2, Ed. by I. Dahl, Reinhold, N. Y., 1962.

Table I.C.2

Mole Fraction of Major Species
in Analytical Flames

Air/Acet. (Fuel Rich)		Air/Acet. (Normal)		Air/H_2 (Normal)	
Species	Mole Fract.	Species	Mole Fract.	Species	Mole Fract.
N_2	.592	N_2	.687	N_2	.405
CO	.215	CO	.081	H_2	.365
H_2	.091	H_2	.016	H_2O	.229
H_2O	.068	H_2O	.113	H	.0010
CO_2	.027	CO_2	.094	OH	.00006
H	.006	H	.0026		
OH	.00087	OH	.0034		
		NO	.0014		
		O_2	.0010		

N_2O/Acet. (Fuel Rich)			N_2O/Acet. (Normal)	
Species	Mole Fract.		Species	Mole Fract.
N_2	.349		N_2	.394
CO	.383		CO	.356
H_2	.188		H_2	.147
H	.058		H_2O	.035
HCN	.018		CO_2	.011
C_2H_2	.0018		H	.051
CN	.0007		OH	.0039
			NO	.00066
			O	.00086
			O_2	.00008

The data for this table was calculated from JANAF Table Data in-
cluding the 1974 supplement (J. Phys. Chem. Ref. Data, 3, 311 (1974)
M. W. Chase et al.) using a free energy minimization computer program supplied by NASA.

sample holder for about 1 to 10 microliter size samples, and is
capable of about 3000°K by resistance heating in an inert or re-
active atmosphere. The sample holder is usually made of graphite
or carbon, and there are three stages of heating; one for driving
off moisture; one for ashing carbonaceous material; and one for
vaporization and atomization of the remaining sample. The
resulting absorption signal is transient, and the peak absorbance,
or perferably the integrated absorbance, is measured. Background
correction is almost always required with this technique, and
normal strip-chart recorders are often too slow to give accurate
response. Matrix effects should always be carefully studied.

    Filaments are of two major types--the carbon rod atomizer or
the metal ribbon atomizer. These are also heated by electrical
resistance to about 3000°K, use sample sizes from 1 to 10 microliters, and have three heating cycles. These are quite similar to
the furnace atomizers and in general have the same advantages
and disadvantages.

    The metal cup technique was first used by Delves and is also
commercially available. This system consists of a small metal
cup (Ni, Mo, and Pt have been used) which is mounted on a stand
which can reproducibly insert the sample (which is in the cup) into a suitable flame (both the air-acetylene and the nitrous oxide-
acetylene have been used). The energy of the flame is used to
atomize the sample. The flame is also used for drying and ashing
steps by judicious placement of the cup with respect to the flame.
This method uses samples up to several hundred microliters and
because of this has certain advantages over the other non-flame
techniques. Once again, the problems of sample matrix must be
studied very closely.

    For volatile species, specifically mercury, the vapor pressure
is high enough that atomic absorption measuremenets can be made at
room temperature. Samples of Hg are reduced chemically in a
closed system, and the absorbance measured through a quartz cell.
Although there are still some matrix problems, this is probably
the most sensitive atomic method for mercury; it is certainly the
simplest.

    2. Atomic Emission. Although a paper or two may have appeared
which used one or another of the cells described above for atomic
emission measurements, this type of cell has not found much use in
analytical atomic emission studies. Of course the field of d.c. arc
or spark emission is large and fruitful. It is much too large to
be included in this Handbook. However, several other atomic emission techniques have been used in analytical emission studies and
are worthy of mention.

Demountable Hollow Cathode Discharge Lamps have been used.
(See Section D of this chapter for a general discussion of these
devices for excitation sources as well as their emission spectra.)
These devices produce relatively simple atomic emission spectra
and are quite sensitive. Their chief disadvantage lies in the fact
that they are rather slow in routine applications; the main
problem being in changing samples.

Flow-through Electrodeless Discharge Lamps have also been used
for analytical emission. (See Section D of this chapter as well.)
These can be much more intense than hollow cathode devices, but
except with special care, often suffer from a lack of stability
and can be easily quenched by sample and/or gas variations. Sample
sizes must therefore be kept very small and matrix effects once
again are a large concern. However, this technique uses relatively
inexpensive equipment and holds much promise with further tech-
nological developments.

Induction Coupled Plasma (ICP) is also a new emission technique worthy of mention. This method was developed into a viable
analytical tool by Fassel's group, and the data presented by his
group looks extremely impressive. The spectra produced by the
high energy ICP is quite complex, however, and the method should be
considered along with the d.c. arc. These authors also feel that
the cost of the ICP devices place them in the realm of the d.c.
arc as well. Finally, it would appear that a computer control of
experimental parameters is a must for accurate results.

3. Atomic Fluorescence. To date no research has been
reported using non-flame cells in atomic fluorescence. (Recently,
some work has appeared using the carbon rod atomizer in AFS, but not
in time for inclusion in this work.)

D.   EXCITATION SOURCES

Hollow Cathode Discharge Lamps (HCL)

These devices are by far the most commonly used excitation
sources in atomic absorption spectrometry. They consist of a
cylindrical cathode made of the element of interest or alloyed
from the element of interest. They are sealed lamps with an inert
gas, usually neon, and are powered by a d.c. source of about 300V
and 1 to 30 (sometimes as high as 50) milliamps, or by an a.c.
supply of similar power. It has been shown that the emission lines
emitted by HCL's are essentially broadened by the Doppler effect
when run at moderate power and operate at a temperature of about
400 to 500 $^\circ$K.

Variation on this basic type have been used, such as demountable
cathodes, pulsed-mode operation, high intensity lamps, but the most
common configuration is described above. The high intensity lamps
have an auxiliary power supply and a second set of electrodes
which are operated at much greater power. The intensity of radiation from these lamps is much greater. If a pulse of power is
added to the low level d.c., a much more intense radiation also
results. These modifications have found some use in atomic fluorescence measurements.

Relative intensities of the transitions from each element with
respect to the most intense transition for the element are detailed
in Table I.D.1. Relative intensities of the transitions all on
the same scale are given in Table I.D.2. The latter table gives an
indication of which lamps cannot be made with sufficient intensity
to be analytically useful.

Because neon is by far the most used fill gas for HCL's, the
relative intensity of Ne transition lines are given in Table I.D.3.
These were obtained by scanning the spectra of specific HCL. It
should be noted that very little difference was found in the
relative Ne intensities by scanning the spectra of several different
HCL's made from different elements.

Electrodeless Discharge Lamps (EDL or EDT)

These devices have found the greatest use in atomic fluorescence
spectrometry; however, in cases where HCL's cannot be made with
sufficient intensity, they are used in atomic absorption spectrometry.
EDL's generally consist of a small diameter quartz tube (8 - 10 mm
i.d.) about 1 to 2 cm in length into which is sealed a small quantity of one of the following: the pure metal, a halide salt of
the metal, or the hydride of the metal in question. Often an ionization suppressant is added and an inert gas at about 1 to 3 torr.

EDL's are powered with a 2450 MHz microwave generator in con-
juction with an 'A' antenna or a focusing cavity. As yet, there
are no commercial units sold for this purpose, but it is thought
that they will appear on the market in the near future. Commercial devices are now available using an rf power supply for a
limited number of elements.

A flow-through EDL system has been used in atomic emission
spectroscopy. In such systems a smaller diameter tube is generally used in conjunction with a focusing cavity. Once again, the
quenching problem is the most trouble.

The reason for the success of the EDL's is the very intense,
narrow line atomic emission which they produce. In atomic fluorescence spectrometry, the analytical signal is directly propor-
tional to the intensity of the excitation source (see Chapter III).
Quite impressive limits of detection have been found using the
more intense EDL's. The main reason that they have not yet found
commercial status is the difficulty in producing working lamps.
Some workers have reported that only one in four or five lamps work
properly. They have also been rather unstable with respect to
time; however, thermostating the lamps seems to cure this problem.

Lasers

Of course, laser emission holds the promise of the ultimate in
excitation sources, and whereas line emission from most lasers do
not correspond to the wavelengths of most analytical transitions,
the advent of the dye laser has provided researchers in the field
of atomic fluorescence a powerful new excitation source.

There are many laser types, and it is not within the realm of
this Handbook to present a discourse on them; however, the dye
laser is generally flash-pumped or nitrogen laser pumped. Flash
pumping provides significantly wider pulses (ca microseconds)
whereas, laser pumping produces nanosecond range pulses. The latter
makes certain electrical problems in detection of the resultant
fluorescent radiation. The dye laser systems can produce very large
intensities within the special range of the dye being irradiated.
A list of useful dyes and their spectral range is presented in
Table I.D.4.

Continuum Sources

These have been used in both atomic absorption and atomic fluorescence spectrometry. Their use in AAS usually requires better
resolution with the monochromator than does the use of line sources;
however, quite good detection limits have been obtained with the
150 watt high pressure xenon arc. Hydrogen and deuterium dis-

charge lamps have been utilized for background correction in AAS techniques, and many newer commercial units now offer an automatic background correction.

In atomic fluorescence spectrometry, the 150 watt Xenon arc has also produced fair results; however, the extra noise from scattered radiation is a limiting factor. Recently, the use of a new high intensity Xenon arc using a smaller arc length and better focusing has provided much better results.

Tungsten ribbon lamps traceable to the NBS are still the most useful means to accurately measure flame temperatures.

Table I.D.1

RELATIVE INTENSITIES OF
HOLLOW CATHODE LINES

(Relative to Most Intense Line
for Each Element)

Element	Fill Gas	Wavelength (Å)	Relative Intensity
Aluminum, Al	Neon	2367.05	1.2
		2373.12 }	2.5
		2373.35	
		2567.98	3.0
		2575.10 }	5.5
		2575.40	
		3092.71 }	95.5
		3092.84	
		3961.52	100 *
Antimony, Sb	Neon	2068.33	40
		2127.39	2.8
		2175.81	70
		2311.47	100 *
Arsenic, As	Argon	1890.00	15
		1936.96	92
		1971.97	100 *
Barium, Ba	Neon	3501.11	30
		5535.48	100 *
Beryllium, Be	Neon	2348.61	100 *
Bismuth, Bi	Neon	2228.25	4
		2230.61	5
		2276.58	3
		3067.72	100 *
Boron, B	Neon	2088.93 }	800
		2089.59	
		2496.78 }	100 *
		2497.73	

Element	Fill Gas	Wavelength (Å)	Relative Intensity
Cadmium, Cd	Neon	2288.02	11
		3261.06	100 *
Calcium, Ca	Neon	2398.56	0.25
		4226.73	100 *
Cerium, Ce	Neon	5200.12 }	65
		5200.42	
		5697.00	100 *
Cesium, Cs	Argon	4555.36	100 * } R213
		4593.18	26
		8521.10	100 * } R406
		8943.50	47
Chromium, Cr	Neon	3578.69	100 *
		4254.35	71
		4274.80	76
		4289.72	59
		5204.52	22
		5208.44	52
Cobalt, Co	Neon	2407.25	35
		3044.00	15
		3465.80	100 *
		3474.02	34
		3909.93	40
Copper, Cu	Neon	2178.94	3.5
		2181.72	1.7
		2225.70	4.5
		2441.64	8.0
		2492.15	20
		3247.54	100 *
		3273.96	65
Dysprosium, Dy	Neon	4191.60	12
		4194.85	53
		4211.72	100 *
		4218.09	15
		4225.14	5

Element	Fill Gas	Wavelength (Å)	Relative Intensity
Erbium, Er	Neon	3892.69	47
		4007.97	100 *
		4020.52	20
		4087.65	25
Europium, Eu	Neon	3334.33	34
		4594.03	100 *
Gadolinum, Gd	Neon	3684.13	47
		4058.22	100 *
		4190.78	85
Gallium, Ga	Neon	2719.65	8
		2874.24	68
		2943.64	100 *
Germanium, Ge	Neon	2651.18 }	100 *
		2651.58	
		2691.34	32
		3039.06	85
Gold, Au	Neon	2427.95	71
		2675.95	100 *
Hafnium, Hf	Neon	3072.88	48
		3682.24	100 *
		3777.64	87
Holmium, Ho	Neon	4103.84	100 *
		4254.43	27
Indium, In	Neon	2710.26	18
		3039.36	100 *
Iridium, Ir	Neon	2088.82	25
		2543.97	56
		2639.71	100 *
		2664.79	92

Element	Fill Gas	Wavelength (Å)	Relative Intensity
Iron, Fe	Neon	2483.27	1.0
		3719.94	100 *
		3859.91	53
		3920.26	4.0
Lanthanum, La	Neon	3574.43	65
		4037.21	20
		5501.34	100 *
Lead, Pb	Neon	2022.02	6
		2053.27	6
		2169.99	12
		2614.18	55
		2833.06	100 *
Lithium, Li	Neon	3232.63	1
		6103.64	12
		6707.84	100 *
Lutetium, Lu	Argon	3312.11	52
		3376.50	51
		3359.56	39
		3567.84	100 *
Magnesium, Mg	Neon	2025.82	0.5
		2852.13	100 *
Manganese, Mn	Neon	2794.82	16
		3216.95	1.5
		4030.76	100 *
Mercury, Hg	Neon	2536.52	100 *
Molybdenum, Mo	Neon	3132.59	100 *
		3208.83	24
Neodymium, Nd	Neon	4866.74	17
		4924.53	100 *

Element	Fill Gas	Wavelength (Å)	Relative Intensity	
Nickel, Ni	Neon	2320.03	8	
		3414.76	56	
		3515.05	43	
		3524.54	100	*
		3624.73	17	
Niobium, Nb	Neon	3349.06	28	
		3580.27	52	
		4058.94	100	*
		4079.73	83	
Osmium, Os	Neon	2909.06	32	
		4260.85	100	*
Palladium, Pd	Neon	2447.91	6	
		2476.42	5	
		3404.58	100	*
Platinum, Pt	Neon	2659.45	100	*
		2997.97	83	
Potassium, K	Neon	4044.14	1	
		7664.91	100	*
		7698.98	80	
Praseodymium, Pr	Neon	4951.36	100	*
		5133.42	89	
Rhenium, Re	Neon	3451.88	30	
		3460.46	100	*
		3464.73	68	
Rhodium, Rh	Neon	3280.55	7	
		3434.89	100	*
Rubidium, Rb	Neon	4201.85	1	
		4215.56	1	
		7800.23	100	*
		7947.60	27	

Element	Fill Gas	Wavelength (Å)	Relative Intensity
Ruthenium, Ru	Neon	3498.94	100 *
		3925.92	64
Samarium, Sm	Neon	4296.74	85
		4760.27	100 *
Scandium, Sc	Neon	3269.91	13
		3273.63	18
		3911.81	100 *
Selenium, Se	Neon	1960.23	77
		2039.85	100 *
Silicon, Si	Neon	2506.90	37
		2514.32	32
		2516.11	100 *
		2524.11	34
		2881.60	94
Silver, Ag	Neon	3280.68	100 *
		3382.89	100
Sodium, Na	Neon	3302.32 }	6
		3302.99	
		5889.95	91
		5895.92	100 *
Strontium, Sr	Neon	4607.33	100 *
Tantalum, Ta	Neon	2714.67	100 *
		2758.31	59
Tellurium, Te	Neon	2142.75	15
		2259.04	100 *
		2385.76	41
Terbium, Tb	Neon	4318.85	82

Element	Fill Gas	Wavelength (Å)	Relative Intensity
Terbium (cont)	Neon	4326.47	100 *
		4338.45	63
Thallium, Tl	Neon	2580.14	9
		2767.87	100 *
Thorium, Th	Neon	3304.24	32
		3719.44	100 *
		3803.07	57
Thulium, Tm	Neon	3717.92	70
		4203.73	64
		4359.93	22
		5307.12	100 *
Tin, Sn	Neon	2246.05	100 *
		2334.80	48
		2661.24	44
Titanium, Ti	Neon	3642.68	100 *
		3653.50	100
		3989.76	98
Tungsten, W	Neon	2551.35	6
		4008.75	100 *
		4074.36	86
Uranium, U	Argon	3489.37	46
		3514.61	68
		3566.60	73
		3584.88	100 *
Vanadium, V	Neon	3066.38	20
		3183.98	100 *
		3185.40	50
		4389.97	29
Ytterbium, Yb	Neon	2464.49	4

Element	Fill Gas	Wavelength (Å)	Relative Intensity
Ytterbium (cont)	Neon	2671.98	3
		3987.98	100 *
Yttrium, Y	Neon	4102.38	100 *
		4142.85	49
Zinc, Zn	Neon	2138.56	29
		3075.90	100 *
Zirconium, Zr	Neon	3601.19	66
		4687.80	100 *

* Intensity reference.

Unless specified spectral response of Hamamatsu R 213 was used.

"Hollow Cathode Lamps," Varian Techtron, 2700 Mitchell Drive, Walnut Creek, California, 94598, 1970.

Table I.D.2

RELATIVE INTENSITIES OF
HOLLOW CATHODE LINES
(All Elements on the Same Scale)

Element	Fill Gas	Wavelength (A)	Relative Emission Intensity
Aluminum, Al	Neon	3092.71 } 3092.84	1200
		3961.52	800
Antimony, Sb	Neon	2175.81	250
		2311.47	250
Arsenic, As	Argon	1936.96	125
		1971.97	125
Barium, Ba	Neon	5535.48	400
		3501.11	200
Beryllium, Be	Neon	2348.61	2500
Bismuth, Bi	Neon	2230.61	120
		3067.72	400
Boron, B	Argon	2497.72	400
Cadmium, Cd	Neon	2288.02	2500
		3261.06	5000
Calcium, Ca	Neon	4226.73	1400
Cerium, Ce	Neon	5200.12 } 5200.42	8
		5697.00	8
Cesium, Cs	Neon	8521.10	15
		4555.36	35

Element	Fill Gas	Wavelength (A)	Relative Emission Intensity
Chromium, Cr	Neon	3578.69	6000
		4254.35	5000
Cobalt, Co	Neon	2407.25	1000
		3453.50	1500
		3526.85	1300
Copper, Cu	Neon	3247.54	7000
		3273.96	6000
Dysprosium, Dy	Neon	4045.99	2000
		4186.78	2000
		4211.72	2500
Erbium, Er	Neon	4007.97	1600
		3862.82	1600
Europium, Eu	Neon	4594.03	1000
		4627.22	950
Gadolinium, Gd	Neon	3684.13	350
		4078.70	700
Gallium, Ga	Neon	2874.24	400
		4172.06	1100
Germanium, Ge	Neon	2651.18 } 2651.58	500
		2592.54	250
Gold, Au	Neon	2427.95	750
		2675.95	1200
Hafnium, Hf	Neon	3072.88	300
		2866.37	200

Element	Fill Gas	Wavelength (A)	Relative Emission Intensity
Holmium, Ho	Neon	4053.93	2000
		4103.84	2200
Indium, In	Neon	3039.36	500
		4101.76	500
Iridium, Ir	Neon	2849.72	400
		2639.71	400
Iron, Fe	Neon	2483.27	400
		3719.94	2400
Lanthanum, La	Neon	5501.34	120
		3927.56	45
Lead, Pb	Neon	2169.99	200
		2833.06	1000
Lithium, Li	Neon	6707.84	700
Lutetium, Lu	Argon	3359.56	30
		3376.50	25
		3567.84	15
Magnesium, Mg	Neon	2852.13	6000
		2025.82	130
Manganese, Mn	Neon	2794.82	3000
		2801.06	2200
		4030.76	14000
Mercury, Hg	Argon	2536.52	1000
Molybdenum, Mo	Neon	3132.59	1500
		3170.35	800

Element	Fill Gas	Wavelength (Å)	Relative Emission Intensity
Neodymium, Nd	Neon	4634.24 4924.53	300 600
Nickel, Ni	Neon	2320.03 3414.76	1000 2000
Niobium, Nb	Neon	4058.94 4079.73	400 360
Osmium, Os	Argon	2909.06 3018.04	400 200
Palladium, Pd	Neon	2447.91 2476.42 3404.58	400 300 3000
Phosphorus, P	Neon	2135.47 } 2136.20 2149.11	30 20
Platinum, Pt	Neon	2659.45 2997.97	1500 1000
Potassium, K	Neon	7664.91 4044.14	6 300
Praseodymium, Pr	Neon	4951.36 5133.42	100 70
Rhenium, Re	Neon	3460.46 3464.73	1200 900
Rhodium, Rh	Neon	3434.89 3692.36 3507.32	2500 2000 200

Element	Fill Gas	Wavelength (Å)	Relative Emission Intensity
Rubidium, Rb	Neon	7800.23 4201.85	1.5 80
Ruthenium, Ru	Argon	3498.94 3925.92	600 300
Samarium, Sm	Neon	4296.74 4760.27	600 800
Scandium, Sc	Neon	3911.81 3907.49 4020.40 4023.69	3000 2500 1800 2100
Selenium, Se	Neon	1960.26 2039.85	50 50
Silicon, Si	Neon	2516.11 2881.60	500 500
Silver, Ag	Argon	3280.68 3382.89	3000 3000
Sodium, Na	Neon	5889.95 3302.32 } 3302.99	2000 40
Strontium, Sr	Neon	4607.33	1000
Tantalum, Ta	Argon	2714.67 2775.88	150 100
Tellurium, Te	Neon	2142.75 2385.76	60 50
Terbium, Tb	Neon	4326.14 } 4326.47	110

Element	Fill Gas	Wavelength (Å)	Relative Emission Intensity
Terbium (cont)	Neon	4318.85	90
		4338.45	60
Thallium, Tl	Neon	2767.87	600
		2580.14	50
Thulium, Tm	Neon	3717.92	40
		4094.19	50
		4105.84	70
Tin, Sn	Neon	2246.05	100
		2863.33	250
Titanium, Ti	Neon	3642.68	600
		3998.64	600
Tungsten, W	Neon	2551.00 } 2551.35	200
		4008.75	1400
Uranium, U	Neon	3584.88	300
		3566.60	200
		3514.61	200
		3489.37	150
Vanadium, V	Neon	3183.41 } 3183.98	600
		3855.37 } 3855.84	200
Ytterbium, Yb	Argon	3987.98	2000
		3464.36	800
Yttrium, Y	Neon	4077.38	500
		4102.38	600
		4142.85	300
Zinc, Zn	Neon	2138.56	2500
		3075.90	2000

Element	Fill Gas	Wavelength (Å)	Relative Emission Intensity
Zirconium, Zr	Argon	3601.19	250
		3519.60	210

From Hollow Cathode Discharge Devices, Westinghouse Electric Corp. Electronic Tube Division, P.O. Box 284, Elmira, NY 14902, Pub. # TD 86-375.

Relative intensity data were obtained on monochromator with a 30,000 groove per inch grating, 52 mm x 52 mm, and having a RLD of 16 Å/mm using 25 mm slits. Most data taken with R136 photomultiplier (S 10) at 800 volts.

The most intense line is the Mn 4030.26 line at 14,000; the least intense is the Rb 7800.23 line at 1.5.

Table I.D.3

THE HOLLOW CATHODE LAMP
SPECTRA OF NEON

Wavelength	Relative Intensity	Origin
3232.38	5.4	Ne II
3311.30	2.8	Ne II
3319.75	8.7	Ne II
3323.75	28.	Ne II
3329.20	1.7	Ne II
3334.87	5.2	Ne II
3344.43	17.	Ne II
3355.05	3.5	Ne II
3360.63	1.7	Ne II
3369.91 } 3369.81	7.8	Ne I
3378.28	17.	Ne II
3392.78	8.3	Ne II
3417.90	16.	Ne I
3447.70	12.	Ne I
3454.19	15.	Ne I
3460.53	6.6	Ne I
3466.58	12.	Ne I
3472.57	12.	Ne I
3498.06	2.9	Ne I
3501.22	3.8	Ne I
3515.19	3.6	Ne I
3520.47	61.	Ne I
3568.53	7.8	Ne II
3574.64	5.9	Ne II
3593.53	19.	Ne I
3600.17	3.5	Ne I
3633.67	3.6	Ne I
3664.11	1.9	Ne II
3694.20	3.5	Ne II
3709.64	4.9	Ne II
3721.86	3.1	Ne II
4042.64	1.4	Ne I

Wavelength	Relative Intensity	Origin
5330.78	1.6	Ne I
5349.20	1.6	Ne I
5400.56	3.3	Ne I
5764.42	2.3	Ne I
5852.49	100.	Ne I
5881.89	8.7	Ne I
5944.83	14.	Ne I
5974.63 } 5975.53	2.6	Ne I
6030.00	2.8	Ne I
6074.34	11.	Ne I
6096.16	15.	Ne I
6143.06	20.	Ne I
6163.59	5.2	Ne I

For the Varian Techtron Copper hollow cathode
lamp operated at 10 mA, the Copper 324.7 nm
line was a factor of 2.9 more intense than
the 5852.49 Å Neon line. Spectra taken on a
IP28 PMT. Relative intensities not corrected
for instrument function or photomultiplier
response.

Table I.D.4

LASER DYES

Dye	Spectral Coverage (nm)
p-terphenyl (p-diphenylbenzene)	343-356
2,5-diphenyl-1,3,4-oxadiazole (PPD)	(348)
2,5 diphenyloxazole (PPO)	358-391
isopropyl PBD	361,370
2,5 diphenylfuran (PPF)	369-380
p-quaterphenyl	(375)
phenylbiphenylyloxadiazole (PBD)	377-415
α-naphtylphenyloxazole (α-NPO)	392-414
p,p'-diphenylstilbene	(408)
dibiphenyloxazole (BBO)	401-420
1,2-di-4-biphenylyl-ethelene	380-440
2-hydroxy-4-methyl-7-amino-quinoline	(413)
1,4-distyrylbenzene	(415)
1,4-di-[2-(5-phenyloxazyl)]-benzene (POPOP)	(418)
4,4'-dichlor-1,4-distyrylbenzene	(420)
p,p'-bis(O-methylstyryl) benzene (bis-MSB)	419-425
2,2'-dimethoxy-1,4-distyryl-benzene	(425)
1,2-di-(α-naphtyl)ethylene	(426)
1,4-di-[2-(4-methyl-5-phenylox-azolyl)] (DIMETHYL POPOP)	424-441
9,10-diphenylanthracene	(430)
1-styryl-4-[ω-vinyl-(n-biphenylyl)]-benzene	(432)
ACRIDONE	(435)
2,5-bis[tert-butylbenzoxazolyl(2)] thiopene (BBOT)	(437)
3-ethylaminopyrene-5,8,10-trisulfonic acid	(441)
7-acetoxy-4-methylcoumarin	441-486
7-hydroxy-4-methylcoumarin (4-methyl-umbelligerone)	420-550
7-methylamino-4,6-dimethyl-coumarin	440-485
7-ethylamino-4,6-dimethyl-coumarin	(457)
9-aminoacridine hydrochloride	457-460
7-hydroxycoumarin	450-470

Dye	Spectral Coverage (nm)
7-diethylamino-4-methyl-coumarin	438-482
7-acetoxy-5-allyl-4,8-dimethylcoumarin	458-515
ESCULIN	450-470
benzyl β methylumbelliferone	464-468
4,8-dimethyl-7-hydroxycoumarin	455-505
2,4,6-triphenylpyrilium-fluorobate	(485)
acriflavin hydrochloride	(510)
1,2-dihydro-4-methoxybenzo-[C]xanthylium	
fluorobate	(539)
EOSIN	(540)
10-methoxy-12H-indono[3,2,6]-naptho-	
[1,2e]pyrlium fluorobate	(544)
Na-fluorescein	530-560
fluorescein diacetate	541-571
2,7-dichlorofluorescein	(550)
pyrlium salt	(551)
8-hydroxy-1,3,6-pyrenetrisulfonic acid	
trisodium salt	(552)
acetamidopyrene-trisulfonate	(565)
RHODAMINE C	(575)
PYRONIN B	580,615-632
RHODAMINE S	578-695
RHODAMINE 6G	570-620
RHODAMINE G	(591)
SAFRANINE T	(610)
XYLENE RED B	584-645
ACRIDINE RED	600-630
KITON RED S	589-642
RHODAMINE B	605-645
3,3'-diethyl-oxadicarbo-cyanine-iodide	(660)
3,3'-diethyl-10-chloro-2,2'-(5,6,5',6'-	
dibenzo)-thiadicarbo-cyanine-iodide	(714)
3,3'-diethyl-2,2'-thiadicarbocyanine iodide	
(DTDC IODIDE)	705-740
3,3'-diethyl-2,2'-(5,5'-dimethyl)-	
thiazolinotri-carbocyanine iodide	(717)

Dye	Spectral Coverage (nm)
3,3'-diethyl-10-chloro-2,2'-(4,5,4',5'-dibenzo) thiadicarbocyanine iodide	720,760
3,3'-diethyl-2,2'-oxytricarbocyanine iodide	718-750
3,3'-diethyl-5,5'-dimetoxy-6,6'-bis(methyl-mercapto)-10-methyl-thiadicarbocyanine bromide	727-739
3,3'-dimethyl-2,2'-oxatricarbocyanine iodide	720-770
1,1'-diethyl-γ-cyano-2,2'-dicarbocyanine tetrafluoroborate	(740)
1,1'-diethyl-4,4'-quinocarbocyanine iodide (CRYPTOCYANINE)	(745)
1,1'-dimethyl-11-bromo-2,2'-quinodicarbocyanine iodide	(749)
1,1'-diethyl-4,4'-quinocarbocyanine bromide	(754)
chloro-aluminum phthalocyanine (CAP)	750-780
3-ethyl-3'-methylthiathiazolino-tricarbo-cyanine iodide	738-801
1,1'-diethyl-2,2'-dicarbocyanine iodide	(771)
3,3'-diethyl-11-metoxythiatricarbocyanine iodide	773-798
1,1'-diethyl-γ-acetoxy-2,2'-dicarbocyanine tetrafluoroborate	(797)
1,3,3,1',3',3'-hexamethyl-2,2'-indotricarbo-cyanine iodide	780-830
1,1'-diethyl-11-bromo-2,2'-quinodicarbocyanine iodide	780-810
1,1'-diethyl-γ-nitro-4,4'-dicarbocyanine tetrafluoroborate	790-830
3,3'-diethylthiatricarbocyanine iodide (DTTC IODIDE)	794-860
3,3'-diethylthiatricarbocyanine bromide (DTTC BROMIDE)	808-840
1,3,3,1',3',3'-hexamethyl-4,5,4',5'-dibenz-oiodotricarbocyanine perchlorate	816-833
3,3'-diethyl-2,2'-selenatricarbocyanine iodide	(826)
1,1'-diethyl-11-bromo-4,4'-quino-dicarbocyanine iodide	(831)
3,3'-diethyl-6,7,6',7'-dibenzothiotricarbo-cyanine iodide	824-854
3,3'-diethyl-2,2'-(5,6,5',6'-tetramethoxy) thiatricarbocyanine iodide	845-875

Dye	Spectral Coverage (nm)
3,3'-diethyl-6,7,6',7'-dibenzo-11-methyl-thiatricarbocyanine iodide	843-869
3,3'-diethyl-2,2'-(4,5,4',5'-dibenzo)thiatricarbocyanine iodide	840-880
3,3'-diethyl-12-ethylthiatetracarbocyanine iodide	916-924
1,1'-diethyl-2,2'-quinotricarbocyanine iodide	890-960+
1,1'-diethyl-4,4'-quinotricarbocyanine iodide	980-1060

Reference: C.F. Dewey, Jr. in <u>Modern Optical Methods in Gas Dynamic Research</u>, Z. Dosanji, ed., Plenum Press, New York, 1971.

E. INTEREFERENCES

Chemical

    A flame cell can be treated just as a dilute solution in terms
of chemical reactions. The analyte species are essentially at
infinite dilution, and many types of chemical reaction can occur.
Of course, any type of reaction which causes the removal of an
analyte's free atom will cause a depression in the analytical
signal, and by the same token, any reaction which produces an
analyte's free atom will enhance the signal. Common reaction
types are listed below:

$$MO = M+O \qquad\qquad\qquad 1)$$

$$MX = M+X \qquad\qquad\qquad 2)$$

$$M = M+e^- \qquad\qquad\qquad 3)$$

$$MOH = M+OH \qquad\qquad\qquad 4)$$

$$MOX = M+OX \qquad\qquad\qquad 5)$$

$$MH = M+H \qquad\qquad\qquad 6)$$

$$MC = M+C \qquad\qquad\qquad 7)$$

Where M is the analyte species, X is any other flame gas species
except oxygen, O, Hydrogen, H, or carbon, C. All of these reactions can be mathematically treated,as in dilute solution equilibrium; with equilibrium constants of the form

$$K_{eq} = \frac{[M][O]}{[MO]} \; , \; \text{for 1) above, etc.} \qquad 8)$$

In the case of the ionization reaction (3 above), an indication of
this tendency is given in Table III.B.4. The lower the ionization
potential, the greater the tendency for ionization. If the analyte
is easily ionized, a species which is even more easily ionized can
be added in a constant way to the analysis matrix (samples and
standards) in order to reduce the problem by making use of the law
of mass action. Solutions of potassium or cesium are often used for
this purpose.

    The strength of the various oxides, halides, hydrides, and oc-
casionally the carbides of many metal species are listed in Table
III.B.5. Here the greater the bond energy, the harder it is to
break the bond, i.e., the more stable the particular molecule is
in the flame. It has been well documented that flame chemistry is
more important than flame temperature in reducing metal compounds
to the atomic state. Flames with carbonaceous fuels have a decided
reducing advantage. In particular, the nitrous oxide-acetylene
flame, whose stoichiometry tends to produce CO rather than CO_2,
has been found to be extremely reducing in nature. (See Table I.C.2

for analytical flame gas compositions.)

Spectral

All atomic spectrometric methods can be plagued with interferences due to unwanted radiation being emitted or absorbed by either the excitation source or sample cell. Such interferences may be defined as spectral.

1. Direct Spectral Overlap.

It is possible for a transition of an interfering matrix element to very nearly coincide with that of the analyte's transition. If the transitions are less than about 0.3 Å, the profile of the two lines will overlap. (See Chapter III.B.8 for half-intensity line width equations and Table III.B.8 for estimated line widths of common transitions of 53 elements.) When such an overlap exists, it becomes impossible to analyze for the analyte if the interfering species is present in the sample matrix unless a double analysis is performed, a chemical separation of the two species is made, or a different analysis line is used. This is true in all three analytical methods. Observed and predicted direct spectral overlaps are listed in Table I.E.1.

2. Unwanted Radiation from the Excitation Source.

In all line sources, there is a fill gas which usually consists of neon or argon; therefore, spectra from both the atomic and ionic forms of these species can also be emitted. Furthermore, ion lines from the analyte species can be present as well as emission from any impurity present in the excitation source. In all of these cases, unless there is a direct spectral overlap, this unwanted radiation can be excluded from the detection system if sufficient monochromator resolution is utilized. The main problem is to know what lines exist in the neighborhood of the analyte transition and adjust the spectral band pass of the monochromator accordingly.

Each element lamp will have its characteristic spectra and unfortunately, there is no complete catalog of either HCL's or EDL's spectra. However, at least for AAS analysis where HCL's are almost always used with Ne fill gas, a table of potentially interfering neon lines and the necessary resolution required to avoid the problem is provided in Table I.E.2.

It should be pointed out that this type of interference will act as stray radiation in AAS analysis causing both curvature of the analytical curve and poorer limits of detection. In AFS this type of interference will result in scattered radiation with essentially the same outcome.

The problem is magnified with a continuum excitation source and generally higher resolution monochromators are required for both AAS and AFS when they are used.

3. Unwanted Emission or Absorption in the Sample Cell.

In AES, unwanted emission in the sample cell results directly in a signal which must be corrected for. If the radiation arises from the flame components, the correction is a simple matter of subtracting a blank signal from the analytical signal; however, if the origin is in the sample matrix, more work is required. The offending species must be chemically removed from the sample; a scanning technique must be used, or the standard matrix must be made identical with the sample matrix. Of course if the unwanted radiation can be eliminated by using a smaller spectral band pass, that is the simplest cure. Flame emission spectra of the common analytical flames are given in Chapter III.B.8.

In AAS, unwanted emission is not recorded by the readout system due to the a.c. amplification, and such emission will only add noise to the detector. If such radiation exists, the best analytical results will be obtained by using the smallest possible spectral band pass that does not cause a significant increase in gain and therefore electronic amplifier noise.

If unwanted absorbers are present in AAS, a direct effect of the absorption signal results. A simple background correction will eliminate the problem if it is due to molecular species (solid particle scattering will have the effect and correction as molecular absorption). Chapter III.B.8 gives the common molecular species which emit in the flames and Chapter I.C gives the major species present in the common flame gases.

If the unwanted absorption is the result of an atomic species, the cure must lie in better resolution of the monochromator or chemical separation.

In AFS, sample cell emission and absorption can result in a direct effect on the analytical signal. Absorption will only result in a signal if it is accompanied by resultant fluorescence radiation. This is the same problem as in AES and has the same cure.

Table I.E.1

Predicted and Observed Direct Spectral
Overlaps in Atomic Spectroscopy

Analyte	Line (Å)	Interferent	Line (Å)	Peak Separation (Å)
*Aluminum	3082.15	Vanadium	3082.11	0.04
*Antimony	2170.23	Lead	2169.99	0.24
*	2311.47	Nickel	2310.97	0.50
Bismuth	2021.21	Gold	2021.38	0.17
Boron	2497.73	Germanium	2497.96	0.23
*Cadmium	2288.02	Arsenic	2288.12	0.10
*Calcium	4226.73	Germanium	4226.57	0.16
Cobalt	2274.49	Rhenium	2274.62	0.13
	2424.93	Osmium	2424.97	0.04
	2521.36	Tungsten	2521.32	0.04
*		Indium	2521.37	0.01
	3465.80	Iron	3465.86	0.06
	3502.28	Rhodium	3502.52	0.24
	3513.48	Iridium	3513.64	0.16
Copper	2165.09	Platinum	2165.17	0.08
*	3247.54	Europium	3247.53	0.01
Gallium	2944.18	Tungsten	2944.40	0.22
*	4032.98	Manganese	4033.07	0.09
Gold	2427.95	Strontium	2428.10	0.15
Hafnium	2950.68	Niobium	2950.88	0.20
	3020.53	Iron	3020.64	0.11
Indium	3258.56	Osmium	3258.60	0.04
	3039.36	Germanium	3039.06	0.30
Iridium	2088.82	Boron	2088.84	0.02
	2481.18	Tungsten	2481.44	0.26
Iron	2483.27	Tin	2483.39	0.12
*	2719.03	Platinium	2719.04	0.01
Lanthanium	3704.54	Vanadium	3704.70	0.16
Lead	2613.65	Tungsten	2613.82	0.17
*Manganese	4033.07	Gallium	4032.98	0.09
*Mercury	2536.52	Cobalt	2536.49	0.03
Molybdenum	3798.25	Niobium	3798.12	0.13

Predicted and Observed Direct Spectral
Overlaps in Atomic Spectroscopy

Analyte	Line (Å)	Interferent	Line (Å)	Peak Separation (Å)
Osmium	2476.84	Nickel	2476.87	0.03
	2644.11	Titanium	2644.26	0.15
	2714.64	Tantalum	2714.67	0.03
	2850.76	Tantalum	2850.98	0.22
	3018.04	Hafnium	3018.31	0.27
Palladium	3634.70	Ruthenium	3634.93	0.23
Platinum	2274.38	Cobalt	2274.49	0.11
Rhodium	3502.52	Cobalt	3502.62	0.10
Scandium	2980.75	Hafinum	2980.81	0.06
	2988.95	Ruthenium	2988.95	0
	3933.38	Calcium	3933.66	0.28
*Silicon	2506.90	Vanadium	2506.90	0
	2524.11	Iron	2524.29	0.18
Silver	3280.68	Rhodium	3280.60	0.08
Strontium	4215.52	Rubidium	4215.56	0.04
Tantalum	2636.90	Osmium	2637.13	0.23
	2661.89	Iridium	2661.98	0.09
	2691.31	Germanium	2691.34	0.03
Thallium	2918.32	Hafinum	2918.58	0.26
	3775.72	Nickel	3775.57	0.15
Tin	2268.91	Aluminum	2269.10	0.19
	2661.24	Tantalum	2661.34	0.10
	2706.51	Scandium	2706.77	0.26
Titanium	2646.64	Platinium	2646.89	0.25
Tungsten	2656.54	Tantalum	2656.61	0.07
	2718.90	Iron	2719.03	0.13
Vanadium	2507.38	Tantalum	2507.45	0.07
	2526.22	Tantalum	2526.35	0.13
*Zinc	2138.56	Iron	2138.59	0.03
Zirconium	3011.75	Nickel	3012.00	0.25
	3863.87	Molybdenum	3864.11	0.24
	3968.26	Calcium	3968.47	0.21

*These overlaps have been reported in the literature.

Reference: R. J. Lovett, D. L. Welch, & M. L. Parsons, submitted,
J. Appl. Spectrosc.

Table I.E.2

Neon Transitions which
should be Resolved
for Best Analytical Results

Analyte	Line (Å)	Neon Line (Å)	Required Spectral Band Pass (Å)
Chromium	3593.49	3593.53	0.02*
	3578.69	3574.64	2
	3605.33	3600.17	2.5
Copper	3247.54	3232.38	7.5
Dysprosium	4045.99	4042.64	1.6
Erbium	3372.76	3369.91	1.4
Gadolinium	3717.48	3721.86	2.8
	3713.57	3709.64	1.9
Lithium	6707.84	3355.05 in 2nd order is 6710.10	1.1
Lutetium	3359.56	3360.63	0.5
Niobium	4058.94	4042.64	8
Rhenium	3460.46	3460.53	0.03*
	3464.73	3466.58	0.9
	3451.88	3454.19	1.1
Rhodium	3434.89	3440.61	2.8
	3692.36	3694.20	0.9
Ruthenium	3728.03	3721.86	3
Scandium	4023.69	4042.64	9.4
Silver	3382.89	3378.28	2.3
Sodium	5889.95	5881.89	4
	5895.92	5881.89	7
Thulium	3717.92	3721.86	1.9
Titanium	3653.50	3664.11	5.3
	3642.68	3633.67	4.5
	3371.45	3369.81 } 3369.91	0.7
Uranium	3584.88	3593.53	4.3
	3566.60	3568.53	0.9
Ytterbium	3464.46	3466.58	1
Zirconium	3601.19	3600.17	0.5
	3519.60	3520.47	0.4

*These lines exhibit direct spectral overlap.

Reference: R. J. Lovett, D. L. Welch, & M. L. Parsons, submitted,
            J. Appl. Spectrosc.

F.   FIGURES OF MERIT

It is fervently hoped that all analytical techniques reported in the scientific literature have certain unique advantages, at least for a specific application. An analytical chemist must be able to judge between many techniques and, hopefully with true scientific objectivity, be able to choose the one most suitable for his particular problem. This is sometimes easy, but often the analyst is confronted with several seemingly suitable choices. Many factors must be considered: the effect of the other elements in the matrix, the applicability of the proposed technique for handling the matrix in question, i.e. is the sample soluble?; is it homogeneous?; must it be ashed?; etc; and the useful concentration range of the proposed method, i.e., will excessive dilution be required?; will the sample "push" the limit of detection? Indeed, many of these questions are unique to the sample in question; however, the analyst needs an indication from method to method as to the overall merit of the technique. It is the authors' opinion that no ideal figure of merit exists; however, one of the many which are currently used stands out above the rest in providing the maximum information with the least amount of work and presents the analyst with one number--the limit of detection. A properly defined limit of detection provides information concerning the statistical limitations of the entire method--including procedure, experimental apparatus, and experimental technique. This, in combination with a fundamental knowledge of the technique being considered, also gives information as to the linear range of the working curve and certainly information about the sample dilution (or concentration) required. What then is a "properly defined" limit of detection? It is one which is statistically defined in such a way that the confidence of being able to detect the signal originating from the analyte can be specified using a stated number of analytical measurements with a known frequency band pass (time constant) for the readout system. Three items must be stated with each reported limit of detection: the confidence level, the number of experimental measurements, and the frequency band pass (time constant) of the readout system.

It is appropriate at this point to emphasize the reasons for not using 'sensitivity' as defined in AAS, i.e., the concentration which produces a 1% absorption signal. This definition simply describes the slope of the analytical curve, and in no way reflects much of the information required for analytical judgments, especially information about the variation of response, i.e. standard deviation. It is useful only in comparison of information from a particular experimental set-up or in observing the effect of changing a parameter in a particular experimental set-up. It is the authors' opinion that the limit of detection gives this information plus much more.

        The signal-to-noise ratio, S/N, on the other hand, is directly
related to the limit of detection as statistically defined. It is
therefore an equally useful measure of a method--if the three items
of information required for the limit of detection as listed above
are stated. The noise as measured at the readout of an instrument is
a direct measure of the standard deviation for the measurement.
Limits of detection based on S/N are therefore just as valid as those
based on the calculation of the standard deviation.

        It is for the reasons stated above that the preferred figure of
merit in this work is the limit of detection. It is used wherever
possible in the tables of experimental results listed in Chapter II.

        Several precautions should be followed in using the limit of
detection: 1) By definition an analytical measurement is inaccurate
at the limit of detection. 2) Analytical measurements (i.e. precise
measurements) can only be made at approximately an order of magnitude
greater than the limit of detection. 3) For atomic spectroscopic
methods the upper limit of the linear working range is usually from
10 to 100 mg/ml.

        The most convenient way to determine the limit of detection is
to calculate it from experimental data taken from samples containing
about an order of magnitude greater concentration. The S/N is experimentally measured for these solutions, and the fact that the S/N
is a linear function of concentration at very low levels is utilized
to extrapolate to the S/N defined as the limit of detection. This
process is mathematically defined by the following equations:

$$S/N_{pp} = \frac{t\sqrt{2}}{5\sqrt{n}} \quad = \quad 0.2828 \, \frac{t}{\sqrt{n}} \qquad\qquad 1)$$

where N_{pp} is the peak-to-peak noise measured on the recorder
        $(N_{pp} \simeq 1/5 \, N_{RMS})$,

    t is the student t value for the selected confidence level,
        usually 95 or 99%   (There is a table of t values in
        Chapter III.B.9)

    n is the number of measurements of the signal, S, i.e., the
        number of $S_{total}-S_{blank}$ measurements).

        The concentration at the limit of detection is then simply
related to the concentration which was experimentally observed by
Equation 2.

$$C_{LOD} = \frac{S/N_{LOD}}{S/N_{Measured}} \, X \, C_{Measured} \qquad\qquad 2)$$

Reference: M. L. Parsons, J. Chem. Ed., 46, 290 (1969).

CHAPTER II

ATLAS OF SPECTRAL LINES

A. "BEST" ANALYSIS LINES

It is theoretically possible using the equations and data in Chapter III to predict the most sensitive transitions for each element. It is unfortunate that much of the information required for this theoretical exercise is woefully lacking in accuracy, (See Table III.B.1 for instance)thus rendering the theoretical prediction of analysis lines of marginal value. For this reason, it was decided to tabulate the most commonly used analytical transitions for the elements. It should be recognized that common is defined as "vulgar" not necessarily "intelligent." The transitions listed in Tables II.A.1-3 are generally the most sensitivity; however, they are not always the "best" for a particular matrix and/or specific experimental set-up or procedure.

It should be noted, however, that these transitions, by and large, correspond to the first resonance transitions listed in Table I.A.2. These transitions are usually found to be satisfactory when high sensitivity is required; however, when samples of high analyte concentration are to be analyzed it is often desirable to use analysis lines of lesser sensitivity in order to minimize sample dilution. This is one of the reasons for listing the relative sensitivities in Table II.B.3-7.

The choice of analysis line should be coupled with a knowledge of the major matrix elements, and therefore the potential spectral interferences (Table I.E.1 and 2 as well as III.B.8 and the Figures of Chapter III.B.8-a through x) as well as potential chemical problems.

Table II.A.1

COMMON ANALYSIS LINES USED
IN FLAME ATOMIC ABSORPTION SPECTROMETRY

Element	Transition (Å)	Element	Transition (Å)
Aluminum	3092.71 }	Gallium	2874.24
	3092.84		2943.64 }
	3961.52		2944.18
Antimony	2175.81	Germanium	2651.18 }
	2068.33		2651.58
	2311.47		2068.65
Arsenic	1936.36	Gold	2427.95
	1971.97		2675.95
	1890.0	Hafnium	3072.88
Barium	5535.48		2866.37
Beryllium	2348.61		
		Holmium	4103.84
Bismuth	2230.61		4053.93
Boron	2496.77 }	Indium	3256.09
	2497.72		3039.36
Cadmium	2288.02	Iodine	1830
Calcium	4226.73		
		Iridium	2088.82
Cerium	5200.12 }		2639.42 }
	5200.42		2639.71
Cesium	8521.10	Iron	2483.27
	8943.50		2489.75
Chromium	3593.49		
	3578.69	Lanthanum	5501.34
Cobalt	2407.25		5455.15
	2424.93		4187.32
Copper	3247.54	Lead	2169.99
	3273.96		2833.06
		Lithium	6707.84
Dysprosium	4211.72	Lutetium	2615.42 II
	4045.99		3359.56
Erbium	4007.97	Magnesium	2852.13
	3372.76 II	Manganese	2794.82
Europium	4594.03		
	4627.22	Mercury	2536.52
		Molybdenum	3132.59
Gadolinium	4078.70		3170.33
	3783.05	Neodymium	4634.24
	3684.13		4924.53

COMMON ANALYSIS LINES USED
IN FLAME ATOMIC ABSORPTION SPECTROMETRY

Element	Transition (Å)	Element	Transition (Å)
Nickel	2320.03	Sulfur	1807.34
	2310.96	Tantalum	2714.67
Niobium	4079.73	Technetium	2614.23 }
	4058.94		2615.87
	3349.06	Tellurium	2142.75
		Terbium	4326.47
Osmium	2909.06	Thallium	3775.72
Palladium	2447.91	Thorium	3244.46 }
	2476.42		3245.78 II
Phosphorus	1774.99		
	1782.86	Thulium	3717.92
			4105.84
Platinum	2659.45	Tin	2246.05
Potassium	7664.91		2863.33
Praseodymium	4951.36		2354.84
	5133.42		
Rhenium	3460.46	Titanium	3642.68
	2287.51		3653.50
	3464.73	Tungsten	2551.35
			4008.75
Rhodium	3434.89		2681.41
	3692.36		
Rubidium	7800.23	Uranium	3584.88
Ruthenium	3728.03		3566.60
	3498.94	Vanadium	3185.40
			3183.41 }
Samarium	4296.74		3183.98
Scandium	3911.81		
	3907.49	Ytterbium	3987.98
Selenium	1960.26	Yttrium	4102.38
	2039.85		4077.38
			4128.31
Silicon	2516.11		
Silver	3280.68	Zinc	2138.56
	3382.89	Zirconium	3601.19
Sodium	5889.95		3547.68
	5895.92		3519.60
Strontium	4607.33		
	4077.71 II		

Table II.A.2

COMMON ANALYSIS LINES USED
IN FLAME ATOMIC EMISSION SPECTROMETRY

Element	Transition (Å)	Element	Transition (Å)
Aluminum	3961.52	Dysprosium	4186.78
	3944.01		4211.72
Antimony	2598.05 }		4045.99
	2598.09	Erbium	4007.97
	2311.47		4151.10
			5826.79
Arsenic	1936.96		
	2349.84	Europium	4594.03
	2860.44		6018.15
Barium	4554.03 II	Gadolinium	4401.86
	5535.48		4519.66
		Gallium	4172.06
Beryllium	2348.61		
Bismuth	3067.72	Germanium	2651.18 }
	4722.19		2651.58
	2230.61	Gold	2675.95
			2427.95
Boron	2496.78 }	Hafnium	3682.24
	2497.73		5311.60 II
Cadmium	3261.06		
	2288.02	Holmium	4053.93
			4103.84
Calcium	4226.73		6604.94
	3933.67 II	Indium	4511.31
Cerium	5699.23		4101.76
	5697.00		
		Iridium	3800.12
Cesium	8521.10	Iron	3719.94
	8943.50		3859.91
Chromium	4254.35	Lanthanum	5791.34
	5206.04		6578.51
Cobalt	4252.31	Lead	3683.48
	3526.85		4057.83
	3453.50	Lithium	6707.84
Copper	3247.54		3232.63
	3273.96	Lutetium	4514.57
			3312.11
			3359.56

COMMON ANALYSIS LINES USED
IN FLAME ATOMIC EMISSION SPECTROMETRY

Element	Transition (Å)	Element	Transition (Å)
Magnesium	2852.13	Scandium	4020.40
Manganese	4030.76		3911.81
	4033.07		6305.67
Mercury	2536.52	Selenium	2039.85
Molybdenum	3798.20	Silicon	2516.11
	3902.96	Silver	3280.68
	3864.11		3382.89
Neodymium	4924.53	Sodium	5889.95
	4883.81		5895.92
Nickel	3524.54	Strontium	4607.33
	3414.76	Tantalum	4740.16
Niobium	4058.94		4812.75
Osmium	4420.47	Tellurium	4866.2
	2909.06		2142.75
Palladium	3634.70	Terbium	4326.47
	3404.58		4318.85
Phosphorous	5408	Thallium	3775.72
	2535.65		5350.46
		Thorium	5760.55
Platinum	2659.45		4919.82 II
	3064.71		
Potassium	7664.91	Thulium	3717.92
	7698.98		4105.84
		Tin	2839.99
Praseodymium	4939.74		3262.34
	4951.36	Titanium	3653.50
Radium	4825.9		3998.64
	3814.4 II		
		Tungsten	4008.75
Rhenium	3460.46		4074.36
Rhodium	3692.36	Uranium	5415.40
	4374.80	Vanadium	4379.24
Rubidium	7800.23		4111.78
	7947.60	Ytterbium	3987.98
		Yttrium	3620.94
Ruthenium	3728.03		4102.38
Samarium	4883.77 }		6435.00
	4883.97	Zinc	2138.56
	4760.27	Zirconium	3601.19
	4728.44		3519.60

Table II.A.3

COMMON ANALYSIS LINES USED IN
FLAME ATOMIC FLUORESCENCE SPECTROMETRY

Element	Transition (Å)	Element	Transition (Å)
Aluminum	3944.01	Lead	2833.06
	3961.53		4057.83
Antimony	2068.33	Magnesium	2852.13
	2175.81	Manganese	2794.82
	2311.47		2798.27 }
	2598.05		2801.06
Arsenic	1936.96		4030.76
	2349.84		4033.07 }
	2492.91		4034.49
	2898.71		
		Mercury	2536.52
Beryllium	2348.61	Molybdenum	3132.59
Bismuth	2061.70	Nickel	2320.03
	3024.64		3524.54
	3067.72	Palladium	3434.58
Cadimum	2288.02		3609.55
Calcium	4226.73		3634.70
Cesium	8521.10	Platinum	2659.45
			3064.71
Chromium	3578.69		
	3593.49	Rhodium	3434.89
Cobalt	2407.25	Rubidium	7947.60
	2424.93	Ruthenium	3728.03
	2521.36	Scandium	3907.48
Copper	3247.54		
	3273.96	Selenium	1960.26
			2039.85
Gallium	4172.06	Silicon	2516.11
Germanium	2651.18 }	Silver	3280.68
	2651.58		3382.89
Gold	2427.95	Sodium	5895.92
	2675.95	Strontium	4607.33
		Tellurium	2142.75
Indium	3039.36		2383.25 }
	4101.76		2385.76
	4511.31		
Iridum	2543.97	Thallium	3775.72
Iron	2483.27		5350.46

COMMON ANALYSIS LINES USED IN
FLAME ATOMIC FLUORESCENCE SPECTROMETRY

Element	Transition (Å)	Element	Transition (Å)
Tin	2863.33		
	3034.12		
	3175.05		
Titanium	3635.46		
	3653.50		
	3998.64		
Vanadium	3183.98		
Zinc	2138.56		

B.   EXPERIMENTALLY OBSERVED TRANSITIONS

This section contains a tabulation of experimentally observed transitions for the flame spectrometric methods plus non-flame atomic absorption techniques. The tables have all been normalized in format to make them more readable and hopefully more useful. This section is an updated and revised version of the Parsons and McElfresh atlas and includes considerably more information especially in the added section on non-flame atomic absorption and in the atomic fluorescence section. The atomic emission section has added information on nitrous oxide-acetylene flame.

All tables are listed alphabetically according to element name. The wavelengths have all been listed to conform to the NBS Tables with the NSRDS-NBS 4, and NSRDS-NBS 22 given first priority and NBS Monograph 53 second. (These are all three tabulations of atomic transition probabilities and are referenced in Chapter III, Part B). Wherever an ion transition is observed, the usual designation, II, is given. If a molecular band has been observed, the molecule is given, i.e. for boron dioxide the listing is BO_2. The relative signal column (relative absorption in the absorption tables) are all based on a scale of 10 with 10 being the strongest intensity or absorption as the case may be. This information has been calculated from the data given in the original paper and reflects the relative signal, absorption sensitivity or limit of detection depending on the authors' data.

The flame, or sample cell, code is given in Table II.B.1. The excitation source code is given in Table II.B.2. The limit of detection column includes the limit of detection in parts per million given by the individual authors and reflects their experimental conditions and definition of detection limit. It should be remembered that quantitative work can not be accomplished at the limits of detection. Generally, for quantitative analysis, work an order of magnitude above the limits is required. If the author did not measure the limit of detection and supplied only sensitivity data, the number is enclosed by parenthesis. Sensitivity in atomic absorption techniques is defined as the concentration giving one percent absorption. References are given at the end of each table.

Special symbols used throughout the tables are given in Table II.B.3. Finally, in Table II.B.7 on atomic fluorescence lines whenever two wavelengths are given separated only by a slash, the first is the excitation wavelength and the second is the fluorescence wavelength.

Table II.B.1

FLAME TYPE OR SAMPLE CELL CODE

FLAME CELLS

Code	Flame Composition
Ac/A	Acetylene – Air
SAc/A	Separated Acetylene – Air
Ac/N	Acetylene – Nitrous Oxide
SAc/N	Separated Acetylene – Nitrous Oxide
PAc/N	Premixed Acetylene – Nitrous Oxide
Ac/O	Acetylene – Oxygen
PAc/O	Premixed Acetylene – Oxygen
Ac/O/N$_2$	Acetylene – Oxygen – Nitrogen
Ac/O/Ar	Acetylene – Oxygen – Argon
Ac/A/O	Acetylene – Air – Oxygen
H/A	Hydrogen – Air
PH/A	Premixed Hydrogen – Air
PH/Ar	Premixed Hydrogen – Argon-Entrained Air
H/Ar	Hydrogen – Argon – Entrained Air
SH/A	Separated Hydrogen – Air
H/EA	Hydrogen – Entrained Air
H/N$_2$	Hydrogen – Nitrogen-Entrained Air
H/N	Hydrogen – Nitrous Oxide
PH/N	Premixed Hydrogen – Nitrous Oxide
SH/N	Separated Hydrogen – Nitrous Oxide
H/O	Hydrogen – Oxygen
PH/O	Premixed Hydrogen – Oxygen
H/O/Ar	Hydrogen – Oxygen – Argon
H/O/N$_2$	Hydrogen – Oxygen – Nitrogen
SH/O/Ar	Separated Hydrogen – Oxygen – Argon
P/A	Propane – Air
PB/A	Propane – Butane – Air
B/A	Butane – Air
NG/A	Natural Gas – Air
MAPP/N	MAPP (Stablized methylacetylene – Propadiene – Propylene) – Nitrous Oxide
PMAPP/N	Premixed MAPP – Nitrous Oxide
MIBK/A	MIBK (Methyl Isobutyl Ketone) – Air

FLAME TYPE OR SAMPLE CELL CODE

NON-FLAME CELLS

Code	Cell Type
GFL	Graphite Furnace (Large)
GFS	Graphite Furnace (Small)
CRA	Carbon Rod Atomizer
Boat	Metal Boat (Delves Cup)
MF	Metal Filament
ICP	Induction Coupled Plasma
DCP	DC Plasma
DHCL	Demountable Hollow Cathode Lamp
QT	Quartz Tube

Table II.B.2

EXCITATION SOURCE CODE

EDT	Electrodless discharge tube
HIHC	High intensity hollow cathode discharge tube
MVL	Metal vapor discharge arc lamp
Xe1.5	150 watt xenon arc lamp
Xe4.5	450 watt xenon arc lamp
Xe9	900 watt xenon arc lamp
DHC	Demountable hollow cathode discharge tube
SHC	Shielded hollow cathode discharge tube
PTDL	Pulsed tunable dye laser
HFHC	High frequency hollow cathode lamp
HCL	Hollow cathode lamp
TL	Tungsten Lamp
PDHC	Pulsed demountable hollow cathode lamp

Table II.B.3

SPECIAL SYMBOLS CODE

NI	No information available in reference cited.
ND	No detected.
( )	In limit of detection column means sensitivity rather than limit of detection.
II	In wavelength column means ion transition rather than atomic transition.
NFO	No fluorescence observed at the wavelength.

Table II.B.4

EXPERIMENTALLY OBSERVED
ATOMIC ABSORPTION LINES
IN ANALYTICAL FLAMES

Element	Line (Å)	Rel. Abs.	Flame	Lamp Type	Limit of Detection (ppm)	Ref.
Actinium Ac	No information					
Aluminum Al	3092.71 } 3092.84	10	Ac/O/N$_2$	HCL	(0.7)	1
	3961.52	8				
	3082.15	7				
	3944.01	5				
	2373.12 } 2373.35	3				
	2367.05	2				
	2575.10 } 2575.40	1				
	3092.71 } 3092.84	10	Ac/A	HCL	0.5	2
	3961.52	9				
	3082.15	7				
	3944.01	5				
	3092.71 } 3092.84	10	Ac/N	HCL	(0.8)	3
	3961.52	7				
	2373.12 } 2373.35	2				
	2367.05	1				
	2575.10 } 2575.41	0.9				
	2567.98	0.6				
	(3092.71) ?} (3092.84)	10	A/N	(HCL)?	0.02	4
	3961.52	10	PH/O	(HCL)?	350	5
	3092.71 } 3092.84	10	PAc/O	HCL	0.1	6

EXPERIMENTALLY OBSERVED
ATOMIC ABSORPTION LINES
IN ANALYTICAL FLAMES

Element	Line (Å)	Rel. Abs.	Flame	Lamp Type	Limit of Detection (ppm)	Ref.
Aluminum Al Continued	3092.71 } 10 3092.84		MAPP/N	HCL	(100)	7
Americium Am	No information					
Antimony Sb	2175.81	10	Ac/A	HCL	(0.35)	3
	2068.33	8				
	2311.47	5				
	2127.39	1				
	2175.81	10	Ac/A	EDT	0.5	8
	2068.33	7				
	2311.47	5				
	2127.39	0.7				
	(2175.81)?	10	(Ac/A)?	HCL	0.04	4
	2175.81	10	PB/A	HCL	0.1	9
	2175.81	10	H/A	HCL	(0.025)	10
	2311.47	10	PH/Ar	HCL	0.004	11
Argon Ar	No information					
Arsenic As	1936.36	10	PH/A	HCL	(0.9)	3
	1971.97	6				
	1890.0	5				
	1936.96	10	H/N_2	EDT	0.5	8
	1890.0	10				
	1971.97	5				
	1936.36	10	H/N_2	HCL	(0.006)	12
	1971.97	6				
	1890.0	6				

EXPERIMENTALLY OBSERVED
ATOMIC ABSORPTION LINES
IN ANALYTICAL FLAMES

Element	Line (Å)	Rel. Abs.	Flame	Lamp Type	Limit of Detection (ppm)	Ref.
Arsenic As Continued	1936.96	10	H/O/Ar	HCL	0.1	9
	1971.97	10	H/A	HCL	(0.25)	10
	1971.97	10	Ac/O	HCL	(1.5)	13
	3944 (1971.97 2nd order)	10				
	1936.96	8			0.4	
	1890.0	2				
	(1936.96)?	10	PH/Ar	(HCL)?	0.0001	14
Astatine At	No information					
Barium Ba	5535.48	10	Ac/N	HCL	(0.22)	3
	3501.11	0.02				
	(5535.48)?	10	Ac/N	(HCL)?	0.008	4
	5535.48	10	PH/O	(HCL)?	0.3	5
	5535.48	10	MAPP/N	HCL	(10)	7
Berkelium Bk	No information					
Beryllium Be	2348.61	10	Ac/N	HCL	0.002	15
Bismuth Bi	2230.61	10	(Ac/A)?	HCL	0.2	16
	2228.25	5				
	3067.72	3				
	2061.70	1				
	1953.89 1959.48 }	0.8				
	2276.58	0.7				
	2110.26	0.4				
	2021.21	0.1				

EXPERIMENTALLY OBSERVED
ATOMIC ABSORPTION LINES
IN ANALYTICAL FLAMES

Element	Line ($\mathring{A}$)	Rel. Abs.	Flame	Lamp Type	Limit of Detection (ppm)	Ref.
Bismuth Bi Continued	2230.61	10	Ac/A	HCL	(0.24)	3
	2228.25	3				
	3067.72	2				
	2276.58	0.3				
	(2230.61)?	10	Ac/A	(HCL)?	0.025	4
	2230.61	10	MAPP/N	HCL	(4)	7
	2230.61	10	PB/A	HCL	0.05	9
	2230.61	10	H/A	HCL	(0.025)	10
	(2230.61)?	10	PH/A	(HCL)?	0.001	14
Boron B	2496.77 } 2497.72	10	Ac/N	HCL	(9.0)	3
	2088.84 } 2089.57	4				
	(2496.77) } (2497.72)	10	Ac/N	(HCL)?	1.5	4
Bromine Br	No information					
Cadmium Cd	2288.02	10	Ac/A	HCL	(0.014)	3
	3261.06	0.02				
	(2288.02)?	10	Ac/A	(HCL)?	0.002	4
	2288.02	10	PH/O	(HCL)?	0.03	5
	2288.02	10	H/O/Ar	HCL	0.001	9
	2288.02	10	H/A	HCL	(0.001)	10
Calcium Ca	4226.73	10	Ac/A	HCL	(0.023)	3
	2398.56	0.05				

EXPERIMENTALLY OBSERVED
ATOMIC ABSORPTION LINES
IN ANALYTICAL FLAMES

Element	Line (Å)	Rel. Abs.	Flame	Lamp Type	Limit of Detection (ppm)	Ref.
Calcium Ca Continued	(4226.73)?	10	Ac/A	(HCL)?	<0.0005	4
	4226.73	10	Ac/A	HCL	(0.5)	17
	3968.47 II	1				
	3933.67 II	1				
	2398.56 ND					
	4226.73	10	PH/O	(HCL)?	0.02	5
	4226.73	10	MAPP/N	HCL	0.2	7
	4226.73	10	PB/A	HCL	0.002	9
	4226.73	10	H/A	HCL	(0.16)	10
Californium Cf	No information					
Carbon C	No information					
Cerium Ce	5200.12 } 5200.42	10	Ac/N	HCL	13	18
Cesium Cs	8521.10	10	P/A	HCL	(0.12)	3
	8543.50	8				
	4555.36	2				
	4593.18	0.5				
	8521.10	10	Ac/A	HCL	0.008	19
	8521.10	10	PB/A	HCL	0.05	9
Chlorine Cl	No information					
Chromium Cr	3593.49	10	Ac/A	HCL	(0.08)	20
	3578.69	7				
	3605.33	6				

EXPERIMENTALLY OBSERVED
ATOMIC ABSORPTION LINES
IN ANALYTICAL FLAMES

Element	Line (Å)	Rel. Abs.	Flame	Lamp Type	Limit of Detection (ppm)	Ref.
Chromium Cr Continued	4254.35	4	Ac/A	HCL		20
	4274.80	3				
	3578.69	10	Ac/A	HCL	(0.07)	3
	4254.35	3				
	4274.80	3				
	4289.72	1				
	5208.44	0.05				
	5204.52	0.02				
	(3593.49)?	10	Ac/A	(HCL)?	0.003	4
	3578.69	10	H/O	HCL	0.5	21
	3605.33	5				
	4254.35	5				
	4274.80	2				
	4289.72	2				
	3578.69	10	PH/O	(HCL)?	0.05	5
	4254.35	4				
	3578.69	10	PB/A	HCL	0.005	9
	3578.69	10	H/A	HCL	(0.2)	10
Cobalt Co	2424.93	10	Ac/A	HCL	(1)	22
	2521.36	9				
	2411.62	8				
	2407.25	8				
	2432.21	6				
	2415.30	4				
	2435.83	4				
	2528.97	4				
	3044.00	2				
	2535.96	2				
	2439.05	2				
	3453.50	1				
	3526.85	1				
	3465.80	0.9				
	3405.12	0.8				

EXPERIMENTALLY OBSERVED
ATOMIC ABSORPTION LINES
IN ANALYTICAL FLAMES

Element	Line (Å)	Rel. Abs.	Flame	Lamp Type	Limit of Detection (ppm)	Ref.
Cobalt Co Continued	2989.59	0.7	Ac/A	HCL		22
	3513.48	0.7				
	2309.02	0.6				
	3506.32	0.6				
	3431.58	0.6				
	3474.02	0.6				
	2295.23	0.5				
	3575.36	0.5				
	3442.93	0.5				
	3502.28 } 3502.62	0.5				
	2987.16	0.4				
	3529.03	0.4				
	2274.49	0.3				
	2419.12	0.2				
	2407.25	10	Ac/A	HIHC	0.005	23
	2424.93	7				
	2521.36	4				
	2411.62	3				
	3526.85	0.4				
	3453.50	0.4				
	2407.25	10	Ac/A	HCL	(0.07)	3
	3044.00	0.6				
	3465.80	0.2				
	3474.02	0.1				
	3909.93	0.005				
	2407.25	10	H/O	HCL	(0.5)	24
	2424.93	8				
	2411.62	6				
	2521.36	4				
	3412.63	0.6				
	3474.02	0.6				
	3405.12	0.5				
	3455.23	0.5				
	3526.85	0.5				
	3462.80	0.4				
	2407.25	10	P/A	HCL	0.04	25

EXPERIMENTALLY OBSERVED
ATOMIC ABSORPTION LINES
IN ANALYTICAL FLAMES

Element	Line (Å)	Rel. Abs.	Flame	Lamp Type	Limit of Detection (ppm)	Ref.
Cobalt Co Continued	2424.93	6	P/A	HCL		25
	2521.36	4				
	2407.25	10	PB/A	HCL	0.005	9
	2407.25	10	H/A	HCL	(0.014)	10
Copper Cu	3247.54	10	Ac/A	HCL	1.1	26
	3273.96	5				
	2178.94	2				
	2165.09	1				
	2181.72	1				
	2225.70	0.6				
	2024.34	0.2				
	2492.15	0.1				
	2244.27	0.05				
	2441.64	0.02				
	3247.54	10	Ac/A	HCL	(0.04)	3
	3273.96	3				
	2178.94	1				
	2181.72	0.9				
	2225.70	0.3				
	2492.15	0.08				
	2441.64	0.04				
	(3247.54)?	10	Ac/A	(HCL)?	0.001	4
	3247.54	10	PAc/O	HCL	0.05	6
	3247.54	10	PH/O	(HCL)?	0.09	5
	3247.54	10	MAPP/N	HCL	(0.4)	7
	3247.54	10	PB/A	HCL	0.05	9
	3247.54	10	H/A	HCL	(0.0075)	10
Curium Cm	No information					

EXPERIMENTALLY OBSERVED
ATOMIC ABSORPTION LINES
IN ANALYTICAL FLAMES

Element	Line (Å)	Rel. Abs.	Flame	Lamp Type	Limit of Detection (ppm)	Ref.
Dysprosium Dy	4211.72 I	10	Ac/N	HCL	0.4	27
	4045.99 I	9				
	4186.78 I	8				
	4194.85 I	6				
	3531.70 II	5				
	3968.42 II	2				
	3645.41 II	2				
	3944.70 II	2				
	4167.99 I	2				
	4077.98 II	0.6				
	4000.48 II	0.3				
	4211.72	10	Ac/N	HCL	(0.67)	3
	4194.85	5				
	4191.60	0.4				
	4225.14	0.3				
	4218.09	0.1				
	(4211.72)?	10	Ac/N	(HCL)?	0.05	4
Einsteinium Es	No information					
Erbium Er	3372.76 II	10	Ac/N	HCL	0.1	27
	4007.97 I	8				
	3862.82 I	6				
	4151.10 I	6				
	3264.79 II	5				
	3499.11 II	4				
	3892.69 II	3				
	3973.04 } 3972.60	2				
	4087.65 I	2				
	3937.02 I	2				
	3810.33 I	2				
	3312.42 II	2				
	3616.58 II	1				
	3692.64 II	0.9				
	3905.44 I	0.8				
	3944.41 I	0.7				

EXPERIMENTALLY OBSERVED
ATOMIC ABSORPTION LINES
IN ANALYTICAL FLAMES

Element	Line (Å)	Rel. Abs.	Flame	Lamp Type	Limit of Detection (ppm)	Ref.
Erbium Er Continued	4606.62 I	0.6	Ac/N	HCL		27
	4409.35 I	0.4				
	4190.71 I	0.3				
	2985.50 I	0.2				
	4426.77 I	0.2				
	3558.02 I	0.05				
	4007.97 I	10	Ac/N	HCL	(0.73)	3
	3892.69 II	2				
	4087.65 I	1				
	4020.52 I	0.3				
	(4007.97)?	10	Ac/N	(HCL)?	0.04	4
Europium Eu	4594.03 I	10	Ac/N	HCL	0.2	27
	4627.22 I	8				
	4661.88 I	7				
	4129.70 II	4				
	4205.05 II	3				
	3724.94 II	2				
	3210.57 I	0.8				
	3212.81 I	0.7				
	3111.43 I	0.7				
	3334.33 I	0.5				
	4594.03	10	Ac/N	HCL	(0.37)	3
	4594.03	10	Ac/N	HCL	0.03	18
Fermium Fm	No information					
Fluorine F	No information					
Francium Fr	No information					

EXPERIMENTALLY OBSERVED
ATOMIC ABSORPTION LINES
IN ANALYTICAL FLAMES

Element	Line (Å)	Rel. Abs.	Flame	Lamp Type	Limit of Detection (ppm)	Ref.
Gadolinium Gd	4078.70 I	10	Ac/N	HCL	4	27
	3783.05 I	10				
	3684.13 I	10				
	4058.22 I	9				
	4053.64 I	8				
	3717.48 I	8				
	3713.57 I	6				
	4346.46 I } 6					
	4346.62 I					
	4190.78 I	4				
	3679.21 I	4				
	4045.01 I	3				
	3362.23 II	2				
	3945.54 I	2				
	3266.73 I	2				
	3513.65 I	1				
	3358.62 II	1				
	3796.37 II	1				
	3768.39 II	1				
	3654.62 II	1				
	3423.90 I } 0.2					
	3423.92 II					
	3684.13	10	Ac/N	HCL	(24)	3
	4058.22	7				
	4190.78	3				
	(4078.70)?	10	Ac/N	(HCL)?	1.2	4
	3684.13	10	PAc/O	HCL	4	6
Gallium Ga	2874.24	10	Ac/O	HCL	0.05	28
	2943.64 } 10					
	2944.18					
	4172.06	6				
	4032.98	4				
	2500.70	1				
	2450.07	0.8				
	2943.64 } 10		Ac/A	HCL	1.3	29
	2944.18					

EXPERIMENTALLY OBSERVED
ATOMIC ABSORPTION LINES
IN ANALYTICAL FLAMES

Element	Line (Å)	Rel. Abs.	Flame	Lamp Type	Limit of Detection (ppm)	Ref.
Gallium Ga Continued	2874.24	10	Ac/A	HCL		29
	4172.06	8				
	4032.98	6				
	2450.07	1				
	2500.17 } 2500.70	0.6				
	2719.65	0.6				
	2943.64 } 2944.18	10	Ac/A	HCL	(1.0)	3
	2874.24	2				
	2719.65	0.3				
	2943.64 } 2944.18	10	Ac/A	HCL	0.3	29
	2874.24	8				
	4172.06	7				
	2719.65	4				
	4032.98	4				
	2500.17 } 2500.70	1				
	2450.07	0.7				
	(2943.64) } (2944.18) ?	10	Ac/N	(HCL)?	0.05	4
	2874.24	10	PB/A	HCL	(0.07)	9
	2874.24	10	H/A	HCL	(1.2)	10
Germanium Ge	2651.18 } 2651.58	10	Ac/N	HCL	1	15
	2592.54	5				
	2709.63	4				
	2754.59	3				
	2691.34	3				
	2041.69	2				
	2068.65	0.5				
	2497.96	0.5				
	2094.23	0.4				

EXPERIMENTALLY OBSERVED
ATOMIC ABSORPTION LINES
IN ANALYTICAL FLAMES

Element	Line (Å)	Rel. Abs.	Flame	Lamp Type	Limit of Detection (ppm)	Ref.
Germanium Ge Continued	2651.18 } 2651.58	10	Ac/N	HCL	2	30
	2068.65	8				
	2041.69	8				
	2043.76	4				
	2198.70	0.8				
	2497.96	0.5				
	2065.20	<0.0004				
	2651.18 } 2651.58	10	Ac/N	HCL	(1.6)	3
	2691.34	2				
	3039.06	0.5				
	2651.18 } 2651.58	10	Ac/N	HCL	0.1	18
	2651.18 } 2651.58	10	Ac/O/N$_2$	HCL	(6.5)	1
	2592.54	4				
	2709.63	3				
	2754.59	3				
	2691.34	2				
Gold Au	2427.95	10	Ac/A	HCL	0.05	31
	2675.95	6				
	3122.78	0.01				
	2748.26	0.009				
	6278.18	0.002				
	2427.95	10	Ac/A	HCL	(0.14)	3
	2675.95	5				
	(2427.95)?	10	Ac/A	(HCL)?	0.01	4
	2427.95	10	PB/A	HCL	0.02	9
	2427.95	10	H/A	HCL	(0.01)	10
	2427.95	10	MIBK/A	HCL	0.002	32

EXPERIMENTALLY OBSERVED
ATOMIC ABSORPTION LINES
IN ANALYTICAL FLAMES

Element	Line (Å)	Rel. Abs.	Flame	Lamp Type	Limit of Detection (ppm)	Ref.
Hafnium Hf	3072.88	10	Ac/O/N_2	HCL	25	1
	2866.37	8				
	2898.26	4				
	2964.88	3				
	3682.24	3				
	2950.68	2				
	3020.53	2				
	2904.41 } 2904.75	2				
	2940.77	2				
	3777.64	2				
	3072.88	10	Ac/N	HCL	(11)	3
	3682.24	2				
	3777.64	1				
	(3072.88)?	10	Ac/N	(HCL)?	2	4
Helium	No information					
Holmium Ho	4103.84 I	10	Ac/N	HCL	0.3	27
	4053.93 I	8				
	4163.03 I	6				
	4173.23 I	2				
	4040.81 I	2				
	4108.62 I	1				
	4127.16 I	0.9				
	3456.00 II	0.9				
	4101.09 I	0.8				
	4227.04 I	0.4				
	4136.22 I	0.3				
	4254.43 I	0.2				
	3955.73 I	0.2				
	5982.90 I	0.1				
	3796.75 I	<.06				
	3810.73 I	<.06				
	2518.73 II	<.06				
	3233.87 II	<.06				
	3398.98 II	<.06				

EXPERIMENTALLY OBSERVED
ATOMIC ABSORPTION LINES
IN ANALYTICAL FLAMES

Element	Line (Å)	Rel. Abs.	Flame	Lamp Type	Limit of Detection (ppm)	Ref.
Holmium Ho Continued	3416.46 II	<.06	Ac/N	HCL		27
	3484.84 II	<.06				
	3891.02 II	<.06				
	4103.84	10	Ac/N	HCL	(0.87)	3
	4254.43	0.1				
	(4103.84)?	10	Ac/N	(HCL)?	0.04	4
Hydrogen H	No information					
Indium In	3256.09	10	Ac/N	HCL	0.7	29
	3039.36	7				
	4511.31	2				
	4101.76	2				
	3258.56	0.7				
	2710.26	0.7				
	2560.15	0.6				
	2932.63	0.2				
	3256.09	10	Ac/A	HCL	0.3	29
	3039.36	8				
	4511.31	3				
	4101.76	3				
	3258.56	1				
	2560.15	0.9				
	2932.63	0.4				
	2710.26	<.06				
	3256.09	10	Ac/A	HCL	.05	28
	3039.36	10				
	4101.76	4				
	4511.31	3				
	2560.15	0.8				
	2753.88	0.3				
	3039.36	10	Ac/A	HCL	(0.41)	3
	2710.26	0.45				
	(3256.09)?	10	Ac/A	(HCL)?	0.02	4

EXPERIMENTALLY OBSERVED
ATOMIC ABSORPTION LINES
IN ANALYTICAL FLAMES

Element	Line (Å)	Rel. Abs.	Flame	Lamp Type	Limit of Detection (ppm)	Ref.
Indium In Continued	3039.36	10	PB/A	HCL	(0.05)	9
	3039.36	10	H/A	HCL	(0.05)	10
Iodine I	1830	10	Ac/N	EDT	8	33
Iridium Ir	2088.82	10	Ac/A	HCL	2	34
	2639.42 } 2639.71	6				
	2664.79	5				
	2849.72	4				
	2372.77	4				
	2502.98	4				
	2092.63	3				
	2924.79	3				
	2475.12	2				
	2543.97	2				
	2481.18	1				
	3513.64	0.7				
	2661.98	0.6				
	3800.12	0.6				
	2088.82	10	Ac/A	HCL	(0.84)	3
	2639.42 } 2639.71	3				
	2664.79	3				
	2543.97	2				
	(2088.82)?	10	Ac/A	(HCL)?	0.6	4
	2088.82	10	P/A	HCL	(15)	35
	2639.42 } 2639.71	8				
	2372.77	6				
	2849.72	5				
	2664.79	5				
	2924.79	3				
	2639.42 } 2639.71	10	PB/A	HCL	(2.0)	9

EXPERIMENTALLY OBSERVED
ATOMIC ABSORPTION LINES
IN ANALYTICAL FLAMES

Element	Line (Å)	Rel. Abs.	Flame	Lamp Type	Limit of Detection (ppm)	Ref.
Iron Fe	2483.27	10	Ac/A	HCL	(0.10)	20
	2522.85	5				
	2719.02	3				
	3020.49 } 3020.64	3				
	2501.13	2				
	2166.77	2				
	3719.94	1				
	2966.90	1				
	3859.91	0.9				
	2983.57	0.8				
	3440.61	0.6				
	2936.90	0.6				
	2462.65	0.4				
	3824.44	0.1				
	3679.92	0.1				
	2483.27	10	Ac/A	HCL	(0.07)	3
	3719.94	1				
	3859.91	0.7				
	3920.26	0.04				
	2483.27	10	Ac/A	HCL	(0.01)	36
	2488.15	5				
	2518.10	3				
	3020.49 } 3020.64	3				
	2491.16	2				
	2719.02	2				
	3719.94	2				
	2966.90	1				
	3859.91	0.8				
	2524.29	0.7				
	2983.57	0.4				
	2994.43	0.4				
	2973.13 } 2973.24	0.4				
	3047.60	0.3				
	3059.09	0.3				
	2479.78	0.3				

EXPERIMENTALLY OBSERVED
ATOMIC ABSORPTION LINES
IN ANALYTICAL FLAMES

Element	Line (Å)	Rel. Abs.	Flame	Lamp Type	Limit of Detection (ppm)	Ref.
Iron Fe Continued	2472.88 } 2472.91	0.2	Ac/A	HCL		36
	2936.90	0.2				
	3000.95	0.2				
	3581.20	0.2				
	2501.13	0.2				
	2947.88	0.2				
	2489.75	10	Ac/A	HIHC	(2)	37
	3745.56 } 3745.90	3				
	2479.78	3				
	3825.88	2				
	3465.86	2				
	3886.28	1				
	3758.24	0.8				
	3618.77	0.7				
	3608.86	0.6				
	3631.46	0.6				
	3570.10	0.5				
	3827.82	0.5				
	3920.26	0.5				
	(2483.27)?	10	Ac/A	(HCL)?	0.005	4
	2483.27	10	MAPP/N	HCL	(1)	7
	2483.27	10	H/O	HCL	(0.8)	24
	2488.15	7				
	2522.85	6				
	3020.64 } 3021.07	3				
	3719.94	2				
	3719.02	2				
	2524.29	2				
	3859.91	1				
	3737.13	1				
	3440.61 } 3440.99	1				
	2966.90	0.9				
	3734.87	0.8				

EXPERIMENTALLY OBSERVED
ATOMIC ABSORPTION LINES
IN ANALYTICAL FLAMES

Element	Line (Å)	Rel. Abs.	Flame	Lamp Type	Limit of Detection (ppm)	Ref.
Iron Fe	3758.24	0.6	H/O	HCL		24
Continued	2720.90	0.5				
	2483.27	10	PB/A	HCL	(0.005)	9
	2483.27	10	H/A	HCL	(0.015)	10
	2483.27	10	PH/O	(HCL)?	0.06	5
	3719.02	1				
Krypton Kr	No information					
Lanthanum La	5501.34 I	10	Ac/N	HCL	3	27
	4187.32 I	7				
	4949.77 I	6				
	4086.72 II	4				
	3574.43 II	3				
	3649.53 I	2				
	3927.56 I	2				
	4037.21 I	2				
	4079.18 I	2				
	3613.08 I	2				
	4766.89 I	0.9				
	3898.60 I	0.6				
	5158.69 I	0.4				
	4662.51 II	<0.2				
	5501.34 I	10	Ac/N	HCL	(55)	3
	4037.21 I	4				
	3574.43 II	2				
	(5501.34)?	10	Ac/N	(HCL)?	2	4
	5455.15	10	PAc/O	TL	20	6
Lead Pb	2169.99	10	Ac/A	HCL	(0.12)	3
	2833.06	5				

EXPERIMENTALLY OBSERVED
ATOMIC ABSORPTION LINES
IN ANALYTICAL FLAMES

Element	Line (Å)	Rel. Abs.	Flame	Lamp Type	Limit of Detection (ppm)	Ref.
Lead Pb Continued	2614.18	0.3	Ac/A	HCL		3
	2022.02	0.2				
	2053.27	0.03				
	2169.99	10	Ac/A	HCL	0.3	38
	2833.06	6				
	2614.18	0.06				
	(2169.99)?	10	Ac/A	(HCL)	0.01	4
	2169.99	10	H/O	HCL	0.45	39
	2833.06	7				
	4057.83	0.03				
	2169.99	10	H/O	HCL	0.03	40
	2833.06	5				
	4057.83	0.003				
	2169.99	10	Ac/O	HCL	0.03	40
	2833.06	5				
	4057.83	0.001				
	2169.99	10	H/A	HCL	0.005	41
	2833.06	7				
	2169.99	10	MAPP/N	HCL	(2)	7
	2833.06	10	PB/A	HCL	(0.01)	9
Lithium Li	6707.84	10	P/A	HCL	(0.02)	3
	3232.63	0.03				
	6103.64	0.001				
	6707.84	10	NG/A	HCL	0.03	38
	3232.63	0.06				
	(6707.84)?	10	Ac/A	(HCL)?	0.0003	4
	6707.84	10	MAPP/N	HCL	(0.06)	7
	6707.84	10	PB/A	HCL	0.005	9

EXPERIMENTALLY OBSERVED
ATOMIC ABSORPTION LINES
IN ANALYTICAL FLAMES

Element	Line (Å)	Rel. Abs.	Flame	Lamp Type	Limit of Detection (ppm)	Ref.
Lutetium Lu	2615.42 II	10	Ac/N	HCL	3	42
	3359.56 I	6				
	3312.11 I	4				
	3376.50 I	3				
	3567.84 I	3				
	3278.97 I }	3				
	3281.74 I					
	3118.43 I	2				
	3396.82 I	0.9				
	2989.27 I	0.7				
	4518.57 I	0.5				
	3507.39 II	0.4				
	3081.47 I	0.3				
	3359.56	10	Ac/N	HCL	(7.5)	3
	3567.84	6				
	3312.11	4				
	3376.50	4				
	(2615.42 II)?	10	Ac/N	(HCL)?	0.7	4
Magnesium Mg	2852.13 I	10	Ac/A	HCL	(.01)	43
	2795.53 II	0.07				
	2802.70 II	0.04				
	2852.13 I	10	Ac/A	HCL	(0.8)	17
	2795.53 II	2				
	2802.70 II	ND				
	2852.13	10	Ac/A	HCL	(0.004)	3
	2025.82	0.3				
	(2852.13)?	10	Ac/A	(HCL)?	< 0.0001	4
	2852.13	10	PH/O	(HCL)?	0.04	5
	2852.13	10	PAc/O	HCL	0.004	6
	2852.13	10	MAPP/N	HCL	(0.04)	7
	2852.13	10	PB/A	HCL	0.0003	9

EXPERIMENTALLY OBSERVED
ATOMIC ABSORPTION LINES
IN ANALYTICAL FLAMES

Element	Line (Å)	Rel. Abs.	Flame	Lamp Type	Limit of Detection (ppm)	Ref.
Magnesium Mg Continued	2852.13	10	H/A	HCL	(0.002)	10
Manganese Mn	2794.82	10	Ac/A	(HCL)?	(1.3)	44
	2798.27	8				
	2801.06	5				
	4030.76	0.8				
	2794.82	10	Ac/A	HCL	(0.03)	3
	4030.76	0.8				
	3216.95	0.004				
	(2794.82)?	10	Ac/A	(HCL)?	0.002	4
	2794.82	10	H/O	HCL	(0.25)	24
	2798.27	8				
	2801.06	5				
	4030.76	0.8				
	2794.82 } 2798.27 } 2801.06	10	MAPP/N	HCL	(0.2)	7
	2794.82	10	PB/A	HCL	0.002	9
	2794.82	10	H/A	HCL	(0.005)	10
Mendelevium Md	No information					
Mercury Hg	2536.52	10	Ac/A	HCL	(2.2)	3
	(2536.52)?	10	Ac/A	(HCL)?	0.25	4
	2536.52	10	MAPP/N	HCL	(40)	7
	2536.52	10	PB/A	HCL	0.5	9
	2536.52	10	H/A	HCL	(0.03)	10

EXPERIMENTALLY OBSERVED
ATOMIC ABSORPTION LINES
IN ANALYTICAL FLAMES

Element	Line (Å)	Rel. Abs.	Flame	Lamp Type	Limit of Detection (ppm)	Ref.
Molybdenum Mo	3132.59	10	Ac/A	HCL	0.5	45
	3170.33	7				
	3798.25	6				
	3193.97	6				
	3864.11	5				
	3902.96	3				
	3158.16	3				
	3208.83	1				
	3112.12	0.4				
	3132.59	10	Ac/A	HCL	(0.4)	3
	3208.83	0.8				
	(3132.59)?	10	Ac/A	(HCL)?	0.02	4
	3132.59	10	MAPP/N	HCL	(2)	7
	3132.59	10	PB/A	HCL	0.03	9
	3132.59	10	H/A	HCL	(> 50)	10
Neodymium Nd	4634.24	10	Ac/N	HCL	(35)	1
	4896.93	8				
	4719.02	5				
	4924.53	10	Ac/N	HCL	(8.8)	3
	4866.74	1				
	4634.24	10	Ac/N	HCL	0.6	18
Neon Ne	No information					
Neptunium Np	No information					
Nickel Ni	2320.03	10	Ac/A	HCL	(0.13)	38
	2310.96	3				
	2345.54	0.6				

EXPERIMENTALLY OBSERVED
ATOMIC ABSORPTION LINES
IN ANALYTICAL FLAMES

Element	Line (Å)	Rel. Abs.	Flame	Lamp Type	Limit of Detection (ppm)	Ref.
Nickel Ni Continued	3524.54	0.5	Ac/A	HCL		38
	3414.76	0.4				
	3050.82	0.4				
	3002.49	0.2				
	3461.65	0.2				
	2320.03	10	(Ac/A)?	(HCL)?	(0.12)	20
	2310.96	7				
	2345.54	3				
	2289.98	2				
	2337.49	0.6				
	3369.57	0.5				
	2347.52	0.4				
	3232.96	0.3				
	3391.05	0.3				
	3437.28	0.2				
	2476.87	0.04				
	2320.03	10	Ac/A	HCL	(0.07)	3
	3414.76	2				
	3524.54	2				
	3515.05	0.9				
	3624.73	0.02				
	(2320.03)?	10	Ac/A	(HCL)?	0.002	4
	2320.03	10	H/O	HCL	(0.4)	24
	(2310.96)?	7				
	3414.76	3				
	2345.54	3				
	3002.49	2				
	3524.54	2				
	3054.32	2				
	3437.28	0.9				
	3101.55 } 3101.88	0.8				
	3433.56	0.8				
	3037.94	0.7				
	3392.99	0.7				
	3012.00	0.5				
	3057.64	0.4				

EXPERIMENTALLY OBSERVED
ATOMIC ABSORPTION LINES
IN ANALYTICAL FLAMES

Element	Line (Å)	Rel. Abs.	Flame	Lamp Type	Limit of Detection (ppm)	Ref.
Nickel Ni continued	3446.26	0.4	H/O	HCL		24
	2320.03	10	PB/A	HCL	0.005	9
	2320.03	10	H/A	HCL	(0.016)	10
Niobium Nb	4079.73 I	10	Ac/N	HIHC	5	15
	4058.94 I	10				
	4100.40 I 4100.92 I}	8				
	4123.81 I	8				
	3726.24 I 3727.23 I}	6				
	3790.15 I 3791.21 I}	5				
	4168.13 I	5				
	3739.80 I	4				
	4163.66 I 4164.66 I}	4				
	3798.12 I	3				
	4152.58 I	3				
	3787.06 I 3787.48 I}	2				
	3759.55 I 3760.64 I}	2				
	3802.92 I	2				
	4137.10 I 4137.59 I}	1				
	2697.06 II	0.8				
	3741.78 I 3742.39 I}	0.6				
	3163.40 II	0.3				
	3130.79 II	0.3				
	3094.18 II	0.3				
	2927.81 II	<0.1				
	2950.88 II	<0.1				
	3349.06	10	Ac/N	HCL	(1)	3
	3580.27	9				

EXPERIMENTALLY OBSERVED
ATOMIC ABSORPTION LINES
IN ANALYTICAL FLAMES

Element	Line (Å)	Rel. Abs.	Flame	Lamp Type	Limit of Detection (ppm)	Ref.
Niobium Nb Continued	4079.73	8	Ac/N	HCL		3
	4058.94	8				
	4058.94	10	PAc/O	HCL	3	6
Nitrogen N	No information					
Nobelium No	No information					
Osmium Os	2909.06	10	Ac/N	HCL	0.5	46
	3058.66	6				
	2637.13	6				
	3018.04	3				
	3301.56	3				
	2714.64	2				
	2806.91	2				
	2644.11	2				
	4420.47	0.6				
	4260.85	0.3				
	2909.06	10	Ac/N	HCL	1	47
	3058.66	6				
	3018.04	4				
	3301.56	3				
	3267.94	2				
	3232.06	0.7				
	3262.29	0.4				
	3030.70	0.1				
	(2909.06)?	10	Ac/N	(HCL)?	0.08	4
	2909.06	10	Ac /A	HCL	(1.2)	3
	4260.85	0.6				
	2909.06	10	PB/A	HCL	(17)	48
	2838.17 } 2838.63	9				
	3058.66	8				
	3232.06	7				

EXPERIMENTALLY OBSERVED
ATOMIC ABSORPTION LINES
IN ANALYTICAL FLAMES

Element	Line (Å)	Rel. Abs.	Flame	Lamp Type	Limit of Detection (ppm)	Ref.
Osmium Os Continued	3262.29 } 3262.75	4	PB/A	HCL		48
Oxygen O	No information					
Palladium Pd	2447.91 2476.42 3404.58 2763.09	10 10 3 3	Ac/A	HCL	0.02	20
	2447.91 2476.42 3404.58	10 7 1	Ac/A	HCL	(0.12)	3
	2447.91 2476.42 2763.09 3404.58 3242.70 3634.70 3609.55	10 8 2 2 1 1 0.6	Ac/A	HIHC	0.03	49
	2447.91 2476.42 2763.09 3404.58 3242.70 3634.70 3609.55	10 6 2 2 1 0.8 0.5	H/A	HIHC	0.15	49
	2476.42	10	H/A	HCL	(0.01)	10
	2476.42	10	PB/A	HCL	0.03	9
	2447.91 2476.42	10 6	P/A	HCL	∿2	50

EXPERIMENTALLY OBSERVED
ATOMIC ABSORPTION LINES
IN ANALYTICAL FLAMES

Element	Line (Å)	Rel. Abs.	Flame	Lamp Type	Limit of Detection (ppm)	Ref.
Phosphorus P	3135.47 } 3136.20	10	Ac/N	HCL	290	51
	2149.11	5				
	1774.99	10	SAc/N	EDT	29	52
	1782.86	9				
	1787.68	5				
	(requires vacuum monochromator)					
Platinum Pt	2659.45	10	Ac/A	HCL	0.1	20
	2174.67	7				
	3064.71	5				
	2628.03	4				
	2144.23	3				
	2830.30	3				
	2929.79	3				
	2733.96	2				
	2702.40	2				
	2705.89	2				
	2487.17	2				
	2646.89	2				
	2997.97	2				
	2067.50	2				
	2719.04	1				
	2677.15	1				
	2467.44	0.9				
	3042.64	0.6				
	2049.37	0.4				
	2450.97	0.4				
	2440.06	0.2				
	2165.17	0.07				
	3315.05	0.06				
	2659.45	10	Ac/A	HCL	0.5	53
	3064.71	4				
	2174.67	1				
	2628.03	0.2				
	2830.30	0.2				

EXPERIMENTALLY OBSERVED
ATOMIC ABSORPTION LINES
IN ANALYTICAL FLAMES

Element	Line (Å)	Rel. Abs.	Flame	Lamp Type	Limit of Detection (ppm)	Ref.
Platinum Pt Continued	2659.45	10	Ac/A	HCL	(1.2)	3
	2997.97	2				
	(2659.45)?	10	Ac/A	(HCL)?	0.05	4
	2659.45	10	Ac/N	(HCL)	2.0	5
	3064.71	5				
	2628.03	4				
	2174.67	4				
	2830.30	3				
	2659.45	10	H/O	HCL	10	21
	2929.79	2				
	3064.71	0.3				
	2997.97	10	P/A	HCL	5	55
	2646.89	10				
	2659.45	10	PB/A	(HCL)	0.1	9
	2659.45	10	H/A	(HCL)	(0.5)	10
Plutonium Pu	No information					
Polonium Po	No information					
Potassium K	7664.91	10	P/A	HCL	(0.011)	3
	7698.98	3				
	4044.15	0.03				
	7664.91	10	PB/A	HCL	0.005	9
	7664.91	10	NG/A	HCL	(0.03)	56
	4044.14	0.06				
	7664.91	10	H/A	HCL	(0.002)	10
	(7664.91)?	10	Ac/A	(HCL)?	<0.002	4

EXPERIMENTALLY OBSERVED
ATOMIC ABSORPTION LINES
IN ANALYTICAL FLAMES

Element	Line (Å)	Rel. Abs.	Flame	Lamp Type	Limit of Detection (ppm)	Ref.
Praseodymium Pr	4951.36	10	Ac/N	HCL	10	27
	4914.03	7				
	5133.42	6				
	5045.53	3				
	4951.36	10	Ac/N	HCL	(72)	1
	5133.42	8				
	4736.69	5				
	4951.36	10	Ac/N	HCL	(20)	3
	5133.42	5				
	(4951.36)?	10	Ac/N	(HCL)?	5	4
	4951.36	10	PH/O	HCL	6	6
Promethium Pm	No information					
Protactinium Pa	No information					
Radium Ra	No information					
Radon Rn	No information					
Rhenium Re	3460.46	10	Ac/N	HCL	(12)	1
	3464.73	6				
	3451.88	4				
	3460.46	10	Ac/N	HCL	(11)	3
	3464.73	6				
	3451.88	4				
	2287.51	10	Ac/N	HCL	(20)	57
	2294.49	8				
	2274.62	7				

EXPERIMENTALLY OBSERVED
ATOMIC ABSORPTION LINES
IN ANALYTICAL FLAMES

Element	Line (Å)	Rel. Abs.	Flame	Lamp Type	Limit of Detection (ppm)	Ref.
Rhenium Re Continued	3460.46	4	Ac/N	HCL		57
	2419.81	0.2				
	2428.58	0.07				
	3460.46	10	Ac/N	HCL	(3.5)	58
	3460.46	10	Ac/N	HCL	0.2	18
	3460.46	10	Ac/O	HCL	25	59
	3464.73	10				
	3451.88	5				
	3460.46	10	Ac/O	HCL	1	6
Rhodium Rh	3434.89	10	Ac/A	HCL	0.03	60
	3692.36	6				
	3396.85	4				
	3502.52	3				
	3657.99	2				
	3700.91	1				
	3507.32	0.4				
	3434.89	10	Ac/A	HCL	(0.17)	3
	3280.55	0.2				
	(3434.89)?	10	Ac/A	(HCL)?	0.004	4
	3434.89	10	Ac/N	HCL	0.7	61
	3692.36	5				
	3396.85	4				
	3502.52	3				
	3507.32	3				
	3657.99	3				
	3700.91	1				
	3434.89	10	PB/A	HCL	0.03	9
	3434.89	10	H/A	HCL	(0.02)	10

EXPERIMENTALLY OBSERVED
ATOMIC ABSORPTION LINES
IN ANALYTICAL FLAMES

Element	Line (Å)	Rel. Abs.	Flame	Lamp Type	Limit of Detection (ppm)	Ref.
Rubidium Rb	7800.23	10	P/A	HCL	(0.04)	3
	7947.60	4				
	4201.85	0.1				
	4215.56	0.03				
	7800.23	10	NG/A	HCL	(0.1)	56
	4201.85	0.1				
	7800.23	10	Ac/A	HCL	0.0003	19
	7800.23	10	PB/A	HCL	0.005	9
Ruthenium Ru	3728.03	10	Ac/A	TL	(0.25)	62
	3498.94	2				
	3498.94	10	Ac/A	HCL	(0.78)	3
	3925.92	0.9				
	?	10	Ac/A	(HCL)?	0.07	4
	3498.94	10	PB/A	HCL	0.3	9
	3498.94	10	P/A	HCL	<2	63
	3728.03	6				
Samarium Sm	4296.74	10	Ac/N	HCL	5	27
	5200.59	6				
	4760.27	4				
	4728.42	4				
	5282.91	3				
	5117.16	3				
	4883.77 } 4883.97	3				
	4783.10	3				
	4581.58	2				
	5271.40	1				
	5403.70	0.8				
	5341.29	0.6				
	5320.60	0.6				
	4296.74	10	Ac/N	HCL	(7.5)	3
	4760.27	6				

EXPERIMENTALLY OBSERVED
ATOMIC ABSORPTION LINES
IN ANALYTICAL FLAMES

Element	Line (Å)	Rel. Abs.	Flame	Lamp Type	Limit of Detection (ppm)	Ref.
Samarium Sm Continued	(4296.74)?	10	Ac/N	(HCL)	2	4
Scandium Sc	3911.81	10	Ac/N	HCL	(0.8)	1
	3907.49	7				
	4023.69	7				
	4020.40	5				
	3269.91	3				
	3911.81	10	Ac/N	HCL	0.2	27
	3907.49	10				
	4023.69	7				
	4020.40	5				
	4054.55	4				
	3269.91	3				
	4082.40	1				
	3273.63	0.8				
	3911.81	10	Ac/N	HCL	(0.38)	3
	3273.63	4				
	(3269.91)?	2				
	3911.81	10	Ac/N	HCL	0.02	18
	3911.81	10	PH/O	TL	0.2	6
	3911.81	10	Ac/O	Xe1.5	(5)	59
	3907.49	10				
	4023.69	10				
	4020.40	10				
	3269.91	5				
	3273.63	5				
	4054.55	1				
	4082.40	1				
	3255.69	1				
	3996.61	1				
	3933.38	0.5				
Selenium Se	1960.26	10	Ac/A	EDT	0.1	8
	2039.85	2				
	2062.79	0.4				

EXPERIMENTALLY OBSERVED
ATOMIC ABSORPTION LINES
IN ANALYTICAL FLAMES

Element	Line (Å)	Rel. Abs.	Flame	Lamp Type	Limit of Detection (ppm)	Ref.
Selenium Se Continued	1960.26	10	Ac/A	HCL	1	64
	2039.85	1				
	2062.79	0.2				
	2074.79	0.1				
	(1960.26)?	10	Ac/A	(HCL)?	0.05	4
	1960.26	10	H/A	HCL	(0.5)	3
	2039.85	0.7				
	2039.85	10	H/A	HCL	(0.25)	10
	1960.26	10	P/A	EDT	(1)	65
	2039.85	5				
	2062.79	1				
	2074.79	0.4				
	2164.00	0.4				
	1960.26	10	PB/A	HCL	0.1	9
	1960.26	10	H/Ar	HCL	(0.02)	66
	(1960.26)?	10	PH/Ar	(HCL)?	0.001	14
Silicon Si	2516.11	10	Ac/N	HCL	0.6	67
	2506.90	4				
	2528.51	3				
	2514.32	3				
	2524.11	3				
	2216.67	3				
	2519.20	2				
	2210.89	2				
	2207.98	0.7				
	2516.11	10	Ac/N	HCL	(1.7)	3
	2506.90	3				
	2514.32	3				
	2524.11	3				
	2881.58	0.6				
	(2516.11)?	10	Ac/N	(HCL)?	0.02	4

EXPERIMENTALLY OBSERVED
ATOMIC ABSORPTION LINES
IN ANALYTICAL FLAMES

Element	Line (Å)	Rel. Abs.	Flame	Lamp Type	Limit of Detection (ppm)	Ref.
Silicon Si Continued	2516.11	10	Ac/O/N$_2$	HCL	10	1
	2514.32	8				
	2519.20	7				
	2506.90	4				
	2528.51	4				
	2524.11	3				
Silver Ag	3280.68	10	Ac/A	HCL	(0.035)	3
	3382.89	5				
	(3280.68)?	10	Ac/A	(HCL)?	0.002	4
	3280.68	10	NG/A	HCL	(.05)	56
	3382.89	3				
	3280.68	10	PH/O	(HCL)?	0.06	5
	3280.68	10	PB/A	HCL	0.005	9
	3280.68	10	H/A	HCL	(0.002)	10
Sodium Na	5889.95 } 5895.92	10	Ac/A	(HCL)?	0.001	68
	3302.37 } 3302.98	0.04				
	(5889.95)?	10	Ac/A	(HCL)?	<0.0002	4
	5889.95	10	P/A	HCL	(0.004)	3
	5895.92	4				
	3302.37 } 3302.98	0.02				
	5889.95	10	PB/A	HCL	0.002	9
	5889.95	10	H/O	HCL	1	21
	5895.92	10				
	3302.32	0.3				

EXPERIMENTALLY OBSERVED
ATOMIC ABSORPTION LINES
IN ANALYTICAL FLAMES

Element	Line (Å)	Rel. Abs.	Flame	Lamp Type	Limit of Detection (ppm)	Ref.
Sodium Na Continued	5889.95	10	PH/O	(HCL)?	0.001	5
	5889.95	10	MAPP/N	HCL	(0.1)	7
	5889.95	10	H/A	HCL	(0.001)	10
Strontium Sr	4607.33 I	10	Ac/A	HCL	(0.2)	17
	4077.71 II	5				
	4215.52 II	0.7				
	2569.47 I	0.5				
	2931.83 I	0.2				
	2428.10 I	0.1				
	6892.59 I	ND				
	(4607.33)?	10	Ac/A	(HCL)?	0.002	4
	4607.33	10	Ac/N	HCL	(0.05)	3
	4607.33	10	PH/O	(HCL)?	0.4	5
	4607.33 I	10	NG/A	HCL	(0.15)	56
	4077.71 II	0.6				
	4607.33	10	MAPP/N	HCL	(0.4)	7
	4607.33	10	PB/A	HCL	0.01	9
	4607.33	10	H/A	HCL	(0.16)	10
Sulfur S	1807.34	10	Ac/N	(HCL)?	30	69
	(requires vacuum monochromator)					
Tantalum Ta	2714.67 I	10	Ac/O/N_2	HCL	(30)	1
	2607.84 II					
	2608.20 I	5				
	2608.63 I					
	2775.88 I	5				
	2559.43 I	4				
	2647.47 I	3				

EXPERIMENTALLY OBSERVED
ATOMIC ABSORPTION LINES
IN ANALYTICAL FLAMES

Element	Line (Å)	Rel. Abs.	Flame	Lamp Type	Limit of Detection (ppm)	Ref.
Tantalum Ta Continued	2661.34 I } 2661.89 I	2	Ac/O/N$_2$	HCL		1
	2758.31 I	2				
	2714.67	10	Ac/N	HCL	(10)	3
	2758.31	2				
	(2714.67)?	10	Ac/N	(HCL)?	1	4
Technetium Tc	2614.23 } 2615.87	10	Ac/A	HCL	0.9	70
	2608.86	3				
	4297.06	2				
	4262.27	1				
	3182.37	1				
	4238.19	1				
	3636.07	1				
	3173.30	0.1				
	3466.28	0.1				
	4031.36	0.1				
Tellurium Te	2142.75	10	Ac/A	HCL	1	71
	2259.04	2				
	2383.25	0.5				
	2530.70	0.06				
	(2769) ?	0.06				
	2142.75	10	Ac/A	HCL	(0.3)	3
	2259.04	0.6				
	2385.76	0.07				
	(2142.75)?	10	Ac/A	(HCL)?	0.05	4
	2142.75	10	H/A	HCL	(0.4)	72
	2259.04	1				
	2385.76	0.05				
	2142.75	10	H/A	HCL	0.025	10
	2142.75	10	PB/A	HCL	0.1	9

EXPERIMENTALLY OBSERVED
ATOMIC ABSORPTION LINES
IN ANALYTICAL FLAMES

Element	Line (Å)	Rel. Abs.	Flame	Lamp Type	Limit of Detection (ppm)	Ref.
Terbium Tb	4326.47	10	Ac/N	HCL	2	27
	4318.85	8				
	3901.35	6				
	4061.59	6				
	4338.45	5				
	4105.37	3				
	4326.47	10	Ac/N	HCL	(11)	3
	4318.85	5				
	4338.45	4				
	(4326.47)?	10	Ac/N	(HCL)?	0.6	4
Thallium Tl	3775.72	10	Ac/A	HCL	(1.2)	17
	2767.87	5				
	2580.14	5				
	2379.69	3				
	2826.16	1				
	5350.46	1				
	2709.23 } 2710.67	0.4				
	2918.32 } 2921.52	0.4				
	2315.98	ND				
	2608.99	ND				
	2665.57	ND				
	3229.75	ND				
	2767.87	10	Ac/A	HCL	(0.30)	3
	2580.14	0.13				
	(2767.87)?	10	Ac/A	(HCL)?	0.03	4
	2767.87	10	NG/A	HCL	(0.8)	56
	3775.72	3				
	2767.87	10	PB/A	HCL	0.03	9
	2767.87	10	H/A	HCL	0.02	10

EXPERIMENTALLY OBSERVED
ATOMIC ABSORPTION LINES
IN ANALYTICAL FLAMES

Element		Line (Å)	Rel. Abs.	Flame	Lamp Type	Limit of Detection (ppm)	Ref.
Thorium	Th	3244.46 I 3245.78 II}	10	Ac/N	HCL	181	18
Thulium	Tm	3717.92 I	10	Ac/N	HCL	0.2	42
		4105.84 I	7				
		3744.07 I	6				
		4094.19 I	6				
		4187.62 I	5				
		4203.73 I	3				
		3883.13 I 3883.44 II}	3				
		3887.45 I					
		4359.93 I	1				
		2973.22 I 2973.39 II}	0.8				
		3410.05 I	0.7				
		5307.12 I	0.5				
		4386.43 I	0.4				
		4733.34 I	0.3				
		3761.91 II	0.3				
		3996.62 I	0.3				
		3291.00 II	0.2				
		5675.85 I	0.2				
		3916.48 I	0.2				
		5631.40 I	0.1				
		3949.28 I	0.1				
		3416.59 I	0.06				
		5964.30 I	0.04				
		3717.91	10	Ac/N	HCL	(0.3)	3
		4203.73	3				
		4359.93	1				
		5307.12	0.6				
		(3717.91)?	10	Ac/N	(HCL)?	0.01	4
Tin	Sn	2246.05	10	H/A	HIHC	0.1	73
		2863.33	6				
		2354.84	5				

EXPERIMENTALLY OBSERVED
ATOMIC ABSORPTION LINES
IN ANALYTICAL FLAMES

Element	Line (Å)	Rel. Abs.	Flame	Lamp Type	Limit of Detection (ppm)	Ref.
Tin Sn Continued	2706.51	2	H/A	HIHC		73
	2546.55	2				
	3034.12	2				
	2839.99	1				
	2429.49	1				
	3009.14	1				
	2268.91	0.8				
	3175.05	0.6				
	2199.34	0.5				
	2334.80	0.5				
	2661.24	0.3				
	2209.65	0.3				
	2483.39	0.2				
	2073.08	0.2				
	2286.68	0.07				
	2246.05	10	H/A	EDT	0.5	8
	2863.33	7				
	2354.84	5				
	2546.55	3				
	2429.49	1				
	2246.05	10	H/A	HCL	(0.4)	3
	2334.80	7				
	2661.24	0.3				
	2246.05	10	H/A	HCL	2	74
	2863.33	4				
	2354.84	3				
	2246.05	10	H/A	HCL	(0.04)	10
	2246.05	10	Ac/A	HIHC	0.4	73
	2354.84	7				
	2706.51	6				
	2863.33	6				
	2839.99	4				
	2429.49	3				
	3034.12	3				
	2268.91	2				
	2546.55	2				

EXPERIMENTALLY OBSERVED
ATOMIC ABSORPTION LINES
IN ANALYTICAL FLAMES

Element	Line (Å)	Rel. Abs.	Flame	Lamp Type	Limit of Detection (ppm)	Ref.
Tin Sn Continued	3175.05	2	Ac/A	HIHC		73
	3009.14	2				
	2199.34	0.8				
	2334.80	0.7				
	2209.65	0.6				
	2661.24	0.4				
	2483.39	0.3				
	2073.08	0.1				
	2286.68	0.1				
	2246.05	10	Ac/A	EDT	0.5	8
	2354.84	7				
	2863.33	6				
	2429.49	3				
	2546.55	2				
	2839.99	--				
	2246.05	10	Ac/A	HCL	2	74
	2354.84	8				
	2863.33	7				
	(2246.05)?	10	Ac/A	(HCL)?	0.01	4
	2246.05	10	Ac/N	HCL	NI	75
	2706.51	4				
	2334.80	2				
	2421.70	0.4				
	2354.84	10	Ac/N	HCL	2	74
	2246.05	7				
	2863.33	6				
	2246.05	10	PH/O	(HCL)?	0.6	5
	2246.05	10	PB/A	HCL	0.03	9
Titanium Ti	3653.50 I	10	Ac/N	HIHC	0.5	76
	3642.68 I	9				
	3199.92 I	8				
	3371.45 I	8				
	3752.86 I	6				
	3191.99 I	6				

EXPERIMENTALLY OBSERVED
ATOMIC ABSORPTION LINES
IN ANALYTICAL FLAMES

Element	Line (Å)	Rel. Abs.	Flame	Lamp Type	Limit of Detection (ppm)	Ref.
Titanium Ti Continued	3741.06 I } 3741.64 II	6	Ac/N	HIHC		76
	3354.64 I	6				
	3186.45 I	5				
	2956.13 I	5				
	2646.64 I	4				
	3341.88 I } 3341.88 II	4				
	3998.64 I	4				
	2948.26 I	4				
	2644.26 I	4				
	2641.10 I	3				
	3989.76 I	3				
	2942.00 I } 2942.90 II	3				
	3635.20 I } 3635.46 I	3				
	3958.21 I	3				
	3729.82 I	3				
	2611.48 I	3				
	3981.76 I } 3982.48 I	3				
	3377.48 I } 3377.58 I	2				
	2605.15 I	2				
	3956.34 I	2				
	3948.67 I	2				
	3947.78 I	0.8				
	3642.68	10	Ac/N	HCL	(1.7)	3
	3653.50	9				
	3989.76	4				
	3642.68	10	Ac/N	HCL	0.01	18
	3642.68	10	PAc/O	HCL	0.5	6
	3653.50 I	10	Ac/O/N$_2$	HCL	(3)	1
	3642.68 I	10				
	3354.64 I	9				
	3199.92 I	9				

EXPERIMENTALLY OBSERVED
ATOMIC ABSORPTION LINES
IN ANALYTICAL FLAMES

Element	Line (Å)	Rel. Abs.	Flame	Lamp Type	Limit of Detection (ppm)	Ref.
Titanium Ti Continued	3635.20 I } 3635.46 I } 8		$Ac/O/N_2$	HCL		1
Tungsten W	2551.35	10	Ac/N	HCL	3	27
	2681.41	8				
	2944.40	7				
	2946.98 } 2947.38 } 5					
	4008.75	4				
	2831.38	4				
	2724.35	4				
	2896.45	4				
	2879.39	3				
	2656.54	3				
	4302.11	3				
	4074.36	2				
	4294.61	2				
	2606.39 } 2606.90 } 2					
	3617.52	1				
	2911.00	1				
	4269.39	0.2				
	2762.34	0.08				
	2452.00	0.08				
	3191.57	0.08				
	3768.45	0.08				
	2551.35	10	Ac/N	HCL	(5.3)	3
	4008.75	3				
	4074.36	1				
	2551.35	10	Ac/N	HCL	0.7	18
	4008.75	10	PAc/O	HCL	3	6
	2551.35	10	$Ac/O/N_2$	HCL	(17)	1
	2681.41	9				
	2724.35	6				
	4008.75	6				
	2946.98 } 2947.38 } 6					

EXPERIMENTALLY OBSERVED
ATOMIC ABSORPTION LINES
IN ANALYTICAL FLAMES

Element	Line (Å)	Rel. Abs.	Flame	Lamp Type	Limit of Detection (ppm)	Ref.
Tungsten W Continued	2831.38 } 2896.45	5	Ac/O/N$_2$	HCL		1
	2656.54	4				
	2718.90	4				
	4074.36	4				
	2944.40	3				
Uranium U	3584.88 I	10	Ac/N	HCL	30	27
	3566.60 I	7				
	3514.61 I	4				
	3943.82 I	4				
	3489.37 I	3				
	4153.97 I	3				
	3659.16 I	3				
	4042.76 I	3				
	3550.82 II	2				
	3500.07 I	2				
	3507.34 I	2				
	3890.36 II	2				
	3638.20 I	1				
	4050.05 II	0.8				
	4341.69 II	0.6				
	3670.07 II	0.4				
	4062.55 II	<0.4				
	4116.10 II	<0.4				
	2889.63 II	<0.4				
	3584.88	10	Ac/N	HCL	120	1
	3514.61	4				
	3489.37	4				
	3466.30	3				
	3584.88	10	Ac/N	HCL	(120)	3
	3566.60	7				
	3514.61	5				
	3489.37	3				
	3584.88	10	PAc/O	HCL	12	6
Vanadium V	3185.40	10	Ac/N	HIHC	0.02	76
	3183.41 } 3183.98	8				
	3060.46	4				
	3066.38	3				

EXPERIMENTALLY OBSERVED
ATOMIC ABSORPTION LINES
IN ANALYTICAL FLAMES

Element	Line (Å)	Rel. Abs.	Flame	Lamp Type	Limit of Detection (ppm)	Ref.
Vanadium V	3056.33	3	Ac/N	HIHC		76
Continued	4379.24	2				
	3703.58	2				
	4384.72	2				
	3855.37 } 3855.84	2				
	3840.44 } 3840.75	2				
	4389.97	2				
	4115.18	1				
	3053.65	1				
	3902.25	1				
	4111.78	1				
	3828.56	1				
	3704.70	1				
	2526.22	0.8				
	2923.62	0.8				
	2519.62	0.6				
	2574.02	0.5				
	2517.14	0.3				
	2530.18	0.3				
	2507.78	0.2				
	2511.65 } 2511.95	0.2				
	3185.40	10	Ac/N	HCL	(1.0)	3
	3183.41 } 3183.98	7				
	3066.38	3				
	4389.97	1				
	3185.40	10	Ac/N	HCL	0.02	9
	3183.41 } 3183.98	10	PAc/O	HCL	0.2	6
	3183.98	10	Ac/O/N$_2$	HCL	(2)	1
	3185.40	8				
	3183.41	8				
	3066.38	2				
	3703.58	2				

EXPERIMENTALLY OBSERVED
ATOMIC ABSORPTION LINES
IN ANALYTICAL FLAMES

Element	Line (Å)	Rel. Abs.	Flame	Lamp Type	Limit of Detection (ppm)	Ref.
Vanadium V Continued	3840.44 } 3840.75	2	Ac/O/N_2	HCL		1
	4379.24	1				
	3855.37 } 3855.84	1				
	4384.72	0.8				
	4389.97	0.8				
Xenon Xe	No information					
Ytterbium Yb	3987.98 I	10	Ac/N	HCL	(0.25)	1
	3464.36 I	3				
	2464.49 I	2				
	2672.65 II	0.2				
	3987.98 I	10	Ac/N	HCL	(0.07)	3
	2464.49 I	0.3				
	2672.65 II	0.03				
	(3987.98)?	10	Ac/N	(HCL)?	0.005	4
Yttrium Y	4102.38	10	Ac/N	HCL	(5)	1
	4128.31	9				
	4077.38	9				
	4142.85	4				
	4102.38	10	Ac/N	HCL	(4.5)	3
	4142.85	7				
	(4102.38)?	10	Ac/N	(HCL)?	0.05	4
	4077.38	10	PAc/O	HCL	0.2	6
	4077.38	10	Ac/O	Xe1.5	(50)	59
	4102.38	5				
	4128.31	5				
	4142.85	5				

EXPERIMENTALLY OBSERVED
ATOMIC ABSORPTION LINES
IN ANALYTICAL FLAMES

Element	Line (Å)	Rel. Abs.	Flame	Lamp Type	Limit of Detection (ppm)	Ref.
Zinc Zn	2138.56	10	Ac/A	HCL	(0.009)	3
	3075.90	0.001				
	(2138.56)?	10	Ac/A	(HCL)?	<0.001	4
	2138.56	10	H/O	HCL	(0.03)	56
	3075.90	0.002				
	2138.56	10	PH/O	(HCL)?	0.03	5
	2138.56	10	H/A	HCL	0.001	10
	2138.56	10	MAPP/N	HCL	0.1	7
	2138.56	10	PB/A	HCL	0.002	9
Zirconium Zr	3547.68	10	Ac/N	HCL	5	27
	3011.75	8				
	2985.39	8				
	3029.52	8				
	3623.86	7				
	3601.19	10	Ac/N	HCL	(10)	3
	4687.80	1				
	(3601.19)?	10	Ac/N	(HCL)?	1	4
	3601.19	10	Ac/O/N_2	HCL	(18)	1
	3519.60	9				
	3011.75	6				
	3863.87	5				
	3547.68	4				
	3623.86	4				
	3029.52	3				
	2985.39	3				
	3890.32	3				
	3509.32	2				

1. M. D. Amos & J. B. Willis, Spectrochim. Acta, 22, 1325 (1966).

2. D. C. Manning, At. Absorption Newslett., 3, 6 (1964).

3. "Hollow Cathode Lamps," Varian Techtron, Walnut Creek, Calif., 1970.

4. S. Slavin, W. B. Barnett, & H. L. Kahn, At. Absorption Newslett., 11, 37 (1972).

5. N. V. Mossholder, V. A. Fassel, & R. N. Kniseley, Anal. Chem., 45, 1614 (1973).

6. J. A. Fiorino, R. N. Kniseley, & V. A. Fassel, Spectrochim, Acta, 23B, 413 (1966).

7. R. E. Mansell, Spectrochim. Acta, 25B, 219 (1970).

8. R. M. Dagnall, K. C. Thompson, & T. S. West, At. Absorption Newslett., 6, 117 (1967).

9. M. E. Britske, Zav. Lab., 35, 1329 (1969).

10. B. Moldan, Pure Appl. Chem., 23, 127 (1966).

11. Y. Yamamoto, T. Kumamaru, & Y. Hayashi, Anal. Letters, 5, 419 (1972).

12. A. Ando, M. Suzuki, K. Fuwa, & B. L. Vallee, Anal. Chem., 41, 1974 (1969).

13. K. E. Smith & C. W. Frank, Appl. Spectrosc., 22, 765 (1968).

14. F. J. Schmidt & J. L. Royer, Anal. Letters, 6, 17 (1973).

15. D. C. Manning, At. Absorption Newslett., 6, 35 (1967).

16. D. C. Manning, J. Vollmer, & F. Fernandez, At. Absorption Newslett., 6, 17 (1967).

17. M. L. Parsons & P. M. McElfresh, "Flame Spectrometry, Atlas of Spectral Lines," Plenum, New York, 1971.

18. N. Shifrin, A. Hell, & J. Ramirez-Muñoz, Appl. Spectrosc. 23, 365 (1969).

19. F. J. Fernandez & D. C. Manning, At. Absorption Newslett., 11, 67 (1972).

20. W. Slavin, "Atomic Absorption Spectroscopy," Interscience, New York, 1968.

21. J. W. Robinson, Anal. Chem., 33, 1067 (1961).

22. W. W. Harrison, _Anal. Chem._, _37_, 1169 (1965).

23. J. S. Cartwright & D. C. Manning, _At. Absorption Newslett._, _5_, 114 (1966).

24. S. L. Sachdev, J. W. Robinson, & P. W. West, _Anal. Chim Acta_, _38_, 499 (1967).

25. B. Fleet, K. V. Liberty, & T. S. West, _Analyst_, _93_, 701 (1968).

26. J. E. Allan, _Spectrochim. Acta_, _17_, 459 (1961).

27. D. C. Manning, _At. Absorption Newslett._, _5_, 127 (1966).

28. C. E. Mulford, _At. Absorption Newslett._, _5_, 28 (1966).

29. R. E. Popham & W. G. Schrenk, _Spectrochim. Acta_, _24B_, 223 (1969).

30. R. E. Popham & W. G. Schrenk, _Spectrochim. Acta_, _23B_, 543 (1968).

31. D. C. Manning & J. Vollmer, _At. Absorption Newslett._, _6_, 38 (1967).

32. M. A. Hildon & G. R. Sully, _Anal. Chim. Acta_, _54_, 245 (1971).

33. G. F. Kirkbright, T. S. West, & J. P. Wilson, _At. Absorption Newslett._, _11_, 53 (1972).

34. D. C. Manning & F. Fernandez, _At. Absorption Newslett._, _6_, 15 (1967).

35. D. F. Makarov & Y. N. Kukushkin, _Anal. Chem. USSR_, _24_, 1118 (1969).

36. K. E. Smith & C. W. Frank, _Anal. Chim. Acta_, _42_, 324 (1968).

37. C. B. Belcher, _Anal. Chim. Acta_, _62_, 87 (1972).

38. W. T. Elwell & J. A. F. Gidley, "Atomic Spectrophotometry," Anchor Press, Ltd., 1966.

39. C. L. Chakrabarti, J. W. Robinson, & P. W. West, _Anal. Chim. Acta_, _34_, 269 (1966).

40. C. L. Chakrabarti, _Appl. Spectrosc._, _21_, 160 (1967).

41. B. Moldan, I. Rubeska, & M. Miksovsky, _Anal. Chim. Acta_, _50_, 342 (1970).

42. F. Fernandez & D. C. Manning, _At. Absorption Newslett._, _7_, 57 (1966).

43. T. L. Chang, T. A. Gover, & W. W. Harrison, _Anal. Chim. Acta_, _34_, 17 (1966).

44. J. E. Allan, Spectrochim. Acta, 15, 800 (1959).

45. D. J. David, Analyst, 86, 730 (1961).

46. F. Fernandez, At. Absorption Newslett., 8, 90 (1969).

47. W. Osolinski & N. H. Knight, Appl. Spectrosc., 22, 532 (1968).

48. D. F. Makarov & Y. N. Kukushkin, Anal. Chem. USSR, 24, 1436 (1969).

49. V. Sychra, P. J. Slevin, J. Matousek, & F. Bek, Anal. Chim
     Acta, 52, 259 (1970).

50. G. Erinc & R. J. Magee, Anal. Chim. Acta, 31, 197 (1964).

51. D. C. Manning & Sabina Slavin, At. Absorption Newslett., 8, 132
     (1969).

52. G. F. Kirkbright & M. Marshall, Anal. Chem., 45, 1610 (1973).

53. A. E. Pitts, J. C. Van Loon & F. E. Beamish, Anal. Chim. Acta, 50,
     181 (1970).

54. A. E. Pitts, J. C. Van Loon, & F. E. Beamish, Anal. Chim. Acta, 50,
     195 (1970).

55. A. Strasheim & G. J. Wessels, Appl. Spectrosc., 17, 65 (1963).

56. B. M. Gatehouse & J. B. Willis, Spectrochim. Acta, 17, 710 (1961).

57. B. W. Bailey & J. M. Rankin, Anal. Chem., 43, 936 (1971).

58. D. G. Biechler & C. H. Long, At. Absorption Newslett., 8, 56 (1969).

59. V. A. Fassell & V. G. Mossotti, Anal. Chem., 35, 252 (1963).

60. P. Heneage, At. Absorption Newslett., 5, 64 (1966).

61. M. G. Atwell & J. Y. Herbert, Appl. Spectrosc., 23, 480 (1969).

62. J. E. Allan, Spectrochim. Acta, 18, 259 (1962).

63. D. F. Makarov & Y. N. Kukushkin, Anal. Chem. USSR, 24, 1191 (1969).

64. C. L. Chakrabarti, Anal. Chim. Acta, 42, 379 (1968).

65. C. S. Rann & A. N. Hambly, Anal. Chim. Acta, 32, 346 (1965).

66. Y. Yamamoto, T. Kumamaru, Y. Hayashi, & M. Kanlse, Anal. Letters,
     5, 717 (1972).

67. J. S. Cartwright, C. Sebens, & D. C. Manning, At. Absorption Newslett., 5, 91 (1966).

68. W. J. Price, "Analytical Atomic Absorption Spectrometry," Heyden & Son, New York, 1972.

69. G. F. Kirkbright & M. Marshall, Anal. Chem., 44, 1288 (1972).

70. W. A. Hareland, E. R. Ebersole, & T. P. Ramachandran, Anal. Chem., 44, 520 (1972).

71. J. Y. L. Wu, A. Droll, & P. F. Lott, At. Absorption Newslett., 7, 90 (1968).

72. C. L. Chakabarti, Anal. Chim. Acta, 39, 293 (1967).

73. L. Capacho-Delgado & D. C. Manning, Spectrochim. Acta, 22, 1505 (1966).

74. B. Molden, I. Rubeska, M. Miksovsky, M. Huka, Anal. Chim. Acta, 52, 91 (1970).

75. J. A. Bowman, Anal. Chim. Acta, 42, 285 (1968).

76. J. S. Cartwright, C. Sebens, & D. C. Manning, At. Absorption Newslett., 5, 91 (1965).

Table II.B.5

EXPERIMENTALLY OBSERVED
ATOMIC ABSORPTION LINES
IN NON-FLAME CELLS

Element	Line (Å)	Rel. Abs.	Cell Type	Purge Gas	Limit of Detection $(10^{-12}g)$	Ref.
Actinium Ac	No information					
Aluminum Al	3092.71 } 3092.84	10	MF	H_2	300	1
	3092.71 } 3092.84	10	GFL	Ar	0.2	2
	3092.71 } 3092.84	10	GFL	Ar	$[0.01]^1$	3
	3092.71 } ? 3092.84	10	GFS	N_2	3	4
	3082.16	10	CRA	Ar	1000	5
	3092.71 } 3092.84	10	CRA	N_2	30	24
	3092.71 } 3092.84	10	DCP	Ar	[2.7]	6
	3961.53	10	ICP	Ar	[0.6]	7
	3092.71 } 3092.84	10				
	3944.03	9				
Americium Am	No information					
Antimony Sb	2175.81	10	MF	Ar	200	1
	(2175.81)?	10	GFS	N_2	20	4
	2311.47	10	GFL	Ar	0.4	2
	2311.47	10	DHCL	?	10^5	8

EXPERIMENTALLY OBSERVED
ATOMIC ABSORPTION LINES
IN NON-FLAME CELLS

Element	Line (Å)	Rel. Abs.	Cell Type	Purge Gas	Limit of Detection (10^{-12}g)	Ref.
Argon Ar	8115.31	10	QT	--	NI	9
	7635.11	10				
	8014.79	7				
	7514.65	6				
	8424.65	5				
	7383.98	5				
	7948.18	4				
	6965.43	3				
	5048.81	3				
	8006.16	3				
	7067.22	3				
	7503.87	2				
	5054.18	2				
	5373.49	1				
	5221.27	1				
	3670.64	0.9				
	3675.22	0.9				
	3834.68	0.5				
	5659.13	0.4				
	8408.21	0.3				
	8103.69	0.2				
	4158.59	0.2				
	8521.44	0.2				
	4879.87	10	DHCL	--	NI	10
	5145.32	2				
	4764.86	2				
	4158.59	0.8				
Arsenic As	1936.96	10	MF	Ar	300	1
	1890.4	10	GFS	Ar	600	11
	1936.96	10	Boat	Ac/A	$<10^5$	12
	1936.96	10	DHCL	Ar	2×10^4	13
	1936.96	10	QT	--	$<5 \times 10^4$	14
	1936.96	10	CRA	H/Ar	400	21

EXPERIMENTALLY OBSERVED
ATOMIC ABSORPTION LINES
IN NON-FLAME CELLS

Element	Line (Å)	Rel. Abs.	Cell Type	Purge Gas	Limit of Detection $(10^{-12}g)$	Ref.
Astatine At	No information					
Barium Ba	5535.48	10	MF	H_2	10	1
	5535.48	10	GFL	Ar	1	2
Berkelium Bk	No information					
Beryllium Be	2348.61	10	MF	Ar	2	1
	2348.61	10	GFL	Ar	0.03	15
Bismuth Bi	2230.61	10	MF	Ar	40	1
	3067.72	10	GFL	Ar	0.06	2
	(2230.61)?	10	GFS	N_2	10	4
Boron B	2497.72	10	GFL	Ar	200	15
Bromine Br	No information					
Cadmium Cd	2288.02	10	MF	Ar	2	1
	2288.02	10	GFL	Ar	0.003	16
	2288.02	10	GFL	Ar	[0.002]	3
	2288.02	10	DHCL	Ar	4×10^4	13
	2288.02	10	DHCL	--	200	8
	(2288.02)?	10	GFS	N_2	0.1	4

EXPERIMENTALLY OBSERVED
ATOMIC ABSORPTION LINES
IN NON-FLAME CELLS

Element	Line (Å)	Rel. Abs.	Cell Type	Purge Gas	Limit of Detection (10^{-12}g)	Ref.
Calcium Ca	4226.73	10	MF	Ar	1	1
	4226.73	10	GFL	Ar	0.4	15
	4226.73	10	GFS	?	50	17
	4226.73 I 3933.67 II	10 6	ICP	Ar	[0.2]	7
	4226.73	10	DHCL	Ar	2×10^4	13
Californium Cf	No information					
Carbon C	No information					
Cerium Ce	No information					
Cesium Cs	8521.10	10	MF	Ar	70	1
	8521.10	10	GFL	Ar	0.4	15
Chlorine Cl	No information					
Chromium Cr	3578.69	10	MF	Ar	20	1
	3578.69	10	MF	Ar	94	18
	4254.35	10	GFL	Ar	1	2
	3578.69	10	GFS	N_2	2	19
	3578.69 3593.49	10 7	CRA	Ar	5	20
	4274.80	10	CRA	N_2	10	24

EXPERIMENTALLY OBSERVED
ATOMIC ABSORPTION LINES
IN NON-FLAME CELLS

Element	Line (Å)	Rel. Abs.	Cell Type	Purge Gas	Limit of Detection (10^{-12}g)	Ref.
Chromium Cr Continued	3578.69	10	CRA	H/Ar	5	21
	3578.69	10	DHCL	--	3×10^{4}	8
Cobalt Co	2407.25	10	MF	H_2	300	1
	2407.25	10	GFL	Ar	2	15
	(2407.25)?	10	GFS	N_2	4	4
Copper Cu	3247.54	10	MF	Ar	10	1
	3247.54	10	MF	N_2	6	22
	3247.54	10	GFL	Ar	0.04	2
	3247.54	10	GFL	Ar	[0.003]	3
	(3247.54)?	10	GFS	N_2	1	4
	3247.54 3273.96	10 5	CRA	Ar	50	5
	3247.54	10	CRA	H/Ar	1	23
	3273.96	10	CRA	N_2	10	24
	3247.54	10	DHCL	--	9000	8

Curium Cm   No information

Dysprosium Dy   No information

Einsteinium Es   No information

Erbium Er   No information

EXPERIMENTALLY OBSERVED
ATOMIC ABSORPTION LINES
IN NON-FLAME CELLS

Element	Line (Å)	Rel. Abs.	Cell Type	Purge Gas	Limit of Detection $(10^{-12}g)$	Ref.
Europium Eu	4594.03	10	MF	H_2	30	1
Fermium Fm	No information					
Fluorine F	No information					
Francium Fr	No information					
Gadolinium Gd	No information					
Gallium Ga	2874.24	10	MF	Ar	100	1
	2874.24	10	GFL	Ar	1	15
Germanium Ge	2651.18 } 2651.58 }	10	GFS	?	300	25
	2589.19 } 2592.54 }	6				
	2709.63	6				
	2094.23	5				
	2754.59	4				
	2691.34	4				
	2497.96	0.4				
Gold Au	2427.95	10	MF	Ar	20	1
	2427.95	10	GFL	Ar	1	15
	(2427.95)	10	GFS	N_2	8	4
	2427.95	10	CRA	Ar	200	26
	2427.95	10	CRA	Ar	0.6	27
	2427.95	10	CRA	H/Ar	20	21

EXPERIMENTALLY OBSERVED
ATOMIC ABSORPTION LINES
IN NON-FLAME CELLS

Element	Line (Å)	Rel. Abs.	Cell Type	Purge Gas	Limit of Detection (10^{-12}g)	Ref.
Hafnium Hf	No information					
Helium He	3888.65	10	QT	--	NI	9
	5875.62	8				
	6678.15	5				
	3187.74	3				
	3964.73	3				
	5015.68	3				
	4471.48	1				
	7281.35	0.8				
	4026.19	0.6				
	4921.93	0.6				
	3819.61	0.4				
Holmium Ho	No information					
Hydrogen H	No information					
Indium In	3039.36	10	MF	Ar	100	1
	3039.36	10	GFL	Ar	0.07	2
	3039.36	10	GFS	Ar	200	11
Iodine I	1830	10	GFL	Ar	40	28, 29
Iridium Ir	No information					
Iron Fe	2483.27	10	MF	Ar	100	1
	2966.90	10	GFL	Ar	0.4	2
	(2483.27)?	10	GFS	N_2	3	4
	2483.27	10	CRA	H/Ar	5	21

EXPERIMENTALLY OBSERVED
ATOMIC ABSORPTION LINES
IN NON-FLAME CELLS

Element	Line (Å)	Rel. Abs.	Cell Type	Purge Gas	Limit of Detection $(10^{-12}g)$	Ref.
Iron Fe	2483.27	10	CRA	N_2	10	30
Continued	2522.85	8				
	2488.15	7				
	2719.02	5				
	3020.64	3				
	3719.94	2				
	3859.91	1				
	2527.43	1				
	2966.90	0.5				
	3047.60	0.5				
	2936.90	0.3				
	2947.88	0.3				
	2994.43	0.3				
	3440.61	0.3				
	3737.13	0.3				
	3745.56	0.2				
	2750.14	0.08				
	2983.57	0.08				
	2756.26	0.07				
	2973.24	0.05				
	2795.01	0.05				
	3679.92	0.05				
	3490.58	0.04				
	3000.95	0.04				
	3886.28	0.04				
	3705.57	0.03				
	3824.44	0.02				
Krypton Kr	8059.50	10	QT	--	NI	9
	8190.05	9				
	7601.54	8				
	7587.41	7				
	8112.90	7				
	8104.36	6				
	7694.54	5				
	7224.10	4				
	7854.82	2				
	7685.24	1				
	5870.91	1				
	4273.97	0.8				
	5570.28	0.8				

EXPERIMENTALLY OBSERVED
ATOMIC ABSORPTION LINES
IN NON-FLAME CELLS

Element	Line (Å)	Rel. Abs.	Cell Type	Purge Gas	Limit of Detection (10^{-12}g)	Ref.
Krypton Kr Continued	7913.44	0.8	QT	--	NI	9
	8298.11	0.4				
	8508.87	0.4				
	5562.22	0.3				
	8281.05	0.2				
Lanthanum La	No information					
Lead Pb	2169.99	10	MF	Ar	10	1
	2169.99	10	MF	Ar	20	23
	2833.06	10	GFL	Ar	0.4	2
	2169.99	10	GFL	Ar	5.5	31
	(2169.99)?	10	GFS	N_2	6	4
	2833.06	10	GFS	Ar	10	11
	2833.06	10	Boat	Ac/A	1.2×10^6	32
	2169.99	10	CRA	Ar	50	5
	2833.06	4				
	2833.06	10	CRA	N_2	5	24
	2833.06	10	CRA	H/Ar	75	21
	2169.99	10	CRA	H_2	2	33
	2833.06	5				
	2169.99	10	DHCL	Ar	6×10^3	13
Lithium Li	6707.84	10	MF	Ar	2	1
	6707.84	10	GFL	Ar	0.5	2

EXPERIMENTALLY OBSERVED
ATOMIC ABSORPTION LINES
IN NON-FLAME CELLS

Element	Line (Å)	Rel. Abs.	Cell Type	Purge Gas	Limit of Detection $(10^{-12}g)$	Ref.
Lutetium Lu	No information					
Magnesium Mg	2852.13	10	MF	Ar	0.1	1
	2852.13	10	GFL	Ar	0.04	2
	2852.13	10	GFS	Ar	0.5	11
	2852.13	10	CRA	H/Ar	1	21
	2852.13	10	CRA	N_2	0.05	24
	2852.13	10	CRA	Ar	100	34
	2852.13	10	ICP	Ar	[0.06]	7
	2852.13	10	DHCL	--	400	8
Manganese Mn	2794.82	10	MF	Ar	7	1
	2794.82	10	GFL	Ar	0.02	28
	2794.82	10	GFL	Ar	[0.002]	3
	(2794.82)?	10	GFS	N_2	1	4
	2794.82 4030.76	10 1	CRA	Ar	50	35
	2794.82	10	DHCL	--	900	8
Mendelevium Md	No information					
Mercury Hg	2536.52	10	MF	Ar	200	1
	2536.52	10	MF	Air	[0.0001]	36
	1849	10	GFL	Ar	1	28, 29

EXPERIMENTALLY OBSERVED
ATOMIC ABSORPTION LINES
IN NON-FLAME CELLS

Element	Line (Å)	Rel. Abs.	Cell Type	Purge Gas	Limit of Detection $(10^{-12}g)$	Ref.
Mercury Hg Continued	2536.52	0.3	GFL	Ar	1	28, 29
	2536.52	10	GFS	Ar	20	11
	2536.52	10	DHCL	Ar	5×10^4	13
	2536.52	10	QT	Air	200	37
	2536.52	10	QT	Air	[<0.0002]	38
	2536.52	10	QT	Air	$[(.001)]^2$	39
	1849	10	QT	Ar	[0.0001]	40
	2536.52	10	QT	Vacuum	10	41
Molybdenum Mo	3132.59	10	GFL	Ar	0.3	2
	3132.59	10	CRA	H/N_2	30	42
Neodymium Nd	No information					
Neon Ne	6402.25	10	QT	--	NI	9
	6143.06	6				
	6382.99	5				
	6334.42	4				
	6096.16	4				
	6266.49	3				
	7032.41	3				
	5944.83	3				
	6074.33	2				
	6929.47	2				
	6678.28	2				
	6163.59	2				
	5881.89	2				
	5852.48	2				
	6598.95	2				
	7245.16	1				
	6217.28	1				

EXPERIMENTALLY OBSERVED
ATOMIC ABSORPTION LINES
IN NON-FLAME CELLS

Element	Line (Å)	Rel. Abs.	Cell Type	Purge Gas	Limit of Detection $(10^{-12}g)$	Ref.
Neon Ne Continued	8377.61	0.9	QT	--	NI	9
	6029.99	0.7				
	7438.90	0.6				
	8300.90	0.5				
	8136.41	0.5				
	7173.93	0.3				
	8418.43	0.2				
	7024.05	0.2				
	2967.18 II	10	DHCL	--	NI	10
	3323.75 II					
	4391.94 II					
	3561.23 II					
Neptunium Np	No information					
Nickel Ni	2320.03	10	MF	Ar	200	1
	3524.54	10	GFL	Ar	4	2
	(2320.03)?	10	GFS	N_2	10	4
	2320.03	10	CRA	H/Ar	60	21
	2320.03	10	CRA	H/Ar	20	43
	2320.03	10	CRA	N_2	10	24
	2320.03	10	CRA	Ar	200	44
	3050.82	2				
	3002.49	2				
	2981.65	2				
	3524.54	2				
	2320.03	10	CRA	Ar	840	45
	2310.96	3				
Niobium Nb	4079.73	10	ICP	Ar	[40]	7

EXPERIMENTALLY OBSERVED
ATOMIC ABSORPTION LINES
IN NON-FLAME CELLS

Element	Line (Å)	Rel. Abs.	Cell Type	Purge Gas	Limit of Detection (10^{-12}g)	Ref.
Niobium Nb Continued	4100.40 } 4100.92	9	ICP	Ar	[40]	7
	4058.94	9				
	3535.30	7				
Nitrogen N	No information					
Nobelium No	No information					
Osmium Os	No information					
Oxygen O	No information					
Palladium Pd	2476.42	10	MF	Ar	700	1
	2476.42	10	GFL	Ar	4	15
Phosphorus P	1774.95	10	GFL	Ar	3	28, 29
	1783	8				
	1788	5				
	2135.47 } 2136.18	10	GFL	Ar	200	46
	2149.14 2152.94 } 2154.08	6				
	2533.99 } 2535.61	0.06				
	2553.24 } 2554.90	0.03				
	1774.95	10	DHCL	--	NI	47

EXPERIMENTALLY OBSERVED
ATOMIC ABSORPTION LINES
IN NON-FLAME CELLS

Element	Line (Å)	Rel. Abs.	Cell Type	Purge Gas	Limit of Detection $(10^{-12}g)$	Ref.
Platinum Pt	2659.45	10	GFL	Ar	10	15
Plutonium Pu	4206.47	vvs[3]	GFL	NI	NI	48, 49
	3878.52	vs				
	4208.24	vs				
	4735.40	vs				
Polonium Po	No information					
Potassium K	7664.91	10	MF	Ar	1	1
	4044.14	10	GFL	Ar	20	2
Praseodymium Pr	No information					
Protactinium Pa	No information					
Radium Ra	No information					
Radon Rn	No information					
Rhenium Re	3460.46	10	ICP	Ar	[30]	7
Rhodium Rh	3434.89	10	GFL	Ar	8	15
Rubidium Rb	7800.23	10	MF	Ar	10	1
	7800.23	10	GFL	Ar	1	15
Ruthenium Ru	No information					

EXPERIMENTALLY OBSERVED
ATOMIC ABSORPTION LINES
IN NON-FLAME CELLS

Element	Line (Å)	Rel. Abs.	Cell Type	Purge Gas	Limit of Detection $(10^{-12}g)$	Ref.
Samarium Sm	No information					
Scandium Sc	No information					
Selenium Se	1960.26	10	MF	H_2	700	1
	1960.26	10	GFL	Ar	9	15
	1960.26	10	GFS	Ar	2000	11
	1960.26	10	CRA	N_2	26	50
	1960.26	10	DHCL	Ar	1×10^4	13
Silicon Si	2516.11	10	MF	H_2	1×10^4	1
	2516.11	10	GFL	Ar	0.05	15
	(2516.11)?	10	GFS	N_2	8	4
Silver Ag	3280.68	10	MF	Ar	4	1
	3280.68	10	MF	N_2	6	23
	3280.68	10	GFL	Ar	0.03	28
	3280.68	10	GFL	(Ar)?	0.5	51
	(3280.68)?	10	GFS	N_2	0.25	4
	3280.68	10	CRA	H/Ar	2	21
	3382.89	10	CRA	N_2	0.2	24
	3280.68	10	CRA	Ar	0.02	27
	3280.68	10	CRA	Ar	100	52

EXPERIMENTALLY OBSERVED
ATOMIC ABSORPTION LINES
IN NON-FLAME CELLS

Element	Line (Å)	Rel. Abs.	Cell Type	Purge Gas	Limit of Detection (10^{-12}g)	Ref.
Silver Ag	3280.68	10	DHCL	--	1000	8
Continued	3280.68	10	DHCL	Ar	1×10^4	13
Sodium Na	5889.95	10	MF	Ar	1	1
	5889.95	10	GFS	Ar	7	11
Strontium Sr	4607.33	10	MF	Ar	10	1
	4607.33	10	GFL	Ar	0.2	2
Sulfur S	1807	10	GFL	Ar	80	28, 29
	1821	8				
	1826	3				
	2169	10	ICP	He	300	53
Tantalum Ta	No information					
Technetium Tc	No information					
Tellurium Te	2142.75	10	MF	H_2	300	1
	2142.75	10	GFL	Ar	1	15
	2142.75	10	Boat	?	[0.005]	54
Terbium Tb	No information					
Thallium Tl	2767.87	10	MF	Ar	70	1
	2767.87	10	GFL	Ar	1	15

EXPERIMENTALLY OBSERVED
ATOMIC ABSORPTION LINES
IN NON-FLAME CELLS

Element	Line (Å)	Rel. Abs.	Cell Type	Purge Gas	Limit of Detection $(10^{-12}g)$	Ref.
Thallium Tl Continued	(2767.87)?	10	GFS	N_2	14	4
	2767.87	10	Boat	Ac/A	2000	55
Thorium Th	No information					
Thulium Tm	No information					
Tin Sn	2863.33	10	MF	Ar	100	1
	2863.33	10	GFL	Ar	2	15
	2246.05	10	GFS	N_2	1500	56
Titanium Ti	3642.68	10	MF	H_2	1000	1
	3653.50	10	GFL	Ar	40	15
	5173.75	10	ICP	Ar	[5]	7
	3635.20 } 3635.46	10				
	3642.68	10				
	3653.50	10				
	4667.59	8				
	4681.92	8				
Tungsten W	4008.75	10	ICP	Ar	[3]	7
	4659.87	9				
Uranium U	No information					
Vanadium V	3183.41 } 3183.98	10	MF	H_2	400	1

EXPERIMENTALLY OBSERVED
ATOMIC ABSORPTION LINES
IN NON-FLAME CELLS

Element	Line ($\overset{\circ}{A}$)	Rel. Abs.	Cell Type	Purge Gas	Limit of Detection (10^{-12}g)	Ref.
Vanadium V Continued	3183.41 } 3183.98	10	CRA	N_2	300	57
	3183.41 3183.98 } 3185.40	10	CRA	N_2	7	58
	3185.40	10	DCP	Ar	[18]	6
	3183.41 } 3183.98	10	ICP	Ar	[2]	7
Xenon Xe	8231.63	10	QT	--	NI	9
	2280.12	9				
	8206.34	5				
	8409.19	5				
	6498.71	3				
	7386.00	2				
	4671.23	2				
	6198.26	2				
	7967.34	2				
	8346.82	1				
	8266.51	1				
	4624.28	0.8				
	4807.02	0.4				
	7887.39	0.3				
	4829.71	0.3				
Ytterbium Yb	No information					
Yttrium Y	3620.94 I	10	ICP	Ar	[10]	7
	4102.38 I	10				
	4643.70 I	10				
	3327.89 II	10				
	4077.38 I	10				
	3710.30 II	8				

EXPERIMENTALLY OBSERVED
ATOMIC ABSORPTION LINES
IN NON-FLAME CELLS

Element	Line (Å)	Rel. Abs.	Cell Type	Purge Gas	Limit of Detection $(10^{-12}g)$	Ref.
Zinc Zn	2138.56	10	MF	Ar	1	1
	2138.56	10	MF	Ar	0.3	23
	2138.56	10	GFL	Ar	0.006	2
	(2138.56)?	10	GFS	N_2	0.06	4
	2138.56	10	CRA	Ar	0.02	42
	2138.56	10	DHCL	--	800	8
	2138.56	10	DHCL	Ar	5000	13

Zirconium Zr   No information

Footnotes

1. Limit of detection in ppm are in brackets.

2. Sensitivity is denoted by parenthesis.

3. In this reference VVS means very very strong, VS means very strong, and many other lines were listed.

1. J. Y. Hwang, C. J. Mokeler, & P. A. Ullucci, Anal. Chem., 44, 2018 (1972).

2. B. V. L'vov, "Atomic Absorption Spectroscopy," Israel Program for Scientific Translations, Jerusalem, 1969.

3. R. Woodriff & G. Ramelow, Spectrochim. Acta, 23B, 665 (1968).

4. S. Slavin, W. B. Barnett, & H. L. Kahn, At. Absorption Newslett., 11, 37 (1972).

5. R. G. Anderson, H. N. Johnson, & T. S. West, Anal. Chim. Acta, 57, 281 (1971).

6. K. E. Friend & A. J. Diefenderfer, Anal. Chem., 38, 1763 (1966).

7. R. H. Wendt & V. A. Fassel, Anal. Chem., 38, 337 (1967).

8. H. Massmann, Spectrochim. Acta, 25B, 393 (1970).

9. J. A. Goleb, Anal. Chem., 38, 1059 (1966).

10. P. M. McElfresh & M. L. Parsons, At. Absorption Newslett., 11, 69 (1972).

11. H. Massman, Spectrochim. Acta, 23B, 215 (1968).

12. G. I. Spielholtz, G. C. Toralballa, & R. J. Steinberg, Mikrochim. Acta, 1971, 918 (1971).

13. B. W. Gandrud & R. K. Skogerboe, Appl. Spectrosc., 25, 243 (1971).

14. R. C. Chu, G. P. Barron, & P. A. W. Baumgarner, Anal. Chem., 44, 1476 (1972).

15. B. V. L'vov, J. Appl. Spectrosc. USSR, 8, 517 (1968).

16. D. A. Katskov & B. V. L'vov, J. Appl. Spectrosc. USSR, 10, 867 (1969).

17. J. H. Cragin & M. M. Herron, At. Absorption Newslett., 12, 37 (1973).

18. T. Maruta & T. Takeuchi, Anal, Chim. Acta, 66, 5 (1973).

19. I. W. F. Davidson & W. L. Secrest, Anal. Chem., 44, 1808 (1972).

20. G. Tessari & G. Torsi, Talanta, 19, 1059 (1972).

21. C. J. Molnar, R. D. Reeves, J. D. Winefordner, M. T. Glenn,
     J. R. Ahlstrom, & J. Savory, Appl. Spectrosc., 26, 606 (1972).

22. J. E. Cantle & T. S. West, Talanta, 20, 459 (1973).

23. M. Glenn, J. Savory, L. Hart, T. Glenn, & J. D. Winefordner,
     Anal. Chim. Acta, 57, 263 (1971).

24. K. G. Brodie & J. P. Matousek, Anal. Chem., 43, 1557 (1971).

25. D. J. Johnson, T. S. West, & R. M. Dagnall, Anal. Chim. Acta,
     67, 79 (1973).

26. J. Aggett & T. S. West, Anal. Chim. Acta, 55, 349 (1971).

27. M. P. Bratzel, Jr., C. L. Chakrabarti, R. E. Sturgeon, M. W.
     McIntyre, & H. Agemian, Anal. Chem., 44, 372 (1972).

28. B. V. L'vov, Pure Appl. Chem., 23, 11 (1966).

29. B. V. L'vov & A. D. Khartsyzov, J. Appl. Spectrosc. USSR, 11,
     413 (1969).

30. L. Ebdon, G. F. Kirkbright, & T. S. West, Anal. Chim. Acta, 61,
     15 (1972).

31. G. K. Pagenkopf, D. R. Neuman, R. Woodriff, Anal. Chem., 44,
     2248 (1972).

32. H. T. Delves, Analyst, 95, 431 (1970).

33. M. P. Bratzel, Jr. & C. L. Chakrabarti, Anal. Chim. Acta, 61, 25
     (1972).

34. T. S. West & X. K. Williams, Anal. Chim. Acta, 45, 27 (1969).

35. L. Ebdon, G. F. Kirkbright, & T. S. West, Anal. Chim. Acta, 58,
     39 (1972).

36. M. J. Fishman, Anal. Chem., 42, 1462 (1970).

37. S. J. Long, D. R. Scott, & R. J. Thompson, Anal. Chem., 45, 2227
     (1973).

38. M. P. Stainton, Anal. Chem., 43, 625 (1971).

39. W. R. Hatch & W. L. Ott, Anal. Chem., 40, 2085 (1968).

40. J. W. Robinson, P. J. Slevin, G. P. Hindman, & D. K. Wolcott,
     Anal. Chim. Acta, 61, 431 (1972).

41. J. A. Goleb, Appl. Spectrosc., 25, 522 (1971).

42. R. M. Dagnall, D. J. Johnson & T. S. West, Anal. Chim. Acta, 66,
     171 (1973).

43. G. Hall, M. P. Bartzel, Jr., & C. L. Chakrabarti, Talanta, 20,
     755 (1973).

44. K. W. Jackson & T. S. West, Anal. Chim. Acta, 59, 187 (1972).

45. J. F. Alder & T. S. West, Anal. Chim Acta, 61, 132 (1972).

46. B. V. L'vov & A. D. Khartsyzov, J. Appl. Spectrosc. USSR, 11, 9
     (1969).

47. W. Slavin, "Atomic Absorption Spectroscopy," Interscience, New
     York, 1968.

48. L. Borey, Spectrochim. Acta, 10, 383 (1957).

49. G. K. T. Conn & C. K. Wu, Trans. Faraday Soc., 34, 1483 (1938).

50. R. B. Baird, S. Pourian & S. M. Gabrielian, Anal. Chem., 44,
     1887 (1972).

51. R. Woodriff, B. R. Culver, D. Shrader, & A. B. Super, Anal. Chem.,
     45, 230 (1973).

52. J. F. Alder & T. S. West, Anal. Chim. Acta., 58, 331 (1972).

53. H. E. Taylor, J. H. Gibson, R. K. Skogerboe, Anal. Chem., 42,
     1569 (1970).

54. R. D. Beaty, Anal. Chem., 45, 234 (1973).

55. A. S. Curry, J. F. Read & A. R. Knott, Analyst, 94, 744 (1969).

56. F. J. Fernandez & D. C. Manning, At Absorption Newslett., 10, 65
     (1971).

57. G. L. Everett, T. S. West & R. W. Williams, Anal. Chim. Acta, 66,
     301 (1973).

58. C. L. Chakrabarti & G. Hall, Spectrosc. Lett., 6, 385 (1973).

Table II.B.6

EXPERIMENTALLY OBSERVED
ATOMIC EMISSION LINES

Element	Line (Å)	Relative Intensity	Flame	Limit of Detection (ppm)	Ref.
Actinium Ac	No information				
Aluminum Al	3961.53	10	PAc/O	0.2	1
	3944.03	7		0.3	
	3961.53	10	SAc/N	.03	10
	3944.03	5		0.2	3
	3092.7	0.8		NI	
	3082.2	0.4		NI	
	3961.53	10	PAc/N	0.003	5
	3961.53	NI	PH /N	7	10
	3961.53	NI	SH/N	5	10
	3961.53	NI	PMAPP/N	2.5	10
	3961.53	NI	SMAPP/N	3	10
	3961.53	10	H/A	NI	38
	3944.03	10		NI	
	3961.53	10	Ac/A	NI	37
	3944.03	1		NI	
	3961.53	10	Ac/O	NI	38
	3944.03	5		0.1	
AlO	4842.3	10	H/O	2	38
	4866.4	10		NI	
	4672.0	7		NI	
	5079	7		NI	
	5102	7		NI	
	5123	7		NI	
	5143	6		NI	
	4648	6		NI	
	4694.6	6		NI	
	4715.6	6		NI	

EXPERIMENTALLY OBSERVED
ATOMIC EMISSION LINES

Element	Line (Å)	Relative Intensity	Flame	Limit of Detection (ppm)	Ref.
Aluminum AlO Continued	4842.3 } 4866.4	10	H/A		38
	4672.0	5			
	5079 } 5102	7			
	5123				
	5143				
	4648.2 } 4684.6	5			
	4715.6				
	4847.3	10	Ac/O		38
	4866.4	8			
	4672.0	6			
	5079	6			
	5102	6			
	5123	6			
	5143	6			
	4648.2	4			
	4684.6	5			
	4715.6	4			
Americium Am	No information				
Antimony Sb	2598.05	10	PAc/O	20	1
	2598.05	2		90	
	2528.52	10	PAc/N	0.6	14
	2598.05	10	H/O	0.2	13
	2311.47	10	H/A	0.4	35
	2598.05	4		NI	
	2528.52	2		NI	
	2311.47	2	Ac/A	3	(15)
	2598.05	10		NI	

EXPERIMENTALLY OBSERVED
ATOMIC EMISSION LINES

Element	Line (Å)	Relative Intensity	Flame	Limit of Detection (ppm)	Ref.
Antimony Sb Continued	2528.52	10	Ac/A	NI	(15)
	2175.81				
	2068.33				
	2311.47	NI	Ac/O/Ar	1	15
	2598.05	NI		NI	
	2528.52	NI		NI	
	2175.81	NI		NI	
	2068.33	NI		NI	
SbO	2574	10	H/A	NI	38
Argon Ar	No information				
Arsenic As	1936.96	NI	PAc/O	NI	1
	2288.12	NI		50	
	2349.84	NI		NI	
	1936.96	10	PAc/N	10	6
	2349.84	NI	Ac/O/Ar	5	15
	1890	NI		NI	
	2288.12	NI		NI	
	1936.96	NI		NI	
	2349.84	NI	H/A	NI	38
	2288.12	NI		NI	
	2492.91	NI		NI	
	2745.00	NI		NI	
	2780.22	NI		NI	
	2860.44	NI		NI	
	2349.84	NI	H/O	2	13
	2288.12	NI		NI	
	2492.91	NI		NI	

EXPERIMENTALLY OBSERVED
ATOMIC EMISSION LINES

Element	Line (Å)	Relative Intensity	Flame	Limit of Detection (ppm)	Ref.
Arsenic As Continued	2745.00	NI	H/O	NI	13
	2780.22	NI		NI	
	2860.44	NI		NI	
	2349.84	10	Ac/O	100	38
	2288.12	NI		NI	
	2492.91	NI		NI	
	2745.00	NI		NI	
	2780.22	NI		NI	
	2860.44	NI		NI	
	1890	NI		NI	
	1936.96	NI		NI	
	2349.84	10		7	
	2285.12	3		NI	15
	2492.91	.1		NI	
	2745.00	1		NI	
	2780.22	1		NI	
	2860.44	10		NI	
	5000	10	H/A	NI	38
Astatine At	No information				
Barium Ba (II)	4554.03	10	PAc/O	0.03	1
Ba (I)	5535.48	6		0.05	
	5535.48	10	PAc/N	10^{-3}	21
	2347.58	10	SAc/N	0.3	7
	5535.48		SAc/N	0.05	16
	5535.48		H/N	0.05	10
	5535.48		SH/N	0.02	10
	5535.48		MAPP/N	0.04	10

EXPERIMENTALLY OBSERVED
ATOMIC EMISSION LINES

Element	Line (Å)	Relative Intensity	Flame	Limit of Detection (ppm)	Ref.
Barium Ba (I) Continued	5535.48		SMAPP/N	0.03	10
	5535.48	10	H/O	0.05	9
Ba(II)	4934.09	6			
	4554.03	2			
	5535.48	10	H/A	NI	38
Ba(II)	4934.09	5		NI	
	4554.03	0.5		NI	
Ba(I)	5535.48	10	Ac/O	NI	38
Ba(II)	4934.09	5		NI	
	4554.03	5		NI	
Ba(I)	5535.48	10	Ac/A	NI	37
Ba(II)	4554.03	0.2		NI	
BaOH	4880	10	H/O	NI	38
	5130	10			
	8300	10			
	8730	10			
	4970	6			
	5020	6			
	5240	6			
	7450	3			
	4880	5	H/A	NI	38
	5130	8			
	8300	10			
	8730	4			
	4970	4			
	5020	4			
	5240	4			
	7450	2			
	5130	2	Ac/O		38
	8300	10		0.03	34
	8730	10			38
	5240	2			
	7450	5			

EXPERIMENTALLY OBSERVED
ATOMIC EMISSION LINES

Element	Line (Å)	Relative Intensity	Flame	Limit of Detection (ppm)	Ref.
Barium Ba OH Continued	5130	10	H/Ar	0.02	22
BaO	5349.7	10	H/O	NI	38
	5492.7	10			
	5644.1	10			
	5701	10			
	5864.5	10			
	6039.6	10			
	5349.7	10	H/A	NI	38
	5492.7	10			
	5644.1	10			
	5701	10			
	5864.5	10			
	6039.6	10			
	5349.7	10	Ac/O	NI	38
	5492.7	8			
	5644.1	8			
	5701	8			
	5864.5	8			
	6039.6	8			
Berkelium Bk	No information				
Beryllium Be	2348.61	10	PAc/O	1	1
	2348.61	10	PAc/N	1	14
	2348.61	10	SAc/N	0.1	17
	2348.61	10	H/O	3	13
	2348.61	NI	H/A	NI	38
	2348.61	10	Ac/O	NI	38
BeO	4708.6	10	PAc/N	0.2	21

EXPERIMENTALLY OBSERVED
ATOMIC EMISSION LINES

Element	Line (Å)	Relative Intensity	Flame	Limit of Detection (ppm)	Ref.
Beryllium BeO	4708.6	10	H/O		38
Continued	4733	7		30	33
	4755	6			38
	5054.4	4			
	5075.7	4			
	5095.1	4			
	4708.6	NI	H/A	NI	38
	4733 }	10			
	4755				
	5054.4 }	5			
	5075.7				
	5095.1	NI			
	4708.6	10	Ac/O	NI	38
	4733	8			
	4755	4			
	5054.4	4			
	5075.7	4			
	5095.1	4			
Bismuth Bi	2230.61	10	PAc/O	40	1
	2228.25	5		80	1
	3067.72	0.05		700	1
	3067.72	10	PAc/N	2	2
	3067.72	10	SAc/N	1.5	10
	3067.72	10	H/N	200	10
	3067.72	10	SH/N	200	10
	3067.72	10	MAAP/N	80	10
	3067.72	10	SMAPP/N	40	10
	2230.61	NI	Ac/O/Ar	1	15

EXPERIMENTALLY OBSERVED
ATOMIC EMISSION LINES

Element	Line (Å)	Relative Intensity	Flame	Limit of Detection (ppm)	Ref.
Bismuth Bi Continued	4722.55	10	H/O	20	33
	3067.72	0.7		NI	
	2230.61	NI		1	13
	2276.58	NI		NI	
	4722.55	10	H/A	NI	38
	3067.72	4			
	2230.61	NI			
	2276.58	NI			
	4722.55	10	Ac/O	NI	38
	3067.72	NI		20	16
	2230.61	NI			38
	2276.58	NI			
	4722.55	NI	Ac/A	NI	37
	3067.72	NI			
	2230.61	10		3	15
	2276.58	NI			
BiH	4424	10	H/O	NI	38
	4394	8			
	4424	10	H/A	NI	38
	4394	10			
	4424	10	Ac/O	NI	38
	4394	8			
	4424	NI	Ac/A	NI	37
	4394	NI			
BiO	5564	10	H/O	NI	38
	5564	10	H/A	NI	38
	5564	10	Ac/O	NI	38
	5564	NI	Ac/A	NI	37

EXPERIMENTALLY OBSERVED
ATOMIC EMISSION LINES

Element	Line (Å)	Relative Intensity	Flame	Limit of Detection (ppm)	Ref.
Boron B	2496.78 } 2497.73	10	PAc/O	30	1
	2496.78	10	H/A	NI	38
	2496.78	10	H/O	NI	38
	2496.78	10	Ac/O	NI	38
	2496.78	NI	Ac/A	NI	37
BO$_2$	5180.7	NI	PAc/N	0.05	6
	5476.0	10	H/O	NI	38
	5180.7	8			
	4941.3	5			
	5790.7	5			
	4719.5	3			
	6030	2			
	5476.0	10	H/A	NI	38
	5180.7	7			
	4941.3	4			
	5790.7	6			
	4719.5	2			
	6030	2			
	5476.0	10	Ac/O	NI	38
	5180.7	4			
	4941.3	3			
	5790.7	7			
	4719.5	2			
	6030	5			
	6202.2	2			
	5476.0	10	Ac/A	NI	37
	5180.7	4			
	5790.7	4			

EXPERIMENTALLY OBSERVED
ATOMIC EMISSION LINES

Element	Line (Å)	Relative Intensity	Flame	Limit of Detection (ppm)	Ref.
Bromine Br	No information				
Cadmium Cd	2288.02	NI	PAc/O	6	1
	3261.06	NI	PAc/N	0.8	6
	3261.06	NI	SAc/N	5	10
	3261.06	NI	SH/N	7	10
	3261.06	NI		5	10
	3261.06	NI	MAPP/N	12	10
	3261.06	NI	SMAPP/N	12	10
	3261.06	10	H/O	0.5	38
	2288.02	5		NI	
	3261.06	10	H/A	NI	38
	2288.02	0.1		0.3	
	3261.06	10	Ac/O	30	38
	2288.02	10		6	
	3261.06	7	Ac/A	NI	37
	2288.02	0.1		0.5	15
	3261.06	NI	H/A/Ar	0.2	22
	2288.02	NI	Ac/O/Ar	0.5	15
Calcium Ca	3933.67 II	10	PAc/O	0.005	1
	4226.73	10		0.005	
	4226.73	NI	SAc/O	0.003	16
	4226.73	NI	PAc/N	10^{-4}	21
	4226.73	NI	SAc/N	0.003	10

EXPERIMENTALLY OBSERVED
ATOMIC EMISSION LINES

Element	Line (Å)	Relative Intensity	Flame	Limit of Detection (ppm)	Ref.
Calcium Ca Continued	4226.73	NI	H/N	0.002	10
	4226.73	NI	SH/N	0.001	10
	4226.73	NI	MAPP/N	0.002	10
	4226.73	NI	SMAPP/N	0.001	10
	4226.73	NI	Ac/A/O	5×10^{-4}	9
	4226.73	10	H/O	NI	38
	4226.73	10	H/A	NI	38
	3933.67 II	NI			
	3968.47 II	NI			
	4226.73	10	Ac/O	NI	38
	3933.67 II	1			
	3968.47 II	0.6			
CaOH	6220	10	H/O	0.004	33
	5540	7		NI	38
	6078	3		NI	
	6220	10	H/A	NI	38
	5540	10			
	6038	2			
	6220	10	Ac/O	NI	38
	5540	3			
	6038	5			

Californium Cf   No information

Carbon C   No information

| Cerium Ce | 5697.00 | NI | PAc/O | 10 | 1 |

EXPERIMENTALLY OBSERVED
ATOMIC EMISSION LINES

Element	Line (Å)	Relative Intensity	Flame	Limit of Detection (ppm)	Ref.
Cerium Ce Continued	5699.23	NI	PAc/O	10	1
	5699.23	NI	PAc/N	16(Alcohol)	18
	5699.23	NI		10	6
CeO	5500-6000	10	H/O	1	34
	4940	7		NI	38
	4684	5			
	4810	5			
	5500-6000	10	H/A	0.15	33
	4940	6		NI	38
	4684	4			
	4810	4			
	5500-6000	10	PA/O	10	1
	4940	10			
	4684	7			
	4810	7			
Cesium Cs	8521.10	10	PAc/O	0.008	1
	4555.36	0.01		8	
	8521.10	NI	PAc/N	2×10^{-5}	2
	4555.36	NI		0.6	14
	8521.10	10	H/O	0.1	33
	8943.50	3			
	4555.36	0.2			
	4593.18	0.1			
	8521.10	10	H/A	NI	38
	8943.50	3			
	4555.36	0.2			
	4593.18	0.05			
	8521.10	10	Ac/A	NI	37
	8943.50	10			
	4555.36	1.0			
	4593.18	0.1			

EXPERIMENTALLY OBSERVED
ATOMIC EMISSION LINES

Element	Line (Å)	Relative Intensity	Flame	Limit of Detection (ppm)	Ref.
Chlorine Cl	4354	10	H/O	700	35
Chromium Cr	3578.69	NI	PAc/O	0.1	1
	3593.49	NI		0.1	
	3605.33	NI		0.1	
	4254.35	NI	PAc/N	0.002	5
	4254.35	NI	SAc/O	0.007	16
	4254.35	NI	Ac/A/O	0.005	9
	4254.35	NI	PAc/N	0.005	9
	4254.35	NI	SAc/N	0.001	10
	4254.35	NI	H/N	0.07	
	4254.35	NI	SH/N	0.04	
	4254.35	NI	MAPP/N	0.02	
	4254.35	NI	SMAPP/N	0.007	
	4254.35	NI	H/A/Ar	0.01	22
	4254.35	10	H/O	0.1	33
	3578.69	8		NI	38
	4274.80	8			
	3593.49	7			
	4289.72	7			
	5206.04	7			
	3605.33	6			
	4254.35	7	H/A	NI	38
	3578.68	3			
	4274.80	6			
	3593.49	1			

EXPERIMENTALLY OBSERVED
ATOMIC EMISSION LINES

Element	Line (Å)	Relative Intensity	Flame	Limit of Detection (ppm)	Ref.
Chromium Cr Continued	4289.72	6	H/A	NI	38
	5206.04	10			
	3605.33	1			
	4254.35	10	Ac/O	NI	34
	3578.69	**10**		0.01	
	4274.80	8			
	3593.49	8			
	4289.72	6			
	5206.04	5			
	3605.33	8			
	4254.35	10	Ac/A	NI	37
	3578.69	7			
	4274.80	10			
	3593.49	7			
	4289.72	10			
	5206.04	0.3			
	3605.33	7			
CrO	5852.1	10	H/O	NI	38
	6051.6	10			
	5356	10			
	5417	10			
	5564.1	10			
	5623	10			
	5794.4	10			
	6394.3	5			
	6850	5			
	5852.1	10	H/A	NI	38
	6051.6	10			
	5356	8			
	5417	8			
	5564.1	8			
	5623	8			
	5794.4	10			
	5852.1	8	Ac/O	NI	38
	6051.6	10			
	5356	5			

EXPERIMENTALLY OBSERVED
ATOMIC EMISSION LINES

Element	Line (Å)	Relative Intensity	Flame	Limit of Detection (ppm)	Ref.
Chromium Cr Continued	5417	5	Ac/O	NI	38
	5564.1	5			
	5623	5			
	5794.4	8			
	6394.3	8			
	6850	8			
Cobalt Co	3453.50	NI	PAc/O	1	1
	3453.50	NI	PAc/N	0.03	2
	4252.31	NI		0.005	9
	3526.85	NI	SAc/O	0.04	16
	3526.85	NI	SAc/N	0.2	10
	3526.85	NI	H/N	0.3	10
	3526.85	NI	SH/N	0.2	10
	3526.85	NI	MAPP/N	0.3	10
	3526.85	NI	SMAPP/N	0.25	10
	3453.50	NI	H/Ar	0.1	22
	3526.85	10	H/O	NI	38
	3412.34	7			
	3453.50	7		0.5	33
	3405.12	6			38
	3873.12	6			
	3502.28	5			
	3513.48	4			
	3575.36	4			
	3449.44	3			
	3465.80	3			
	3845.47	3			
	4121.32	3			

EXPERIMENTALLY OBSERVED
ATOMIC EMISSION LINES

Element	Line (Å)	Relative Intensity	Flame	Limit of Detection (ppm)	Ref.
Cobalt Co Continued	3443.64	3	H/O	NI	38
	3594.87	3			
	3894.08	3			
	3506.32	3			
	3995.31	3			
	3569.38	2			
	3474.02	2			
	3433.04	1			
	3526.85	8	H/A	NI	38
	3412.34	9			
	3453.50	10			
	3405.12	9			
	3873.12	7			
	3502.28	7			
	3513.48	4			
	3575.36	4			
	3449.44	6			
	3465.80	4			
	3845.47	5			
	4121.32	6			
	3443.64	4			
	3594.87	4			
	3894.08	5			
	3506.32	7			
	3995.31	6			
	3569.38	4			
	3474.02	4			
	3433.04	2			
	3526.85	10	Ac/O	NI	38
	3412.34	4			
	3453.50	10			
	3405.12	7			
	3873.12	6			
	3502.28	4			
	3513.48	4			
	3575.36	4			
	3449.44	4			
	3465.80	4			

EXPERIMENTALLY OBSERVED
ATOMIC EMISSION LINES

Element	Line (Å)	Relative Intensity	Flame	Limit of Detection (ppm)	Ref.
Cobalt Co	3845.47	3	Ac/O	NI	38
Continued	4121.32	3			
	3443.64	4			
	3594.87	3			
	3894.08	3			
	3506.32	3			
	3995.31	3			
	3569.38	2			
	3474.02	3			
	3433.04	2			
	3526.85	7	Ac/A	NI	37
	3412.34	7			
	3453.50	7			
	3405.12	7			
	3873.12	7			
	3502.28	10			
	3513.48	5			
	3575.36	7			
	3449.44	2			
	3465.80	7			
	3845.47	2			
	4121.32	2			
	3443.64	2			
	3594.87	2			
	3894.08	1			
	3506.32	7			
	3995.31	2			
	3569.38	2			
	3474.02	5			
	3433.04	2			
CoO	5635	10	H/O	NI	38
Copper Cu	3247.54	NI	PAc/O	0.1	1
	3273.96			0.2	
	3273.96	NI	PAc/N	0.003	2
	3247.54	NI		0.08	12

EXPERIMENTALLY OBSERVED
ATOMIC EMISSION LINES

Element	Line (Å)	Relative Intensity	Flame	Limit of Detection (ppm)	Ref.
Copper Cu Continued	3273.96	NI	SAc/O	0.04	16
	3273.96	NI	Ac/A/O	0.1	9
	3273.96	NI	SAc/N	0.03	10
	3273.96	NI	H/N	0.13	10
	3273.96	NI	SH/N	0.06	10
	3273.96	NI	MAPP/N	0.09	10
	3273.96	NI	SMAPP/N	0.06	10
	3247.54	10	Ac/O	NI	38
	3273.96	10			
	5105.54	0.7			
	3247.54	10	Ac/A	NI	37
	3273.96	10			
	5105.54	0.01			
	3247.54	10	H/O	0.01	33
	3273.96	10			38
	5105.54	0.8			
	3247.54	10	H/A	NI	38
	3273.96	7			
CuOH	5370	NI	H/Ar	0.05	22
	5370	10	H/A	NI	38
	5240	7			
	5050	5			
Dysprosium Dy	4186.78	NI	PAc/O	0.3	1
	4211.72	NI		0.1	

EXPERIMENTALLY OBSERVED
ATOMIC EMISSION LINES

Element	Line (Å)	Relative Intensity	Flame	Limit of Detection (ppm)	Ref.
Dysprosium Dy Continued	4045.99	NI	PAc/N	0.05	2
	4045.99	NI	PAc/N	0.02(Alcohol)	18
	4045.99	NI	H/O	0.4	14
	3531.71	0.7	Ac/O	NI	30
	3757.06	0.7			
	3868.81	0.8			
	4013.83	0.7			
	4045.99	8			
	4103.88	6			
	4130.42	0.5			
	4146.07	0.8			
	4167.97	2			
	4183.61	0.8			
	4186.81	6			
	4191.63	1			
	4194.83	5			
	4197.93	0.6			
	4202.25	0.8			
	4211.72	10			
	4213.18	2			
	4215.17	1			
	4218.09	2			
	4221.10	2			
	4222.22	1			
	4225.15	1			
	4232.03	0.5			
	4239.87	0.5			
	4565.12	0.6			
	4577.80	1			
	4589.38	4			
	4612.27	3			
	4880.17	0.7			
	5236.25	0.5			
	5301.59	2			
	5395.58	0.5			
	5423.32	0.7			

EXPERIMENTALLY OBSERVED
ATOMIC EMISSION LINES

Element	Line (Å)	Relative Intensity	Flame	Limit of Detection (ppm)	Ref.
Dysprosium Dy Continued	5451.09	2	Ac/O	NI	30
	5547.27	0.7			
	5634.50	2			
	5652.01	2			
	5832.03	2			
	5974.50	2			
	5988.59	2			
	6088.26	2			
	6168.43	1			
	6259.09	5			
	6421.93	2			
	6579.38	5			
	6835.44	4			
DyO	5204.4	10	H/O	NI	38
	5160	6			
	6057	3			
	6091	3			
	6057	10	Ac/A	NI	38
	4574.2	6			
Einsteinium Es	No information				
Erbium Er	4007.97	NI	PAc/O	0.3	1
	4151.10	NI		1	
	5826.79	NI		1	
	4007.97	NI	PAc/N	0.02	18
	3558.02	0.7	Ac/O	NI	30
	3570.76	0.5			
	3638.68	0.5			
	3746.06	0.5			
	3810.33	3			
	3855.9	0.5			
	3863.08	7			
	3892.69	6			
	3902.77	0.5			

EXPERIMENTALLY OBSERVED
ATOMIC EMISSION LINES

Element	Line (Å)	Relative Intensity	Flame	Limit of Detection (ppm)	Ref.
Erbium Er	3905.41	2	Ac/O	NI	30
Continued	3937.02	3			
	3944.42	2			
	3956.43	0.8			
	3964.51	0.7			
	3973.04	6			
	3977.02	0.8			
	3982.33	0.8			
	3987.66	0.8			
	4007.97	10			
	4012.58	0.7			
	4020.52	1			
	4021.96	0.5			
	4077.97	0.5			
	4087.64	3			
	4098.10	0.7			
	4131.5	0.5			
	4151.11	7			
	4185.72	0.5			
	4190.70	1			
	4218.43	1			
	4286.56	0.6			
	4298.91	0.5			
	4386.40	0.5			
	4409.36	2			
	4418.70	0.5			
	4424.57	1			
	4426.77	2			
	4606.61	5			
	4673.16	1			
	5172.76	0.8			
	5206.52	0.8			
	5762.79	2			
	5826.79	6			
	5855.29	1			
	5221.01	5			
	6308.79	2			
	6583.46	2			

EXPERIMENTALLY OBSERVED
ATOMIC EMISSION LINES

Element		Line (Å)	Relative Intensity	Flame	Limit of Detection (ppm)	Ref.
Erbium ErO		5520	10	H/O	0.2	34
		5459.3	8			
		5600	7			
		5041.3	6			
		5660	5			
		5150.9	3			
Europium Eu		4594.03	NI	PAc/O	0.003	1
		4594.03	NI	PAc/N	2×10^{-4}	5
		4594.03	NI	SAc/N	0.01	10
		4594.03	NI	H/N	0.1	10
		4594.03	NI	SH/N	0.06	10
		4594.03	NI	MAPP/N	0.06	10
		4594.03	NI	SMAPP/N	0.03	10
		6018.15	10	H/O	0.1	34
		4594.03	2			
		4627.22	2			
		4661.88	2			
		3334.33	2	Ac/O	NI	30
	Eu(II)	3724.99	1			
	Eu(I)	3819.66	2			
	Eu(II)	3907.11	0.8			
		3930.50	1			
		3971.99	1			
	Eu(I)	4129.62	2			
	Eu(II)	4129.74	2			
		4205.05	2			
		4435.53	1			
	Eu(I)	4594.02	10			
		4627.12	9			

EXPERIMENTALLY OBSERVED
ATOMIC EMISSION LINES

Element	Line (Å)	Relative Intensity	Flame	Limit of Detection (ppm)	Ref.
Europium Eu(I) Continued	4661.88	8	Ac/O	NI	30
	5645.80	2			
	5765.20	6			
	5831.05	1			
	5967.16	1			
	6029.01	8			
	6262.28	0.6			
	6266.95	1			
	6291.34	2			
	6299.76	0.5			
	6864.55	7			
	4594.03	10	Ac/A	NI	38
	4627.22	10			
	4661.88	10			
EuOH	5980	10	H/O	0.1	34
	6230	7			38
	7020	5			
	6470	2			
	6840	2			
	5980	10	Ac/O	NI	38
	6230	10			
	7020	10			
	6470	10			
	6840	10			
	5980	10	Ac/A	NI	37
	6230	10			
	7020	10			
	6470	10			
	6840	10			
Fermium Fm	No information				
Fluorine F	No information				

EXPERIMENTALLY OBSERVED
ATOMIC EMISSION LINES

Element	Line (Å)	Relative Intensity	Flame	Limit of Detection (ppm)	Ref.
Francium Fr	No information				
Gadolinium Gd	4346.46 } 4346.62	NI	PAc/O	4	1
	4519.66	NI		2	
	4401.86	NI	PAc/N	1.0	2
	4401.86	NI	PAc/N	1.0 (Alcohol)	18
Gd(I)	3513.66	1	Ac/O	NI	30
	3605.25	2			
	3674.06	1			
	3684.12	2			
	3757.95	1			
	3783.06	3			
	3843.26	2			
	3866.98	2			
	3934.80	1			
	3945.54	1			
	3969.00	1			
	4023.35	5			
	4045.01	3			
	4053.65	6			
	4054.73	3			
	4058.23	6			
	4078.71	6			
	4090.42	2			
	4092.71	3			
	4100.26	3			
	4134.17	3			
	4150.61	1			
	4157.79	1			
	4167.27	1			
	4175.54	1			
	4190.78	3			
	4191.62	2			
	4225.85	8			

EXPERIMENTALLY OBSERVED
ATOMIC EMISSION LINES

Element	Line (Å)	Relative Intensity	Flame	Limit of Detection (ppm)	Ref.
Gadolinium Gd(I)	4260.11	3	Ac/O	NI	30
Continued	4266.60	2			
	4267.02	2			
	4274.17	1			
	4306.35	3			
	4313.85	5			
Gd(II)	4321.11	1			
	4325.57	5			
Gd(I)	4327.10	4			
	4346.63	1			
	4373.84	4			
	4401.85	10			
	4403.14	1			
	4411.16	5			
	4414.73	5			
	4422.41	8			
	4430.63	8			
	4476.14	7			
	4486.91	4			
	4497.33	5			
Gd(II)	4506.35	5			
Gd(I)	4519.66	8			
	4537.82	7			
	4542.03	2			
	4581.30	2			
	4821.71	3			
	5015.06	1			
	5103.46	1			
	5617.91	2			
	5643.25	2			
	5696.22	3			
	5911.42	4			
	6114.08	1			
	6730.76	1			
	6828.25	1			
GdO	5910.8	10	H/O	NI	38
	5987.9	10		0.12	34
	6120	9			38
	6221.0	9			

EXPERIMENTALLY OBSERVED
ATOMIC EMISSION LINES

Element	Line (Å)	Relative Intensity	Flame	Limit of Detection (ppm)	Ref.
Gadolinium GdO	5807.5	6	H/O	NI	38
Continued	5698.3	4			
	4615.8	2			
	4633.7	2			
	5450.7	2			
	4892.0	2			
	4909.5	2			
	5405.1	2			
	5910.8	10	H/A	NI	38
	5987.9	10			
	6120	10			
	6221.0	10			
	5807.5	5			
	5698.3	3			
	5910.8	10	Ac/O	NI	38
	5987.9	10			
	6120	10			
	6221.0	10			
	5807.5	5			
	5910.8	3	Ac/A	NI	37
	5987.9	3			
	6120	7			
	6221.0	10			
	5807.5	2			
	4615.8	7			
	4633.7	5			
Gallium Ga	4172.06	NI	PAc/O	0.07	1
	4172.06	NI	SAc/N	0.08	10
	4172.06	NI	H/N	0.07	10
	4172.06	NI	SH/N	0.04	10
	4172.06	NI	MAPP/N	0.07	10
	4172.06	NI	PAc/N	0.005	2

EXPERIMENTALLY OBSERVED
ATOMIC EMISSION LINES

Element	Line (Å)	Relative Intensity	Flame	Limit of Detection (ppm)	Ref.
Gallium Ga Continued	4172.06	NI	SMAPP/N	0.05	10
	4172.06	NI	H/Ar	0.02	22
	4172.06	10	H/O	0.05	33
	4032.98	5		NI	38
	4172.06	10	H/A	NI	38
	4032.98	5			
	4172.06	10	Ac/O	NI	38
	4032.98	5			
	4172.06	10	Ac/A	NI	37
	4032.98	10			
	2944.18	2			
Germanium Ge	2651.18	NI	PAc/O	0.6	1
	2651.18	NI	PAc/N	0.4	6
	2651.58	NI	Ac/O/Ar	3	15
	2592.54	NI			
	2691.34	NI			
	2709.63	NI			
	2754.59	NI			
	2651.18	NI	SAc/N	1.3	20
	2651.18	NI	SAc/N	1.1	
	2651.18	10	H/O	NI	38
	2651.18	10	H/A	5	35
	2651.58	10	Ac/A	NI	37
	2592.54	4			
	2709.63	2			

EXPERIMENTALLY OBSERVED
ATOMIC EMISSION LINES

Element	Line (Å)	Relative Intensity	Flame	Limit of Detection (ppm)	Ref.
Germanium Ge	2651.18	NI	Ac/A	7	15
Continued	2592.54	NI			37
	2691.34	NI			
	2709.63	NI			
	2754.59	NI			
Gold Au	2675.95	NI	PAc/O	7	1
	2675.95	NI	PAc/N	0.5	2
	2675.95	10	H/O	5	33
	2427.95	5		NI	38
	2675.95	10	H/A	NI	38
	2427.95	3			
	2675.95	10	Ac/O	NI	
	2427.95	6			
	2675.95	10	Ac/A	NI	37
	2427.95	10			
Hafnium Hf	3682.24	NI	PAc/O	75	1
	5311.60 II	NI	PAc/N	20	14
Helium He	No information				
Holmium Ho	4103.84	NI	PAc/O	0.1	1
	4053.93	NI	PAc/N	0.02	2
	4053.93	NI	PAc/N	0.01 (Alcohol)	18
	3510.75	0.8	Ac/O	NI	30
	3570.4	0.5			
	3666.65	0.5			

EXPERIMENTALLY OBSERVED
ATOMIC EMISSION LINES

Element	Line (Å)	Relative Intensity	Flame	Limit of Detection (ppm)	Ref.
Holmium Ho Continued	3667.97	0.7	Ac/O	NI	30
	3682.65	0.5			
	3852.4	0.5			
	3857.65	0.5			
	3862.6	0.5			
	3891.02	0.7			
	5955.74	1			
	3998.28	1			
	4031.75	0.5			
	4037.60	0.7			
	4053.92	9			
	4101.09	0.7			
	4103.84	10			
	4108.63	1			
	4120.20	2			
	4125.65	0.7			
	4127.16	2			
	4136.24	0.7			
	4164.03	8			
	4173.42	3			
	4194.34	1			
	4254.43	2			
	4624.07	1			
	4350.73	3			
	4939.01	3			
	4979.97	2			
	5359.99	1			
	5860.28	2			
	5921.76	2			
	5973.52	1			
	5982.90	5			
	6081.79	2			
	6305.36	1			
	6604.94	8			
	6007.47	1			
	6028.99	4			
HoO	5659	10	H/O	0.08	34
	5157.0	4		NI	38
	5270.2	4			

EXPERIMENTALLY OBSERVED
ATOMIC EMISSION LINES

Element	Line (Å)	Relative Intensity	Flame	Limit of Detection (ppm)	Ref.
Holmium HoO Continued	5319.6	4	H/O	NI	34
	5104.8	3			38
	5849.4	3			
	5659	10	Ac/A	NI	37
	5157.0	2			
	5270.2	2			
	5319.6	2			
	5849.4	2			
Hydrogen H	No information				
Indium In	4511.31	NI	PAc/O	0.03	1
	4511.31	NI	PAc/N	0.001	2
	4511.31	NI	Ac/N	0.04	10
	4511.31	NI	H/N	0.05	10
	4511.31	NI	SH/N	0.03	10
	4511.31	NI	MAPP/N	0.08	10
	4511.31	NI	SMAPP/N	0.07	10
	4511.31	NI	H/Ar	0.005	22
	4511.31	10	H/O	0.03	33
	4101.76	6			
	4511.31	10	H/A	NI	38
	4101.76	6			
	3256.09	0.6			
	4511.31	10	Ac/O	NI	38
	4101.76	7			

EXPERIMENTALLY OBSERVED
ATOMIC EMISSION LINES

Element	Line (Å)	Relative Intensity	Flame	Limit of Detection (ppm)	Ref.
Indium In Continued	4511.31	10	Ac/A	NI	33
	4101.76	10			37
	3256.09	3			
	3039.36	3			
Iodine I	2062	NI	Ac/A	2500	15
	2062	NI	Ac/O/Ar	1000	15
IO	4694	10	H/O	NI	38
	4845	10			
	4964	10			
	5131	10			
	5209	10			
	5308	10			
	5533	10			
	5730	10			
Iridium Ir	3513.64	NI	PAc/O	200	1
	3800.12			100	
	3800.12	NI	PAc/N	3	14
	5500 (band)	NI		0.4	
Iron Fe	3719.94	NI	PAc/O	0.7	1
	3719.94	NI	PAc/N	0.01	2
	3719.94	NI	SAc/O	0.03	6
	3719.94	NI	Ac/A/O	0.2	9
	3719.94	NI	SAc/N	0.05	10
	3719.94	NI	H/N	0.09	10
	3719.94	NI	SH/N	0.07	10

EXPERIMENTALLY OBSERVED
ATOMIC EMISSION LINES

Element	Line (Å)	Relative Intensity	Flame	Limit of Detection (ppm)	Ref.
Iron Fe Continued	3719.94	NI	MAPP/N	0.12	10
	3719.94	NI	SMAPP/N	0.08	10
	3719.94	10	H/O	0.2	33
	3859.91	9		NI	38
	3737.13	8			
	3747	6			
	3719.94	10	H/A	NI	38
	3859.91	4			
	3737.13	8			
	3747	4			
	3719.94	10	Ac/O	0.5	6
	3859.91	7		NI	38
	3737.13	5			
	3747	4			
	3719.94	10	Ac/A	NI	37
	3859.91	10			
	3737.13	10			
	3747	9			
FeO	5646.6	10	H/O	NI	38
	5789.8	10			
	5819.2	10			
	5868.1	10			
	5614.0	9			
	5531.4	5			
	6097.3	5			
	6180.5	5			
	6218.9	5			
Fe	5269.54	4			
	5328.05	4			
	3886.28	3			
	3825.88	3			
	3930.30	2			
	3878.58	2			
	3440.61	2			
	3581.20	2			

EXPERIMENTALLY OBSERVED
ATOMIC EMISSION LINES

Element	Line (Å)	Relative Intensity	Flame	Limit of Detection (ppm)	Ref.
Iron FeO	5646.6	5	H/A	NI	38
Continued	5789.8	10			
	5819.2 }	10		0.06	34
	5868.1				
	5614.0	9		NI	38
	5531.4	6			
	6097.3	6			
	6180.5	6			
	6218.9	6			
Fe	5269.54	4			
	5328.05	4			
	3886.28	4			
	3825.88	4			
	3930.30	4			
	3878.58	3			
	3440.61	4			
	3581.20	3			
	3899.71	3			
	4383.55	3			
FeO	5646.6	10	Ac/O	NI	38
	5789.8	10			
	5819.2	10			
	5868.1	10			
	5614.0	8			
	5531.4	5			
	6097.3	10			
	6180.5	10			
	6218.9	10			
Fe	5269.54	3			
	5328.05	3			
	3886.28	2			
	3886.28	10	Ac/A	NI	37
	3825.88	10			
	3930.30	5			
	3878.58	3			
	3440.61	10			
	3581.20	3			

EXPERIMENTALLY OBSERVED
ATOMIC EMISSION LINES

Element	Line (Å)	Relative Intensity	Flame	Limit of Detection (ppm)	Ref.
Iron Fe	3899.71	3	Ac/A	NI	37
Continued	3020.64	10			
Krypton Kr	No information				
Lanthanum La	5791.34	NI	PAc/O	1	1
	5791.34	NI	PAc/N	1	2
	5501.34	NI		4	18
	3574.43	2	Ac/O	NI	30
	3927.56	2			
La(II)	4007.66	2			
La(I)	4015.39	3			
	4037.21	3			
	4060.33	2			
	4064.79	2			
	4079.18	2			
	4089.61	1			
	4104.87	2			
	4137.04	1			
	4160.26	1			
	4187.32	4			
	4280.27	5			
	4549.50	1			
	4567.91	1			
	4570.02	1			
	4766.89	2			
	4850.82	1			
	4949.77	3			
	5145.42	3			
	5158.69	3			
	5177.31	3			
	5211.86	2			
	5234.27	2			
	5253.46	2			
	5271.19	3			

EXPERIMENTALLY OBSERVED
ATOMIC EMISSION LINES

Element	Line (Å)	Relative Intensity	Flame	Limit of Detection (ppm)	Ref.
Lanthanum La	5455.15	6	Ac/O	NI	30
	5501.34	8			
	5740.66	2			
	5761.84	1			
	5769.34	2			
	5789.24	4			
	5791.34	4			
	5930.62 } 5930.67	7			
	6249.93	6			
	6325.91	3			
	6394.23	8			
	6410.99	5			
	6454.52	3			
	6455.99	8			
	6543.16	5			
	6578.51	10			
	6616.59	1			
	6644.41	1			
	6650.81	4			
	6661.40	1			

EXPERIMENTALLY OBSERVED
ATOMIC EMISSION LINES

Element	Line (Å)	Relative Intensity	Flame	Limit of Detection (ppm)	Ref.
Lanthanum LaO	5602.4	10	H/O	NI	38
Continued	5628.6	7			
	4379.7	4			
	4418.2	NI		6	13
	4423.2	4		NI	38
	5406	3			
	5431	3			
	7430	3			
	7920	3		0.005	33
	5457	3		NI	38
	5382.5	2			
	5602.4 }	6	H/A	NI	38
	5628.6				
	4379.7	2			
	4423.2	2			
	7430	10			
	7920	8			
	7430	10	Ac/O	NI	38
	7920	10			
	5602.4	2	Ac/A	NI	37
	4379.7	7			
	4423.2	4			
	7430	10			
	7920	10			
	4400.0	7			
	7691.1	10			
	7725.5	10			
	8159.0	10			
	8233.1	10			
	8270.7	10			
Lead Pb	3683.48	NI	PAc/O	3	1
	4057.83	NI	PAc/N	0.1	6
	3683.48	NI		2×10^{-4}	9

EXPERIMENTALLY OBSERVED
ATOMIC EMISSION LINES

Element	Line (Å)	Relative Intensity	Flame	Limit of Detection (ppm)	Ref.
Lead Pb Continued	2614.18	NI	Ac/O/Ar	1	15
	2169.99	NI			
	2833.06	NI			
	3639.58	NI			
	4057.83	NI	SAc/O	0.5	16
	4057.83	NI	Ac/A/O	2	16
	3683.48	NI	Ac/A/O	NI	16
	4057.83	NI	SAc/N	0.4	10
	4057.83	NI	H/N	1.0	10
	4057.83	NI	SH/N	0.8	10
	4057.83	NI	MAPP/N	0.8	10
	4057.83	NI	SMAPP/N	0.7	10
	3683.48	10	H/O	NI	
	4057.83	10		1	33
	3639.58	5			
	3683.48	10	H/A	NI	38
	4057.83	10			
	3639.58	5			
	3683.48	7	Ac/O	3	9
	4057.83	10		10	16
	3639.58	3			
	3683.48	10	Ac/A	NI	37
	4057.83	10			
	3639.58	3			
	2833.06	3			

EXPERIMENTALLY OBSERVED
ATOMIC EMISSION LINES

Element	Line (Å)	Relative Intensity	Flame	Limit of Detection (ppm)	Ref.
Lithium Li	3232.63	NI	PAc/O	0.1	1
	6103.64	NI		0.001	
	6707.84	NI		3×10^{-6}	
	6707.84	NI	PAc/N	1×10^{-6}	2
	3232.63	NI		> 10ppm	5
	4602.86	NI		> 10ppm	5
	6103.64	NI		0.1	5
	6707.84	10	H/O	2×10^{-4}	33
	6707.84	10	H/A	NI	38
	6707.84	10	Ac/O	0.05	9
	6707.84	10	Ac/A	NI	37
Lutetium Lu	3312.11	NI	PAc/O	0.2	1
	4518.57	NI	PAc/N	0.4	18
	2989.27	0.8	Ac/O	NI	30
	3081.47	2			
	3118.43	2			
	3171.36	1			
	3278.97	2			
	3281.74	4			
	3312.11	6			
	3359.56	10			
	3376.50	5			
	3385.50	1			
	3396.82	2			
	3508.42	2			
	3567.84	7			
	3636.25	0.8			
	3647.77	2			
	3841.18	2			
	3868.46	1			

EXPERIMENTALLY OBSERVED
ATOMIC EMISSION LINES

Element	Line (Å)	Relative Intensity	Flame	Limit of Detection (ppm)	Ref.
Lutetium Lu Continued	4054.45	0.8	Ac/O	NI	30
	4124.73	2			38
	4518.57	4			
	5001.14	0.8			
	5402.57	2			
	6004.52	2			
	6055.03	2			
LuO	5170.3	10	H/O	0.2	34
	4661.7	6			38
	6000	4			
	6750	2			
	5170.3	10	Ac/O	NI	38
	4661.7	0.5			
	4661.7	0.5	Ac/A	NI	37
	6000	0.5			
	6750	10			
Magnesium Mg	2852.13	10	PAc/O	0.2	1
	2852.13	10	PAc/N	0.005	21
	2852.13	10	SAc/O	0.3	16
	2852.13	10	Ac/A/O	.05	9
	2852.13	10	SAc/N	0.001	10
	2852.13	10	H/N	0.08	10
	2852.13	10	SH/N	0.07	10
	2852.13	10	MAPP/N	0.05	10
	2852.13	10	SMAPP/N	0.04	10
	2852.13	10	H/O	0.1	34
	5172.68 } 5183.60	5		NI	38

EXPERIMENTALLY OBSERVED
ATOMIC EMISSION LINES

Element	Line (Å)	Relative Intensity	Flame	Limit of Detection (ppm)	Ref.
Magnesium Mg Continued	2852.13	10	H/A	NI	38
	2852.13	10	Ac/O	0.2	9
	2852.13	10	Ac/A	NI	37
MgOH	3702	NI	H/Ar	0.02	22
	3702	10	H/O	NI	38
	3810-3830	8			
	3877	5			
	3912	5			
	3624	2			
	3702	10	H/A	0.02	33
	3810-3830	10		NI	38
	3877	4			
	3912	4			
	3624	2			
	3702	10	Ac/A	NI	37
	3810-3830	10			
	3877	3			
	3912	3			
	3624	0.5			
MgO	5007	10	H/O	NI	38
	5007	10	H/A	NI	38
Manganese Mn	4030.76	NI	PAc/O	0.1	1
	4030.76	NI	PAc/N	0.001	5
	4033.07	NI		0.008	6
	4033.07	NI	SAc/N	0.005	10
	4033.07	NI	H/N	0.01	10

EXPERIMENTALLY OBSERVED
ATOMIC EMISSION LINES

Element	Line (Å)	Relative Intensity	Flame	Limit of Detection (ppm)	Ref.
	4033.07	NI	SH/N	0.004	10
Manganese Mn Continued	4033.07	NI	MAPP/N	0.03	10
	4033.07	NI	SMAPP/N	0.02	10
	4032	NI	H/Ar	0.01	22
	4030.76 }				
	4033.07 }	10			
	4034.49 }				
	4033.07	10	H/O	0.01	33
	4033.07	10	H/A	NI	38
	4033.07	10	Ac/O	0.1	9
	4033.07	10	Ac/A	NI	38
	2794.82	2			
	2798.27	2			
	2801.06	2			
MnO	5586.5	10	H/A	NI	38
	5609.3	10			
	5389.5	5			
	5880.3	7			
	5359.4	5			
	5860	7			
	5423.7	5			
	5192.5	5			
	5228.4	5			
	6154.7	3			
	5158.0	5			
	5586.5	10	Ac/A	NI	37
	5609.3	6			
	5389.5	3			
	5880.3	10			
	5359.4	2			
	5860	3			
	5423.7	2			

EXPERIMENTALLY OBSERVED
ATOMIC EMISSION LINES

Element	Line (Å)	Relative Intensity	Flame	Limit of Detection (ppm)	Ref.
Mendelevium Md	No information				
Mercury Hg	2536.52	10	PAc/O	40	1
	2536.52	10	PAc/N	10	14
	2536.52	10	Ac/O/Ar	1	15
	2536.52	10		0.15	14
	2536.52	10	H/O	6	33
	2536.52	10	H/A	4	35
	2536.52	10	Ac/O	NI	38
	2536.52	10	Ac/A	NI	37
Molybdenum Mo	3798.25	NI	PAc/O	0.03	1
	3902.96	NI	PAc/N	0.01	2
	3798.25	NI	SAc/N	0.3	10
	3208.83	NI			
	3193.97	NI		0.3	8
	3170.35	NI		0.5	11
	3158.16	NI		1.5	11
	3132.59	NI		0.5	11
	3902.96	NI		0.2	7
	3798.25	NI	H/N	19.0	10
	3798.25	NI	SH/N	12.0	10
	3798.25	NI	MAPP/N	9	10
	3798.25	NI	SMAPP/N	8	10
	3798.25	10	H/O	NI	38

EXPERIMENTALLY OBSERVED
ATOMIC EMISSION LINES

Element	Line (Å)	Relative Intensity	Flame	Limit of Detection (ppm)	Ref.
Molybdenum Mo Continued	3864.11	10	H/O	NI	38
	3902.96	10			
	3798.25	10	H/A	NI	38
	3864.11	10			
	3902.96	10			
	3798.25	10	Ac/O	NI	29
	3864.11	5			38
	3902.96	5			
	3798.25	10	Ac/A	80	15
	3864.11	7		NI	37
	3902.96	7		NI	
MoO$_2$	5500–6000	10	H/O	1	33
	5500–6000	10	H/A	0.4	34
	5500–6000	10	Ac/A	NI	37
Neodymium Nd	4883.81	NI	PAc/O	1	19
	4924.53	NI		2	1
	4924.53	NI	PAc/N	0.2	18
	4444.98	0.9	Ac/O	NI	30
	4456.13	0.9			
	4477.88	1			
	4480.97	2			
	4481.89	1			
	4527.24	0.6			
	4529.76	0.9			
	4542.05	1			
	4548.24	0.6			
	4559.67	1			
	4560.42	1			
	4561.85	0.6			

EXPERIMENTALLY OBSERVED
ATOMIC EMISSION LINES

Element	Line (Å)	Relative Intensity	Flame	Limit of Detection (ppm)	Ref.
Neodymium Nd	4586.61	1	Ac/O	NI	30
Continued	4609.87	1			
	4621.94	1			
	4624.20	0.6			
	4626.50	0.6			
	4627.99	0.9			
	4634.23	4			
	4637.21	0.9			
	4639.14	1			
	4641.10	1			
	4646.40	2			
	4649.67	2			
	4651.02	1			
	4652.39	1			
	4654.73	2			
	4671.09	0.8			
	4683.44	4			
	4684.04	1			
	4690.34	1			
	4696.44	1			
	4706.96	2			
	4719.03	3			
	4726.56	0.6			
	4731.78	2			
	4734.91	0.6			
	4749.75	1.4			
	4755.85	0.6			
	4758.50	0.6			
	4759.34	0.6			
	4760.46	0.6			
	4770.19	1			
	4772.26	0.6			
	4778.40	0.6			
	4779.46	2			
	4806.62	1			
	4835.66	0.6			
	4836.62	1			
	4853.34	0.8			
	4855.32	0.9			

EXPERIMENTALLY OBSERVED
ATOMIC EMISSION LINES

Element	Line (Å)	Relative Intensity	Flame	Limit of Detection (ppm)	Ref.
Neodymium Nd	4859.59	0.6	Ac/O	NI	30
Continued	4866.73	1			
	4879.79	0.6			
	4883.81	2			
	4885.01	0.6			
	4891.06	3			
	4893.22	1			
	4896.93	4			
	4901.54	1			
	4901.85	4			
	4910.06	1			
	4913.42	3			
	4924.53	10			
	4944.83	4			
	4952.51	0.9			
	4954.78	4			
	4963.33	0.9			
	4969.75	0.6			
	4975.49	0.9			
	5029.45	0.9			
	5040.19	0.6			
	5056.89	1			
	5071.87	0.6			
	5074.5	0.6			
	5103.11	0.9			
	5105.35	0.6			
	5149.56	0.6			
	5198.07	0.6			
	5199.73	0.6			
	5204.38	0.9			
	5377.79	0.9			
	5529.07	0.6			
	5561.17	1			
	5611.18	3			
	5669.77	0.9			
	5675.97	4			
	5729.29	2			
	5749.66	0.9			
	5772.16	0.6			
	5776.12	0.9			

EXPERIMENTALLY OBSERVED
ATOMIC EMISSION LINES

Element	Line (Å)	Relative Intensity	Flame	Limit of Detection (ppm)	Ref.
Neodymium Nd Continued	5784.96	0.9	Ac/O	NI	30
	5788.22	0.6			
	5800.09	0.6			
	5826.74	0.6			
	5887.91	1			
	5921.22	0.6			
	6385.20	0.9			
	6485.69	0.9			
NdO	5990	10	H/O	NI	38
	6600	10			
	6910	10			
	7020	10		2	34
	7120	10		NI	38
	5314	5			
	4620	3			
	6220	2			
	6360	2			
	6430	2			
	6600	10	Ac/O	1	33
	6910	5		NI	38
	7020	10			
	7120	10			
	6360	2			
	6430	3			
	6600	10	Ac/A	NI	37
	5910	10			
	7020	10			
	7120	10			
	6360	2			
	6430	7			
	6623.9	10			
	6942.6	10			

Neon Ne   No information

Neptunium Np   No information

EXPERIMENTALLY OBSERVED
ATOMIC EMISSION LINES

Element	Line (Å)	Relative Intensity	Flame	Limit of Detection (ppm)	Ref.
Nickel Ni	3414.76	NI	PAc/O	1.0	1
	3524.54	NI		0.6	
	3414.76	NI	PAc/N	0.01	2
	3524.54	NI		0.02	6
	3524.54	NI	SAc/N	0.05	10
	3524.54	NI	H/N	0.08	10
	3524.54	NI	SH/N	0.06	10
	3524.54	NI	MAPP/N	0.2	10
	3524.54	NI	SMAPP/N	0.1	10
	3524.54	NI	H/Ar	0.02	22
	3414.76	10	H/O	NI	38
	3524.54	10		0.3	33
	3515.05	4		NI	38
	3619.39	4			
	3461.65	4			
	3392.99	3			
	3492.96	3			
	3446.26	2			
	3433.56	2			
	3566.37	2			
	3369.57	1			
	3380.57	1			
	3510.34	1			
	3610.46	0.8			
	3472.54	0.8			
	3858.30	0.8			
	3002.49	0.6			
	3414.76	10	H/A	0.12	34
	3524.54	10		NI	38
	3515.05	4			
	3619.39	4			
	3461.65	4			

EXPERIMENTALLY OBSERVED
ATOMIC EMISSION LINES

Element	Line (Å)	Relative Intensity	Flame	Limit of Detection (ppm)	Ref.
Nickel Ni	3392.99	3	H/A	NI	34
Continued	3492.96	3			38
	3446.26	2			
	3433.56	2			
	3566.37	2			
	3369.57	2			
	3380.57	2			
	3510.34	6			
	3610.46	2			
	3472.54	0.9			
	3858.30	1			
	3002.49	0.5			
	3414.76	7	Ac/O	NI	38
	3524.54	10			
	3515.05	6			
	3619.39	3			
	3461.65	5			
	3392.99	3			
	3492.96	5			
	3446.26	3			
	3433.56	2			
	3566.37	2			
	3369.57	2			
	3380.57	2			
	3510.34	3			
	3610.46	2			
	3472.54	2			
	3858.30	1			
	3002.49	0.6			
	3414.76	10	Ac/A	NI	37
	3524.54	10			
	3515.05	10			
	3619.39	10			
	3461.65	10			
	3392.99	10			
	3492.96	10			
	3446.26	10			
	3433.56	10			

EXPERIMENTALLY OBSERVED
ATOMIC EMISSION LINES

Element	Line (Å)	Relative Intensity	Flame	Limit of Detection (ppm)	Ref.
Nickel Ni	3566.37	3	Ac/A	NI	37
Continued	3369.57	10			
	3380.57	3			
	3510.34	3			
	3610.46	0.5			
	3472.54	3			
	3858.30	3			
	3002.49	3			
NiO	5200-6000	10	H/O	NI	38
	5174	9			
	5024	6			
Niobium Nb	4058.94	NI	PAc/O	1	1
	4058.94	NI	Ac/N	0.5	8
	4058.94	NI	SAc/N	0.06	8
	3535.30	2	Ac/O	NI	31
	3580.27	3			
	3697.85	2			
	3713.01	2			
	3726.24	2			
	3739.80	2			
	3742.39	2			
	3787.06	2			
	3790.15	2			
	3791.21	2			
	3798.12	2			
	3802.92	2			
	4058.94	10			
	4079.73	8			
	4100.92	6			
	4123.81	5			
	4137.10	3			
	4139.71	2			
	4152.58	4			
	4163.66	4			

EXPERIMENTALLY OBSERVED
ATOMIC EMISSION LINES

Element	Line (Å)	Relative Intensity	Flame	Limit of Detection (ppm)	Ref.
Niobium Nb Continued	4164.66 4168.13	4 4	Ac/O	NI	31
NbO	4500 5500	10 10	H/O	NI	38
	4500 5500	10 10	H/A	2	33
	4500 5500	6 10	Ac/A	NI	37
Nitrogen N	No information				
Nobelium No	No information				
Osmium Os	4420.47	NI	PAc/O	10	1
	4420.47	NI	PAc/N	2	6
	2909.06 3018.04 3058.66 3301.56	10 5 5 5	Ac/A	NI	37
Oxygen O	No information				
Palladium Pd	3634.70	NI	PAc/O	1	1
	3634.70	NI	PAc/N	0.05	2
	3634.70 3404.58 3609.55 3242.70	10 9 5 3	SH/N	0.04 0.09 0.18 2.0	23

EXPERIMENTALLY OBSERVED
ATOMIC EMISSION LINES

Element	Line (Å)	Relative Intensity	Flame	Limit of Detection (ppm)	Ref.
Palladium Pd	3421.24	2	SH/N	0.70	23
Continued	3516.94	2		0.70	
	3460.77	2		1.5	
	3481.15	0.7		5.0	
	3553.08	5		1.5	
	2447.91	2		2	
	3634.70	NI	SAc/N	0.04	10
	3634.70	NI	H/N	0.13	10
	3634.70	NI	MAPP/N	0.08	10
	3634.70	NI	SMAPP/N	0.05	10
	3634.70	10	H/O	0.2	33
	3404.58	9			38
	3609.55	6			
	3516.94	2			
	3421.24	1			
	3460.77	1			
	3634.70	10	H/A	NI	38
	3404.58	8			
	3609.55	5			
	3516.94	1			
	3421.24	2			
	3460.77	1			
	3634.70	10	Ac/O	NI	38
	3404.58	10			
	3609.55	7			
	3516.94	2			
	3421.24	2			
	3460.77	2			
	3634.70	10	Ac/A	NI	37
	3404.58	10			
	3609.55	10			
	3516.94	7			

EXPERIMENTALLY OBSERVED ATOMIC EMISSION LINES

Element	Line (Å)	Relative Intensity	Flame	Limit of Detection (ppm)	Ref.
Palladium Pd Continued	3421.24	7	Ac/A	NI	37
	3460.77	5			
Phosphorous P	2535.65	NI	Ac/O	400	24
	2464.0	NI	H/O	900	24
	2464.0	0.2	H/A	5	24
	5408	10		500	38
	5280	NI	H/N$_2$	0.1	25
	5408	NI	H/O	100	24
	5200	NI		NI	38
	2464.0	NI		400	24
	2375	NI		NI	38
	2385	NI			
	2478	NI			
	2396	NI			
	25 29	NI			
	2540	NI			
Platinum Pt	2659.45	NI	PAc/O	40	1
	2659.45	NI	PAc/N	2	4
	3064.71	NI	SAc/O	2	16
	3064.71	10	H/O	10	33
	2659.45	8			
	3064.71	10	H/A	NI	38
	2659.45	5			
	3064.71	10	Ac/O	NI	38
	2659.45	7			
	3064.71	10	Ac/A	NI	37

EXPERIMENTALLY OBSERVED
ATOMIC EMISSION LINES

Element	Line (Å)	Relative Intensity	Flame	Limit of Detection (ppm)	Ref.
Platinum Pt Continued	2659.45	10	Ac/A	NI	37
Plutonium Pu	No information				
Polonium Po	No information				
Potassium K	7664.91	NI	PAc/O	0.003	1
	7698.98	NI		0.02	
	7664.91	NI	PAc/N	10^{-5}	2
	7664.91 } 7698.98	10	H/O	3×10^{-4}	34
	7664.91 } 7698.98	10	H/A	NI	38
	7664.91	NI	Ac/O	NI	38
	7698.98	NI			
	7664.91	10	Ac/A	NI	37
	7698.98	10			
	4044.15	2			
Praseodymium Pr	4939.74	NI	PAc/O	2	1
	4939.74	NI	PAc/N	0.5	2
	4951.36	NI		0.5	18
	4163.01	0.8	Ac/O	NI	30
	4552.26	0.8			
	4632.28	0.8			
	4635.59	1			
	4639.55	1			
	4639.88	1			
	4674.80	0.8			

EXPERIMENTALLY OBSERVED
ATOMIC EMISSION LINES

Element	Line (Å)	Relative Intensity	Flame	Limit of Detection (ppm)	Ref.
Praseodymium Pr Continued	4687.81	1	Ac/O	NI	30
	4695.77	1			
	4713.10	0.8			
	4730.89	1			
	4744.16	0.8			
	4896.13	1			
	4906.98	1			
	4914.03	1			
	4924.59	6			
	4936.00	0.8			
	4939.73	6			
	4940.30	2			
	4951.36	10			
	4976.40	0.8			
	5018.58	2			
	5019.75	2			
	5026.97	3			
	5033.38	1			
	5043.83	2			
	5045.53	2			
	5053.40	2			
	5087.11	1			
	5133.42	6			
	5194.41	0.8			
	5228.00	2			
PrO	5763	10			
	5691	8			
	5610	7			
	5380	7			
	6030	7			
	6950	7			
	7095	7		10	33
	7321 }	7		NI	
	7376				
	5157	5			
	6298	3			
	6363.1	3			
	6481.2	3			

EXPERIMENTALLY OBSERVED
ATOMIC EMISSION LINES

Element	Line (Å)	Relative Intensity	Flame	Limit of Detection (ppm)	Ref.
Praseodymium PrO	8050	3	Ac/O	NI	38
Continued	8494	1			
	6950	7	Ac/A	NI	37
	7095	10			
	7321	7			
	7376	2			
	8050	5			
	8494.0	7			
Promethium PmO	6400	10	H/O	NI	38
	6800	10			
	6400	10	Ac/O	NI	38
	6800	10			
Protactinium Pa	No information				
Radium Ra	4825.9	10	H/O	NI	38
Ra (II)	3814.4	6			
	4682.3	2			
RaOH	6270	10	H/O	NI	38
	6650	10			
	6020	2			
Radon Rn	No information				
Rhenium Re	3460.46	NI	PAc/O	1	1
	3464.73	6		4	26
	3460.46	10	PAc/N	0.2	4
	3451.88	3		4	26
	4889.14	6		1.5	26
	5776.83	2		4	26

EXPERIMENTALLY OBSERVED
ATOMIC EMISSION LINES

Element	Line (Å)	Relative Intensity	Flame	Limit of Detection (ppm)	Ref.
Rhenium Re Continued	4889.14	10	H/O	100	33
	5275.56	10		NI	
	3451.88	4	Ac/O	NI	38
	3460.46	10			
	3464.73	8			
	4889.14	3			
	5275.56	2			
Rhodium Rh	3692.36	NI	PAc/O	0.3	1
	3692.36	NI	PAc/N	0.03	6
	3692.36	NI	SAc/N	0.01	10
	3692.36	NI	H/N	0.03	10
	3692.36	NI	SH/N	0.01	10
	3692.36	NI	MAPP/N	0.05	10
	3692.36	NI	SMAPP/N	0.04	10
	3692.36	10	H/O	0.7	33
	3434.89	4		NI	38
	3657.99	3			
	3700.91	3			
	3396.85	2			
	3502.52	2			
	4374.80	2			
	3528.02	2			
	3596.19	1			
	4211.14	1			
	3583.10	1			
	3856.52	1			
	3323.09	0.7			
	3692.36	5	H/A	NI	38

EXPERIMENTALLY OBSERVED
ATOMIC EMISSION LINES

Element	Line (Å)	Relative Intensity	Flame	Limit of Detection (ppm)	Ref.
Rhodium Rh	3434.89	2	H/A	NI	38
Continued	3657.99	4			
	3700.91	4			
	3396.85	3			
	3502.52	2			
	4374.80	10			
	3528.02	2			
	3596.19	3			
	4211.14	7			
	3583.10	3			
	3856.52	5			
	3323.09	1			
	3692.36	10	Ac/O	NI	38
	3434.89	10			
	3657.99	4			
	3700.91	3			
	3502.52	3			
	4374.80	1			
	3528.02	3			
	3596.19	1			
	4211.14	1			
	3583.10	1			
	3856.52	1			
	3323.09	1			
	3396.85	3			
	3692.36	10	Ac/A	NI	37
	3434.89	10			
	3657.99	10			
	3700.91	7			
	3396.85	7			
	3502.52	7			
	4374.80	2			
	3528.02	7			
	3596.19	7			
	4211.14	2			
	3583.10	5			
	3856.52	5			
	3323.09	5			

EXPERIMENTALLY OBSERVED
ATOMIC EMISSION LINES

Element	Line (Å)	Relative Intensity	Flame	Limit of Detection (ppm)	Ref.
Rhodium RhO	5425	10	H/O	NI	38
	5425	10	H/A	NI	38
Rubidium Rb	7800.23	NI	PAc/O	0.002	1
	7981	NI	PAc/N	2×10^{-5}	2
	7947.60	NI		3	14
	7800.23	NI		0.008	14
	7800.23	10	H/O	0.003	34
	7947.60	7.2		NI	
	7800.23	10	H/A	NI	38
	7947.60	7			
	7800.23	10	Ac/O	NI	38
	7947.60	8			
	7800.23	10	Ac/A	NI	37
	7947.60	10			
	4201.85	0.3			
	4215.56	1			
Ruthenium Ru	3728.03	NI	PAc/O	0.3	1
	3728.03	NI	PAc/N	0.3	14
	3728.03	10	H/O	0.5	33
	3799.35	5		NI	38
	3498.94	2			
	3428.31	1			
	3661.35	1			
	3436.74	0.5			
	3593.02	0.5			
	5699.05	2			
	3728.03	10	Ac/O	NI	38

EXPERIMENTALLY OBSERVED
ATOMIC EMISSION LINES

Element	Line (Å)	Relative Intensity	Flame	Limit of Detection (ppm)	Ref.
Ruthenium Ru Continued	3799.35	8	Ac/O	NI	38
	3498.94	1			
	3428.31	0.5			
	3661.35	0.7			
	3436.74	0.3			
	3593.02	0.3			
	5699.05	0.3			
	3728.03	10	Ac/A	NI	37
	3799.35	10			
	3498.94	10			
	3661.35	7			
	3436.74	7			
	3593.02	5			
Samarium Sm	4883.77 } 4883.97	NI	PAc/O	0.6	1
	5175.42	NI		0.8	19
	4760.27	NI	PAc/N	0.05	2
	4760.27	NI		0.05	18
	4783.10	NI		0.06	18
	4883.77 } 4883.97	NI		0.05	18
	4690.08	0.8	Ac/O	NI	30
	3745.46	0.6			
	3748.51	0.6			
	3756.41	0.6			
	3773.34	0.6			
	3803.94	1			
	3834.41	1			
	3853.29	1			
	3854.56	1			
	3860.14	0.6			
	3877.47	0.8			
	3881.79	1			
	3925.20	2			
	3951.88	2			

EXPERIMENTALLY OBSERVED
ATOMIC EMISSION LINES

Element	Line (Å)	Relative Intensity	Flame	Limit of Detection (ppm)	Ref.
Samarium Sm	3962.14	0.8	Ac/O	NI	30
Continued	3974.66	2			
	3990.00	2			
	3991.02	0.6			
	3998.35	1			
	4079.83	0.6			
	4135.50	0.6			
	4145.24	0.6			
	4183.33	2			
	4205.77	1			
	4219.30	0.6			
	4226.17	2			
	4240.45	0.6			
	4266.31	0.8			
	4271.86	1			
	4282.20	3			
	4282.83	3			
	4283.50	3			
	4296.75	4			
	4299.14	1			
	4312.85	1			
	4319.53	3			
	4324.46	2			
	4330.01	3			
	4331.44	3			
	4334.15	1			
	4336.13	4			
	4339.35	1			
	4350.81	0.6			
	4362.91	3			
	4365.95	0.6			
	4380.42	2			
	4386.22	1			
	4397.35	2			
	4401.17	4			
	4403.12	8			
	4411.58	6			
	4419.34	9			
	4423.38	1			
Sm (II)	4424.34	0.6			

EXPERIMENTALLY OBSERVED
ATOMIC EMISSION LINES

Element	Line (Å)	Relative Intensity	Flame	Limit of Detection (ppm)	Ref.
Samarium Sm (I)	4429.64	8	Ac/O	NI	30
Continued	4433.07	8			
Sm (II)	4433.88	0.6			
	4434.32	0.6			
Sm (I)	4441.80	8			
	4442.27	8			
	4443.27	1			
	4445.15	4			
	4452.95	2			
	4459.29	4			
	4470.89	6			
	4477.49	1			
	4480.31	6			
	4499.10	7			
	4503.38	7			
	4511.31	4			
	4522.54	5			
	4523.18	3			
	4527.42	1			
	4532.44	1			
	4533.80	4			
	4550.04	0.6			
	4566.78	1			
	4569.58	0.6			
	4581.58	4			
	4596.75	2			
	4598.36	1			
	4611.26	1			
	4629.43	0.2			
	4645.40	4			
	4648.08	1			
	4649.49	4			
	4663.55	2			
	4670.77	7			
	4681.56	2			
	4688.74	4			
	4716.11	8			
	4717.09	8			
	4718.67	0.6			
	4728.44	10			
	4750.73	2			

EXPERIMENTALLY OBSERVED
ATOMIC EMISSION LINES

Element	Line (Å)	Relative Intensity	Flame	Limit of Detection (ppm)	Ref.
Samarium Sm	4760.26	10	Ac/O	NI	30
Continued	4770.19	1			
	4783.10	10			
	4770.19	1			
	4783.10	10			
	4785.87	9			
	4789.96	2			
	4841.70	7			
	4843.33	6			
	4883.78 } 4883.98	8			
	4905.98	6			
	4910.41	8			
	4918.98	7			
	4924.06	2			
	4946.31	3			
	4975.99	6			
	5044.28	6			
	5049.50	0.6			
	5060.92	0.6			
	5088.32	0.8			
	5117.16	7			
	5122.14	5			
	5157.22	1			
	5172.74	3			
	5175.42	4			
	5187.08	0.6			
	5200.59	10			
	5251.89	3			
	5265.65	0.6			
	5271.40	8			
	5282.91	3			
	5320.60	3			
	5341.29	3			
	5348.09	0.6			
	5349.14	0.8			
	5350.62	0.6			
	5368.36	3			
	5403.69	3			
	5405.24	4			

EXPERIMENTALLY OBSERVED
ATOMIC EMISSION LINES

Element	Line (Å)	Relative Intensity	Flame	Limit of Detection (ppm)	Ref.
Samarium Sm	5416.35	1	Ac/O	NI	30
Continued	5421.57	1			
	5453.02	4			
	5466.73	4			
	5485.42	3			
	5493.72	7			
	5498.21	4			
	5511.10	0.6			
	5512.10	3			
	5516.14	6			
	5548.95	2			
	5550.40	3			
	5573.43	0.6			
	5574.91	1			
	5626.01	2			
	5659.86	3			
	5725.59	1			
	5779.25	1			
	5788.39	0.8			
	5802.82	2			
	5867.79	2			
	5868.62	1			
	5871.06	1			
	5874.19	1			
	5875.93	0.6			
	5902.60	1			
	5906.07	1			
	5909.04	1			
	5918.37	1			
	5979.39	1			
	6004.20	1			
	6027.16	1			
	6044.99	2			
	6070.07	2			
	6084.13	2			
	6159.56	0.6			
	6528.02	0.8			
	6588.92	1			
	6671.48	2			
	6725.90	1			

EXPERIMENTALLY OBSERVED
ATOMIC EMISSION LINES

Element	Line (Å)	Relative Intensity	Flame	Limit of Detection (ppm)	Ref.
Samarium Sm	6779.16	0.6	Ac/O	NI	30
Continued	68 61.06	0.8			
SmO	6140	10	H/O	NI	38
	5950	9			
	6030	9			
	5870	8			
	6240	8			
	5820	5			
	6380	5			
	6420.9	5			
	6520	5	H/A	0.5	34
	4720	2		NI	38
	6140	7	Ac/O	NI	38
	6030	2			
	6240	8			
	6380	7			
	6420.9	7			
	6520	10			
	6140	2	Ac/A	NI	37
	6240	7			
	6380	7			
	6420.9	10			
	6520	10			
	6680	7			
	6830	7			
Scandium Sc	3907.49	NI	PAc/O	0.2	1
	3911.81	NI		0.07	
	4020.40	NI		0.1	
	4020.40	NI	PAc/N	0.03	4
	3907.49	NI		0.03	18
	3911.81	NI		0.01	18
	4020.40	NI		0.05	18

EXPERIMENTALLY OBSERVED
ATOMIC EMISSION LINES

Element	Line (Å)	Relative Intensity	Flame	Limit of Detection (ppm)	Ref.
Scandium Sc	3019.34	0.5	Ac/O	NI	36
Continued	3030.76	0.5			38
	3255.69	0.5			
	3269.91	2			
	3273.63	3			
	3907.49	8		2	19
	3911.81	8		NI	38
	3933.38	3			
	3996.61	2			
	4020.40	6			
	4023.69	6			
	4047.79	2			
	4054.55	2			
	4082.40	4			
	4741.02	0.7			
	4743.81	1			
	4753.16	2			
	4779.35	1			
	5081.56	0.8			
	5083.72	0.5			
	5342.96	0.5			
	5349.71	2			
	5671.81	2			
	5686.84	2			
	5700.21	2			
	6210.68	8			
	6239.41	2			
	6239.78	6			
	6258.96	5			
	6276.31	1			
	6305.67	10			
	6378.82	2			
	6413.35	2			
	3911.81	NI	H/O	2	13
ScO	6073	10	H/O	0.04	34
	6110	8			38
	6017 }	7			
	6036				

EXPERIMENTALLY OBSERVED
ATOMIC EMISSION LINES

Element	Line (Å)	Relative Intensity	Flame	Limit of Detection (ppm)	Ref.
Scandium ScO Continued	6154	4	H/O	NI	34
	6190	2			38
	5812	2			
	6073	10	Ac/O	NI	38
	6110	8			
	6017 } 6036	7			
	6154	3			
	6190	2			
	5812	2			
	5849	2			
	6073	10	Ac/A	NI	37
	6110	10			
	6017 } 6036	10			
	6154	10			
	6190	10			
	5812	3			
	5849	3			
	5887	3			
	6230	10			
	4673.1	3			
Selenium Se	1960.26	NI	PAc/N	100	14
	2039.85	NI	H/O	3	13
	1960.26	NI	Ac/A	NI	38
	2039.85	NI		50	15
	2062.79	NI			
	1960.26	NI	Ac/O/Ar	NI	15
	2039.85	NI		25	
	2062.79	NI		NI	

EXPERIMENTALLY OBSERVED
ATOMIC EMISSION LINES

Element	Line (Å)	Relative Intensity	Flame	Limit of Detection (ppm)	Ref.
Silicon Si	2516.11	NI	PAc/O	7	1
	2516.11	NI	PAc/N	3	6
	2516.11	NI	SAc/N	6	27
	2516.11	NI	SAc/N	5	27
	2516.11	NI	H/O	8	13
Silver Ag	3280.68	NI	PAc/O	0.3	1
	3280.68	NI	PAc/N	0.002	5
	3280.68	NI	SAc/N	0.02	10
	3280.68	NI	H/N	0.07	10
	3280.68	NI	SH/N	0.06	10
	3280.68	NI	MAPP/N	0.08	10
	3280.68	NI	SMAPP/N	0.05	10
	3382.89	NI	H/Ar	0.007	22
	3382.89	10	Ac/A	NI	37
	3280.68	10			
	3382.89	10	H/O	0.06	33
	3280.68	6		NI	
	3382.89	10	H/A	0.04	34
	3280.68	4			
	3382.89	10	Ac/O	0.03	9
	3280.68	10			
Sodium Na	3302.37	NI	P Ac/O	5×10^{-4}	1
	5889.95	NI		1×10^{-4}	
	5895.92	NI		2×10^{-4}	
	5889.95	NI	PAc/N	10^{-5}	2
	5889.95 }	10	H/O	NI	38
	5895.92				

EXPERIMENTALLY OBSERVED
ATOMIC EMISSION LINES

Element	Line (Å)	Relative Intensity	Flame	Limit of Detection (ppm)	Ref.
Sodium Na Continued	5889.95 } 5895.92	10	H/A	10^{-4}	33
	5889.95 } 5895.92	10	Ac/O	NI	38
	5889.95 } 5895.92	10	Ac/A	NI	37
	8194.82	7			
Strontium Sr	4607.33	NI	PAc/O	0.004	1
	4607.33	NI	PAc/N	10^{-4}	21
	4607.33	NI	SAc/O	0.002	16
	4607.33	NI	SAc/N	0.02	10
	4607.33	NI	H/N	0.009	10
	4607.33	NI	SH/N	0.007	10
	4607.33	NI	MAPP/N	0.05	10
	4607.33	NI	SMAPP/N	0.02	10
	4607.33	NI	H/Ar	0.001	22
	4607.33	10	H/O	0.01	33
	4607.33	10	H/A	NI	38
Sr (II)	4607.33 4077.71 4215.52	10 0.9 0.6	Ac/O	NI	38
Sr (II)	4607.33 4077.71	10 0.3	Ac/A	NI	37

EXPERIMENTALLY OBSERVED
ATOMIC EMISSION LINES

Element	Line (Å)	Relative Intensity	Flame	Limit of Detection (ppm)	Ref.
Strontium SrOH	6050	10	H/O	NI	38
	6820	10			
	6660	7			
	6590	2			
	6050	10	H/A	NI	38
	6050	10	Ac/O	NI	38
	6820	3			
	6660	4			
	6590	2			
	6050	10	Ac/A	NI	37
	6820	10			
Sulfur	4310	3	H/N_2	1.6	28
(S Molecular)	4270	4		NI	
	4195	4			
	4150	6			
	4050	9			
	3940	10			
	3840	10			
	3740	9			
	3645	7			
	3590	3			
	3555	4			
	3500	3			
	3470	2			
Tantalum Ta	4812.75	NI	PAc/O	18	1
	4740.16	NI	PAc/N	4	6
Technetium Tc	No information				

EXPERIMENTALLY OBSERVED
ATOMIC EMISSION LINES

Element		Line (Å)	Relative Intensity	Flame	Limit of Detection (ppm)	Ref.
Tellurium Te		2383.25	NI	PAc/O	200	1
		2385.76	NI		600	
		4866.2	NI	PAc/N	2	14
		2383.25 }	NI	H/O	NI	38
		2385.76				
		2142.75	NI		0.6	13
		2142.75	NI	Ac/A	5	15
	TeO	3714	10	H/O	NI	38
		3827	8			
		3884	8			
		3954	8			
		4131	8			
		4007	7			
		4075	7			
		4205	7			
		4268	7			
		4343	7			
		4487	7			
		3607	7			
		3662	7			
		3773	7			
		4640	6			
		3561	5			
		3714	10	H/A	NI	38
		3827	10			
		3884	10			
		3954	10			
		4131	10			
		4007	10			
		4075	10			
		4205	10			
		4268	9			
		4343	10			
		4487	9			

EXPERIMENTALLY OBSERVED
ATOMIC EMISSION LINES

Element	Line (Å)	Relative Intensity	Flame	Limit of Detection (ppm)	Ref.
Tellurium TeO Continued	3607	9	H/A	NI	38
	3662	10			
	3773	10			
	4640	7			
	3561	7			
	3714	9	Ac/O	NI	38
	3827	10			
	3884	10			
	3954	10			
	4131	10			
	4007	10			
	4075	10			
	4205	10			
	4268	10			
	4487	10			
	3607	9			
	3662	9			
	3773	9			
	4640	10			
	3561	9			
Terbium Tb	3901.35	NI	PAc/O	4	1
	4326.47	NI		1	
	4318.85	NI	PAc/N	0.2	18
	4326.47	NI		0.4	
	3745.07	0.8	Ac/O	NI	30
	3753.35	2			
	3761.12	0.8			
	3783.54	0.8			
	3789.92	2			
	3830.29	4			
	3833.40	0.8			
	3901.35	2			
	3907.91	0.8			
	3915.45	2			

EXPERIMENTALLY OBSERVED
ATOMIC EMISSION LINES

Element	Line (Å)	Relative Intensity	Flame	Limit of Detection (ppm)	Ref.
Terbium Tb Continued	4060.38	0.8	Ac/O	NI	30
	4061.57	2			
	4105.38	0.8			
	4143.53	2			
	4158.28	2			
	4203.71	2			
	4215.13	0.8			
	4235.34	0.8			
	4266.35	2			
	4318.85	8		10	19
	4320.28	0.8		NI	30
	4322.24	0.8			
	4326.47	10			
	4332.13	0.8			
	4336.50	4			
	4338.45	6			
	4340.63	5			
	4342.50	0.8			
	4356.84	4			
	4372.04	0.8			
	4388.25	0.8			
	4417.99	0.8			
	4420.20	0.8			
	4430.14	0.8			
	4434.06	0.8			
	4491.02	3			
Thallium Tl	3775.72	NI	PAc/O	0.09	1
	5350.46	NI	PAc/N	0.002	2
	3775.72	NI		0.05	4
	3775.72	NI	SAc/N	0.08	10
	3775.72	NI	H/N	0.03	10
	3775.72	NI	SH/N	0.025	10

EXPERIMENTALLY OBSERVED
ATOMIC EMISSION LINES

Element	Line (Å)	Relative Intensity	Flame	Limit of Detection (ppm)	Ref.
Thallium Tl Continued	3775.72	NI	MAPP/N	0.1	10
	3775.72	NI	SMAPP/N	0.08	10
	3775.72	NI	SAc/O	0.04	16
	3775.72	10	H/O	0.1	34
	5350.46	7		NI	
	3519.24	0.3			
	3775.72	10	H/A	NI	38
	5350.46	6			
	3519.24	0.8			
	3775.72	10	Ac/O	0.2	16
	5350.46	5		NI	38
	3519.24	0.7			
	2767.87	0.5			
	3775.72	10	Ac/A	NI	37
	5350.46	7			
	3519.24	3			
	2767.87	1			
Thorium Th	5760.55	NI	PAc/O	150	1
Th (II)	4919.82	NI	PAc/N	10	14
Th	5760.55	10	H/O	NI	38
	5760.55	10	Ac/O	NI	38
Thulium Tm	4094.19	NI	PAc/O	0.3	1
	4105.84	NI		0.2	1
	3717.90	NI	PAc/N	0.08	6
	3717.90	NI		0.01	18
	3717.90	NI		0.004	5

EXPERIMENTALLY OBSERVED
ATOMIC EMISSION LINES

Element	Line (Å)	Relative Intensity	Flame	Limit of Detection (ppm)	Ref.
Thulium Tm	4105.84	NI	H/O	NI	38
Continued	4094.19	NI		0.4	13
Tm (II)	3172.82	0.5	Ac/O	NI	30
Tm (I)	3233.75	0.5			
	3410.05	2			
	3416.00	0.6			
	3563.88	1			
	3567.36	0.8			
	3717.92	6			
	3744.07	7			
	3751.82	5			
	3781.14	0.5			
	3807.72	0.5			
	3826.39	2			
	3883.13	8			
	3887.35	6			
	3896.62	2			
	3916.47	2			
	3949.27	0.8			
	4094.18	9		0.5	19
	4105.84	10		NI	30
	4138.36	0.8			
	4187.62	7			
	4203.73	6			
	4222.67	0.5			
	4359.93	5			
	4386.42	3			
	4599.00	1			
	4724.25	0.9			
	4733.32	4			
	5060.89	2			
	5113.96	0.6			
	5307.11	8			
	5631.41	5			
	5675.83	9			
	5764.29	3			
	5971.27	2			

EXPERIMENTALLY OBSERVED
ATOMIC EMISSION LINES

Element		Line (Å)	Relative Intensity	Flame	Limit of Detection (ppm)	Ref.
Thulium	TmO	5380	10	H/O	0.35	34
		5415	10		NI	30
		5530	10			
		4910	9			
		4830	6			
		5230	5			
Tin	Sn	3034.12	NI	PAc/O	9	1
		2839.99	NI	PAc/N	0.1	6
		2354.85	NI	Ac/O/Ar	1	15
		2839.99	NI	SAc/N	0.8	10
		2839.99	NI	H/N	250	10
		2839.99	NI	SH/N	100	10
		2839.99	NI	MAPP/N	14	10
		2429.49	NI	H/O	0.5	13
		2354.84	NI		NI	38
		3262.34	10			
		3034.12	6			
		3009.14	4			
		2863.33	2			
		2839.99	1			
		2706.51	1			
		3262.34	2	H/A	NI	38
		3034.12	2			
		3009.14	0.8			
		3801.02	10			
		3175.05	4			
		3262.34	2	Ac/O	NI	38
		3034.12	2			

EXPERIMENTALLY OBSERVED
ATOMIC EMISSION LINES

Element	Line (Å)	Relative Intensity	Flame	Limit of Detection (ppm)	Ref.
Tin Sn Continued	3009.14	2	Ac/O	NI	38
	2863.33	1			
	2839.99	1		20	33
	2706.51	10		NI	38
	3801.02	2			
	3175.05	1			
	3330.62	2			
	2354.85	NI	Ac/A	2	15
	3262.34	10		NI	37
	3034.12	10			
	3009.14	7			
	2863.33	7			
	2839.99	10			
	2706.51	7			
	3801.02	7			
	3175.05	9			
	3330.62	5			
SnO	3585.4	10	H/O	NI	38
	3388.3	8			
	3415.8	8		25	33
	3691.4	8		NI	30
	3721.2	8			
	3864.9	8			
	3833.2	7			
	3983.9	6			
	3323.5	4			
	3485.5	3			
	4850	2			
	3585.4	7	H/A	NI	38
	3388.3	5			
	3415.8	6			
	3691.4	5			
	3721.2	5			
	3864.9	7			
	3833.2	5			
	3983.9	6			

EXPERIMENTALLY OBSERVED
ATOMIC EMISSION LINES

Element	Line (Å)	Relative Intensity	Flame	Limit of Detection (ppm)	Ref.
Tin SnO	3323.5	3	H/A	NI	38
Continued	3484.5	3			
	4850	10			
Titanium Ti	3653.50	NI	PAc/O	0.5	1
	3998.64	NI		0.5	1
	3653.50	NI	PAc/N	0.03	5
	3998.64	NI	SAc/N	0.5	7
	3989.76	NI			
	3981.76	NI			
	3958.21	NI			
	3956.34	NI			
	3948.67	NI			
	3752.86	NI			
	3741.06	NI			
	3729.82	NI			
	3653.50	NI		1	7
	3642.68	NI			
	3635.46	NI			
	3998.64	NI	H/O	NI	38
	3653.50	NI		2	13
	3186.45	10	Ac/O	NI	31
	3191.99	10			38
	3199.92	3			
	3341.88	3			
	3354.64	3			
	3370.44	2			
	3371.45	3			
	3377.48	3			
	3385.95	3			
	3635.46	6			
	3642.68	7			
	3653.50	8			
	3658.10	2			

EXPERIMENTALLY OBSERVED
ATOMIC EMISSION LINES

Element	Line (Å)	Relative Intensity	Flame	Limit of Detection (ppm)	Ref.
Titanium Ti Continued	3668.97	2	Ac/O	NI	31
	3671.67	2			
	3729.82	5			
	3741.06	6			
	3752.86	8			
	3753.64	2			
	3771.66	2			
	3924.53	3			
	3929.88	2			
	3947.78	3			
	3948.67	4			
	3956.34	5			
	3958.21	8			
	3962.85	3			
	3964.27	3			
	3981.76	6			
	3982.48	2			
	3989.76	8			
	3998.64	10		5	29
	4008.93	3		NI	31
	4024.57	3			
	4300.56	2			
	4301.09	2			
	4305.92	3			
	4533.24	2			
	5039.95	3			
	5064.66	3			
	5173.75	2			
	5192.98	2			
	5210.39	2			
TiO	5449	10	H/O	NI	38
	5759	10			
	5167	9		0.2	33
	4955	8		NI	38
	5003	8			
	4805	7			
	4848	7			
	6730	4			

EXPERIMENTALLY OBSERVED
ATOMIC EMISSION LINES

Element	Line (Å)	Relative Intensity	Flame	Limit of Detection (ppm)	Ref.
Titanium TiO Continued	7150	4	H/O	NI	38
	5449	2	Ac/O	NI	38
	5759	2			
	5167	2			
	6730	10			
	7150	10			
Tungsten W	4008.75	NI	PAc/O	4	1
	4008.75	NI	PAc/N	0.5	4
	4008.75	NI	SAc/N	0.2	11
	4008.75	NI	H/O	NI	38
	4008.75	NI	H/A	NI	38
	3617.52	3	Ac/A	NI	37
	3867.98	3			
	4008.75	10		90	29
	4045.60	3		NI	37
	4074.36	8			
	4269.39	3			
	4294.61	6			
	4302.11	4			
Uranium U	5915.40	NI	PAc/O	10	1
(U molecular species)	5448	NI	PAc/N	5	14
	5915.40	NI	H/O	NI	38
	5915.40	NI	H/A	NI	37
	5915.40	NI	Ac/O	NI	38
	5915.40	NI	Ac/A	NI	37

EXPERIMENTALLY OBSERVED
ATOMIC EMISSION LINES

Element	Line (Å)	Relative Intensity	Flame	Limit of Detection (ppm)	Ref.
Uranium UO$_2$	5500	10	H/O	NI	38
	5500	10	H/A	0.1	33
Vanadium V	4379.24	NI	PAc/O	0.3	1
	4408.20 }	NI		0.3	1
	4408.51				
	4379.24	NI	PAc/N	0.007	5
	3183.41	NI		NI	
	3183.98	NI	SAc/N	0.2	7
	3185.40	NI		0.2	10
	4379.24	NI		0.5	7
	4384.72	NI		NI	
	4389.97	NI			
	4395.23	NI			
	4400.58	NI			
	4406.64	NI			
	4408.51	NI			
	3185.40	NI	H/N	125	10
	3185.40	NI	SH/N	100	10
	3185.40	NI	SMAPP/N	400	10
	3183.98	NI	Ac/O/Ar	25	15
	4111.78	NI	H/O	2	13
	2923.62	1	Ac/A	NI	29
	3043.12	2			
	3043.56	2			
	3044.94	1			
	3050.89	1			
	3053.65	2			
	3056.33	3			
	3060.46	3			

EXPERIMENTALLY OBSERVED
ATOMIC EMISSION LINES

Element	Line (Å)	Relative Intensity	Flame	Limit of Detection (ppm)	Ref.
Vanadium V Continued	3066.38	5	Ac/A	NI	29
	3183.41	6			
	3183.98	6			
	3185.40	5			
	3198.01	3			
	3202.38	3			
	3207.41	3			
	3212.43	1			
	3263.24	1			
	3271.64	1			
	3283.31	1			
	3298.14	1			
	3675.70	2			
	3703.58	5			
	3704.70	5			
	3705.04	4			
	3778.68	2			
	3790.32	2			
	3793.61	1			
	3794.96	4			
	3799.91	2			
	3803.47	2			
	3807.50	2			
	3808.52	2			
	3809.60	2			
	3813.49	3			
	3817.84	1			
	3818.24	4			
	3819.96	2			
	3821.49	2			
	3822.01	4			
	3822.89	2			
	3823.21	2			
	3828.56	4			
	3840.75	5			
	3844.44	2			
	3847.33	2			
	3855.37	6			
	3855.84	6			

EXPERIMENTALLY OBSERVED
ATOMIC EMISSION LINES

Element	Line (Å)	Relative Intensity	Flame	Limit of Detection (ppm)	Ref.
Vanadium V	3864.86	3	Ac/A	NI	29
Continued	3867.60	4			
	3875.08	3			
	3876.09	3			
	3890.18	3			
	3892.86	3			
	3902.25	5			
	3909.89	2			
	4090.58	2			
	4092.69	4			
	4095.49	2			
	4099.80	5			
	4102.16	2			
	4105.17	4			
	4109.79	4			
	4111.78	8			
	4115.18	5			
	4116.47	3			
	4123.57	4			
	4128.07	4			
	4132.02	4			
	4134.49	4			
	4179.42	1			
	4182.59	1			
	4189.84	1			
	4209.86	1			
	4306.21	1			
	4307.18	1			
	4309.80	1			
	4330.02	3			
	4332.82	3			
	4341.01	3			
	4355.94	2			
	4368.04	2			
	4379.24	10		0.3	29
	4384.72	6		NI	
	4389.97	5			
	4395.23	4			
	4400.58	3			

EXPERIMENTALLY OBSERVED
ATOMIC EMISSION LINES

Element	Line (Å)	Relative Intensity	Flame	Limit of Detection (ppm)	Ref.
Vanadium V Continued	4406.64	3	Ac/A	NI	29
	4407.64	6			
	4408.20	6			
	4408.51	6			
	4416.47	2			
	4419.94	1			
	4421.57	2			
	4426.00	1			
	4428.52	1			
	4429.80	1			
	4436.14	2			
	4437.84	2			
	4441.68	2			
	4444.21	2			
	4457.48	2			
	4459.76	3			
	4460.29	3			
	4577.17	2			
	4580.40	2			
	4586.36	2			
	4594.11	4			
	4827.45	2			
	4831.64	2			
	4832.43	2			
	4851.48	2			
	4864.74	2			
	4875.48	2			
	4881.56	2			
	5627.64	1			
	5670.85	1			
	5727.03	1			
	6039.73	1			
	6081.44	1			
	6090.22	2			
	6111.67	1			
	6119.52	1			
	6199.19	2			
	6213.87	1			
	6216.37	2			
	6224.50	1			

EXPERIMENTALLY OBSERVED
ATOMIC EMISSION LINES

Element	Line (Å)	Relative Intensity	Flame	Limit of Detection (ppm)	Ref.
Vanadium V Continued	6230.74	2	Ac/A	NI	29
	6233.20	1			
	6242.81	1			
	6243.10	3			
	6251.82	2			
	6274.65	2			
	6285.16	2			
	6292.83	2			
	6296.49	2			
VO	5469.3	10	H/O	NI	38
	5736.7	10		3	33
	5228.3	8		NI	38
	5275.8	8			
	5058	6			
	6086.4	5			
	7100	2			
	8000	2			
	7470	2			
	5469.3	9			
	5736.7	10	H/A	0.02	34
	5228.3	7		NI	38
	5275.8	7			
	5058	6			
	6086	4			
	7100	7	Ac/A	NI	37
	8000	10			
	7470	10			
Xenon Xe	No information				
Ytterbium Yb	3987.98	NI	PAc/O	0.05	1
	3987.98	NI	PAc/N	0.001	2
	3987.98	NI		3×10^{-4} (Alcohol)	18

EXPERIMENTALLY OBSERVED
ATOMIC EMISSION LINES

Element		Line (Å)	Relative Intensity	Flame	Limit of Detection (ppm)	Ref.
Ytterbium Yb Continued		3987.98	NI	PAc/N	2×10^{-4}	5
		3987.98	NI	H/O	0.001	13
		3987.98	10	H/A	NI	38
		3464.37	4	Ac/O	NI	30
Yb (II)		3694.20	1			38
Yb (I)		3987.99	10		0.1	19
		5556.48	8		NI	38
YbOH		5725	10	H/O	NI	38
		5325	7		0.15	33
		5550				38
	}	5556.48	6		NI	
		4981	4			
		5174	4			
		5443	3			
		4850	3			
		5870	3			
		4778	2			
		5725	10	A/H	0.03	34
		5325	8		NI	38
		5550	8			
		4981	7			
		5174	5			
		5443	4			
		4850	3			
		5870	3			
		4778	3			
		5725	10	Ac/O	NI	38
		5325	5			
		5550	10			
		4981	5			
		5174	3			
		5443	3			
		4850	2			
		5870	2			

EXPERIMENTALLY OBSERVED
ATOMIC EMISSION LINES

Element	Line (Å)	Relative Intensity	Flame	Limit of Detection (ppm)	Ref.
Ytterbium YbOH Continued	4778	2	Ac/O	NI	38
	5725	10	Ac/A	NI	37
	5325	7			
	5550	10			
	4981	3			
	4778	3			
Yttrium Y	4077.38	NI	PAc/O	0.3	1
	4102.38	NI		0.3	
	4128.31	NI		0.5	
	3620.94	NI	PAc/N	0.04	18
	3620.94	NI		0.1	5
	4102.38	NI	H/O	2	13
	2984.26	2	Ac/O	NI	30
	3552.69	1			
	3592.92	2			
	3620.94	4			
Y (II)	3710.30	0.5			
	3774.33	0.5			
Y (I)	4039.83	2			
	4047.63	3			
	4077.38	9			
	4083.71	2			
	4102.38	9			
	4128.31	8		3	19
	4142.85	6		NI	
	4167.52	2			
	4174.13	2			
	4235.94	2			
	4527.25	0.5			
	4643.70	8			
	4674.85	5			
	4760.98	1			
	5466.46	0.5			

EXPERIMENTALLY OBSERVED
ATOMIC EMISSION LINES

Element	Line (Å)	Relative Intensity	Flame	Limit of Detection (ppm)	Ref.
Yttrium Y Continued	5527.54	0.8	Ac/O	NI	19
	6222.59	2			
	6402.01	0.5			
	6435.00	10			
	6557.39	0.5			
	6687.58	2			
	6793.71	1			
YO	5990	10	H/O	0.5	33
	6150	10		NI	38
	5990	10	Ac/O	NI	38
	6150	10			
	5990	10	Ac/A	NI	37
	6150	10			
	4818.2	7			
	4842.0	2			
	4676.3	2			
	4650.2	2			
Zinc Zn	2138.56	NI	PAc/O	50	1
	4810.53	NI		1500	
	2138.56	NI	PAc/N	10	14
	2138.56	NI	Ac/O/Ar	1	15
	2770.86 } 2770.98	NI		NI	
	2800.87	NI			
	3345.02	NI			
	4810.55	NI			
	2138.56	10	H/O	NI	38
	4810.53	10			
	3075.90	3			

EXPERIMENTALLY OBSERVED
ATOMIC EMISSION LINES

Element	Line (Å)	Relative Intensity	Flame	Limit of Detection (ppm)	Ref.
Zinc Zn Continued	2138.56	1	H/A	NI	38
	4810.53	10		NI	
	3075.90	3		NI	
	2138.56	NI	Ac/A	7	37
	2771	NI			
	2801.87	NI			
	3075.90	NI			
	3345.02	NI			
	4810.53	NI			
Zirconium Zr	3519.60	NI	PAc/O	50	1
	3601.19	NI		75	1
	3601.19	NI	PAc/N	3	4
	3519.60	NI	SAc/N	1.2	8
	3519.60	NI	SAc/N	2	8
ZrO	5640	10	H/O	8	33
	5740	10		NI	38
	5640	10	Ac/O	NI	38
	5740	10			

1. V. A. Fassel, D. W. Golightly, Anal. Chem., 39, 466-76 (1967).

2. E. D. Prudnikov, Zh. Anal. Khim., 27, 2327-32 (1972).

3. E. E. Pickett, S. R. Koirtyohann, Spectromchim. Acta, 24B,
    325-33 (1969).

4. E. E. Pickett, S. R. Koirtyohann, Spectrochim. Acta, 23B,
    235-44 (1968).

5. P. W. J. M. Boumans, F. J. deBoer, Spectrochim. Acta, 27B,
    391-414 (1972).

6. G. D. Christian, F. J. Feldman, Appl. Spectrosc., 25, 660-3
    (1971).

7. G. F. Kirkbright, A. Semb, & T. S. West, Talanta, 15, 441-50
    (1968).

8. G. F. Kirkbright, M. Sargent, Talanta, 16, 245-53 (1969).

9. J. F. Chapman, L. S. Dale, Analyst (London), 94, 1969 (1969).

10. M. S. Cresser, P. B. Joshipura, & P. N. Kelihev, Spectrosc
    Letters, 3, 267-275 (1970).

11. G. F. Kirkbright, & T. S. West, Appl. Opt., 7, 1305-11 (1968).

12. V. G. Mossotti, & M. Duggan, Appl. Opt., 7, 1325-30 (1968).

13. R. K. Skogerboe, A. T. Heybey, & G. H. Morrison, Anal. Chem.,
    38, 1821-4 (1966).

14. G. D. Christian, F. J. Feldman, Anal. Chem., 43, 611-13 (1971).

15. J. F. Alder, K. C. Thompson, & T. S. West., Anal. Chim. Acta,
    50, 383-397 (1970).

16. R. S. Hobbs, G. F. Kirkbright, & T. S. West, Analyst, 94, 554-62
    (1969).

17. D. N. Hingle, G. F. Kirkbright, & T. S. West, Analyst, 93,
    522-27 (1968).

18. R. N. Kniseley, C. C. Butler, & V. A. Fassel, Anal. Chem., 41,
    1494-96 (1969).

19. A. P. D'Silva, R. N. Kniseley, & V. A. Fassel, Anal. Chem.,
    36, 1287-9 (1964).

20. R. M. Dagnall, G. F. Kirkbright, T. S. West, & R. Wood, Analyst, 95, 425-30 (1970).

21. S. R. Koirtyohann, E. E. Pickett, Spectrochim. Acta, 23B, 673-85 (1968).

22. K. Zacha, & J. D. Winefordner, Anal. Chem. 38, 1537-9 (1966).

23. V. Sychra, P. J. Slevin, J. Matousek, & F. Bek, Anal. Chim. Acta, 52, 259-73 (1970).

24. R. K. Skogerboe, A. S. Gravatt, & G. H. Morrison, Anal. Chem., 39, 1602-05 (1967).

25. R. M. Dagnall, K. C. Thompson, & T. S. West, Analyst, 93, 72-78 (1968).

26. R. Smith, & A. E. Lawson, Analyst, 96, 631-39 (1971).

27. R. M. Dagnall, G. F. Kirkbright, T. S. West, & R. Wood, Anal. Chim. Acta, 47, 407-13 (1969).

28. R. M. Dagnall, K. C. Thompson, & T. S. West, Analyst, 92, 506-12 (1967).

29. V. A. Fassel, R. B. Myers, & R. N. Kniseley, Spectrochim. Acta, 19, 1187-94 (1963).

30. V. A. Fassel, R. H. Curry, & R. N. Kniseley, Spectrochim. Acta, 18, 1127-53 (1962).

31. V. A. Fassel, R. B. Myers, & R. N. Kniseley, Spectrochim. Acta, 19, 1187-94 (1963).

32. T. C. Rains, H. P. House, & O. Menis, Anal. Chim. Acta, 22, 315-27 (1960).

33. P. T. Gilbert, Jr., Beckman Bulletin, 753 (1959).

34. P. T. Gilbert, Jr., Symposium on Spectroscopy, ASTM Spec. Tech. Publ. (1960).

35. P. T. Gilbert, Jr., x^{th} Colloquium Spectrocopium Internationale, Spartan Press (1967), Washington, D. C. pp. 171ff.

36. M. Honman, Anal. Chem., 27, 165 6 (1955).

37. R. Mavrodineanu and H. Boiteux, "Flame Spectroscopy," John Wiley and Sons, 1965.

38. Handbook of Chemistry and Physics, 1967-68, Chemical Rubber Co., R. C. Weast, Ed. pp E 136 ff.

Table II.B.7

EXPERIMENTALLY OBSERVED
ATOMIC FLUORESCENCE LINES

Element	Line (Å)	Rel. Int.	Flame	Excit. Source	Limit of Detection (ppm)	Ref.
Actinium Ac	No Information					
Aluminum Al	3961.53	10	H/N	EDT	2	2
	3961.53	10	PAc/N	EDT	0.2	42
	3944.01	8	SAc/N		0.12	
	3092.71 } 3092.84	4				
	3082.15	0.9				
	2575.10	NFO				
	3961.53	10	SAc/N	HIHC	1.5	44
	3944.01	5			2.5	
	3092.71 } 3092.84	3			3.5	
	3082.15	1			5.5	
	2575.10 } 2575.40	0.002				
	2567.98	0.002				
	2373.12 } 2373.35	0.002				
	2367.05	0.001				
	3944.01/ 3961.53	10	Ac/N	PTDL	0.005	45
	3961.53	2			0.03	
	3944.01	0.5			0.1	
Americium Am	No Information					
Antimony Sb	2311.47	10	H/A	EDT	0.05	3
	2068.33	5				
	2598.05 } 2598.09	5				
	2877.92	3				
	2769.95	2				
	2670.64	0.9				
	2510.54	NFO				
	2175.81	10	H/N$_2$	EDT	0.08	2

EXPERIMENTALLY OBSERVED
ATOMIC FLUORESCENCE LINES

Element	Line (Å)	Rel. Int.	Flame	Excit. Source	Limit of Detection (ppm)	Ref.
Antimony	2175.81	10	Ac/A	HCL	50	43
Continued	2175.81	10		HIHC	1.5	
	2175.81	10	SAc/A	HCL	20	
	2175.81	10		HIHC	0.6	
	2311.47	NFO	Ac/A	HCL		55
	2311.47	NFO	SAc/A			
	2311.47	10	H/A	Xe1.5	100	56
	2311.47	10	H/A	EDT	0.4	14
Argon Ar	No Information					
Arsenic As	2349.84	10	H/Ar	EDT	0.25	4,5
	2898.71	4				
	2492.91	4				
	2456.53	3				
	2860.44	3				
	2381.18	3				
	1971.97	3				
	1936.96	2				
	1890.41	2				
	2437.23	0.8				
	3032.85	0.7				
	2288.12	0.2				
	2780.22	0.04				
	2745.00	0.02				
	2349.84	10	H/N$_2$	EDT	0.5	4,5
	2492.91	5				
	2381.18	4				
	2456.53	4				
	1971.97	3				
	2860.44	3				
	1936.96	3				
	1890.41	3				
	2437.23	1				
	3032.85	0.9				
	2898.71	0.4				

EXPERIMENTALLY OBSERVED
ATOMIC FLUORESCENCE LINES

Element	Line (Å)	Rel. Int.	Flame	Excit. Source	Limit of Detection (ppm)	Ref.
Arsenic Continued	2288.12	0.3	H/N$_2$	EDT	0.5	4,5
	2780.22	0.04				
	2745.00	0.04				
	2349.84	10	Ac/A	EDT	2	4,5
	2288.12	3				
	2456.53	3				
	2492.91	3				
	1936.96	2				
	1971.97	2				
	2381.18	2				
	1890.41	2				
	2437.23	0.5				
	1936.96	10	SAc/A	HCL	ND	46
	1936.96	10	H/A		1.0	
	1936.96	10	Ac/A	HCL	ND	43
	1936.96	10	SAc/A		ND	
	1936.96	10	SAc/A	HCL	ND	55
	1936.96	10	Ac/A		ND	
Astatine At	No Information					
Barium Ba	No Information					
Berkelium Bk	No Information					
Beryllium Be	2348.61	10	PAc/N	EDT	0.04	6
	2348.61	10	SAc/N	EDT	0.01	7
	2348.61	10	Ac/N	HIHC	0.5	8
	2348.61	10	Ac/O	HIHC	10	8

EXPERIMENTALLY OBSERVED
ATOMIC FLUORESCENCE LINES

Element	Line (Å)	Rel. Int.	Flame	Excit. Source	Limit of Detection (ppm)	Ref.
Beryllium	2348.61	10	SAc/A	HIHC	50	44
Continued	2348.61	10	SAc/N	HIHC	0.02	
Bismuth	3024.64	10	H/Ar	EDT	0.5	9
Bi	3067.72	5				
	2061.70	1				
	2696.76	0.6				
	2897.98	0.5				
	4722.19					
	4722.55}	0.3				
	4722.83					
	2938.30	0.3				
	2230.61	0.3				
	2989.03	0.3				
	2993.34	0.3				
	4121.53	0.1				
	2228.25	0.1				
	3397.21	0.1				
	3510.85	0.07				
	2627.91	0.07				
	2276.58	0.05				
	2780.52	0.04				
	3024.64	10	H/N$_2$	EDT	1	9
	3067.72	6				
	2061.70	1				
	2696.76	0.6				
	2897.98	0.4				
	4722.19					
	4722.55}	0.3				
	4722.83					
	2230.61	0.2				
	2989.03	0.2				
	2938.30	0.2				
	2228.25	0.1				
	4121.53	0.07				
	3397.21	0.07				
	2627.91	0.04				
	3510.85	0.04				
	2276.58	0.03				
	2780.52	0.03				

EXPERIMENTALLY OBSERVED
ATOMIC FLUORESCENCE LINES

Element	Line (Å)	Rel. Int.	Flame	Excit. Source	Limit of Detection (ppm)	Ref.
Bismuth Continued	3067.72	10	H/A	EDT	0.005	1
	3067.72	10	Ac/A	HCL	33	43
	2230.61	8			40	
	3067.72	10	SAc/A	HCL	5	
	2230.61	4			14	
	3067.72	10	Ac/A	HCL	220	55
	3067.72	10	SAc/A		55	
	3067.72	10	H/Ar	Xe1.5	10	56
	3067.72	10	H/Ar	EDT	0.7	14
Boron B	No Information					
Bromine Br	No Information					
Cadmium Cd	2288.02	10	SAc/A	MVL	0.0005	10
	3261.06	0.005				
	2288.02	10	H/Ar	Xe9	0.08	11
	2288.02	10	P/A	Xe1.5	0.2	12
	2288.02	10	H/A/Ar	EDT	NI	13
	2288.02	10	H/A/N$_2$		NI	
	2288.02	10	PH/A	EDT	NI	6
	2288.02	10	PH/Ar			
	2288.02	10	PH/N			
	2288.02	10	PH/O			
	2288.02	10	PAc/N			
	2288.02	10	Ac/A	HCL	0.06	43
	2288.02	10		MVL	0.002	
	2288.02	10	SAc/A	HCL	0.02	

EXPERIMENTALLY OBSERVED
ATOMIC FLUORESCENCE LINES

Element	Line (Å)	Rel. Int.	Flame	Excit. Source	Limit of Detection (ppm)	Ref.
Cadmium	2288.02	10	SAc/A	HCL	0.1	55
Continued	2288.02	10	Ac/A		0.15	
	2288.02	10	H/A	Xel.5	0.08	56
	2288.02	10	H/O	EDT	0.000001	14
	2288.02	10	H/A	PDHC	0.003	60
	2288.02	10	H/A	EDT	0.00002	61
Calcium Ca	4226.73	10	H/A	EDT	0.02	14
	4226.73	10	H/EA	Xe9	0.02	15
	4226.73	10	Ac/O	Xe9	NI	16
	4226.73	10	Ac/A	HCL	0.1	55
	2398.56	NFO	Ac/A		ND	
	4226.73	10	SAc/A		0.05	
	2398.56	NFO	SAc/A		ND	
	4226.73	10	H/A	Xel.5	10	56
	4226.73	10	H/A	PTDL	0.005	45
	No Fluorescence Observed		H/O H/Ar			
Californium Cf	No Information					
Carbon C	No Information					
Cerium Ce	3716.37 II/ 3942.15 II	10	Ac/N	PTDL	0.5	57
	3716.37 II/ 3716.37 II	NI				

EXPERIMENTALLY OBSERVED
ATOMIC FLUORESCENCE LINES

Element	Line (Å)	Rel. Int.	Flame	Excit. Source	Limit of Detection (ppm)	Ref.
Cerium Continued	3716.37 II/ 3952.54 II	NI	Ac/N	PTDL	0.5	57
	3716.37 II/ 3999.24 II	NI				
	3716.37 II/ 4289.94 II	NI				
	3716.37 II/ 4296.67 II	NI				
	5699.23	NFO			ND	
Cesium Cs	8521.10	10	H/O/N$_2$	MVL	NI	17
Chlorine Cl	No Information					
Chromium Cr	3578.69	10	H/N$_2$	EDT	0.05	2
	3578.69	10	H/A	DHC	1	18
	3593.49	10	H/A	MVL	5	19
	3578.69	10	SAc/A	EDT	0.005	50
	3593.49	7.7				
	3605.32	3.3				
	4254.33	1				
	4274.81	2			0.1	
	4289.73	1.8				
	5204.51 5206.02} 5208.42	0.15				
	3593.49	10	H/A	EDT	0.5	
	3593.49	10	SH/A	EDT	0.1	
	3593.49	10	Ac/A	EDT	0.05	
	4254.33	10	Ac/A	HCL	1.0	55
	3578.69	0.7			14	
	4254.33	10	SAc/A		0.5	
	3578.69	0.9			5.7	

EXPERIMENTALLY OBSERVED
ATOMIC FLUORESCENCE LINES

Element	Line (Å)	Rel. Int.	Flame	Excit. Source	Limit of Detection (ppm)	Ref.
Chromium Continued	3593.49	10	H/A	EDT	10	14
	3578.69	10	H/A	PTDL	0.02	45
	3593.49	10	H/A	PTDL	0.02	
Cobalt Co	2407.25	10	P/A	EDT	0.005	20
	2407.25	10	H/A		0.005	
	2407.25	10	Ac/A		0.01	
	2407.25	10	PH/A	HIHC	0.02	21
	2424.93	5				
	2521.36	4				
	2411.62	4				
	2432.21	2				
	2414.46 }	2				
	2415.30					
	2528.97	1				
	2407.25	10	PH/O/Ar	HIHC	0.01	21
	2407.25	10	PAc/A		0.05	
	2407.25	10	H/O/Ar		0.01	
	2407.25	10	H/N$_2$	EDT	0.04	2
	2407.25	10	H/Ar/EA	Xe9	NI	16
	2407.25	10	H/O	EDT	2	22
	2407.25	10	H/Ar	EDT	0.1	14
	2407.25	10	Ac/A	HCL	1.0	43
	2407.25	10		HIHC	0.04	
	2407.25	10	SAc/A	HCL	0.15	
	2407.25	10		HIHC	0.015	
	2407.25	10	Ac/A	HCL	2.5	55
	2407.25	10	SAc/A		1.7	
	2407.25	10	H/Ar	Xe1.5	1.0	56
	3474.53/					
	3575.36	0.04	Ac/N	PTDL	0.2	45
	3575.36	10			50	

EXPERIMENTALLY OBSERVED
ATOMIC FLUORESCENCE LINES

Element	Line (Å)	Rel. Int.	Flame	Excit. Source	Limit of Detection (ppm)	Ref.
Copper Cu	3247.54	10	H/A	HIHC	0.001	23
	3247.54	10	H/Ar	EDT	0.005	14
	3247.54	10	H/O	HIHC	0.1	8
	3247.54	10	H/EA		0.5	
	3247.54	10	H/N$_2$	EDT	0.03	2
	3247.54	10	P/A	Xe1.5	1	12
	3247.54	10	SH/O/Ar	HIHC	0.0015	53
	3273.96	5			0.0025	
	2178.94 } 2181.72	0.13			0.06	
	2165.09	0.034			0.3	
	2225.70	0.024			0.5	
	2492.15	NI			5	
	5105.54	weak			NI	
	5782.13	weak			NI	
	3247.54	10	H/A	HIHC	0.003	53
	3273.96	5			0.005	
	2178.94 } 2181.72	0.152			0.06	
	2165.09	0.045			0.3	
	2225.70	0.026			0.5	
	2492.15	NI			7	
	5105.54	weak			NI	
	5782.13	weak			NI	
	3247.54	10	Ac/A	HIHC	0.006	53
	3273.96	5.1			0.01	
	2178.94 } 2181.72	0.15			0.1	
	2165.09	0.045			0.5	
	2225.70	0.03			0.7	
	2492.15	NI			20	
	5105.54	weak			NI	
	5782.13	weak			NI	
	3247.54	10	SAc/A	HCL	0.05	55
	3247.54	10	Ac/A		0.14	

EXPERIMENTALLY OBSERVED
ATOMIC FLUORESCENCE LINES

Element	Line (Å)	Rel. Int.	Flame	Excit. Source	Limit of Detection (ppm)	Ref.
Copper	3247.54	10	SH/O/Ar	HFHC	0.0005	47
Continued	3247.54	NI	SH/O/Ar	HIHC	NI	
	3247.54	10	H/Ar	Xe1.5	0.2	56
	3247.54	10	H/O	PDHC	0.02	60
	3273.96	6				
	2165.09					
	2178.94}	0.02				
	2181.72					
	3247.54	10	H/A	EDT	0.001	61
Curium Cm	No Information					
Dysprosium Dy	3645.41 II/ 3536.03 II	10	Ac/N	PTDL	0.3	57
	3645.41 II/ 3531.70 II	NI				
	3645.41 II/ 3534.96 II	NI				
	3645.41 II/ 3538.50 II	NI				
	3645.41 II/ 3645.41 II	NI				
	4045.99/ 4045.99	NI				
	4045.99/ 4186.78	NI				
	4045.99/ 4194.85	NI				
	4045.99/ 4215.15	NI				
	4045.99/ 4218.09	NI				
	4186.78	5			0.6	
Einsteinium Es	No Information					

EXPERIMENTALLY OBSERVED
ATOMIC FLUORESCENCE LINES

Element	Line (Å)	Rel. Int.	Flame	Excit. Source	Limit of Detection (ppm)	Ref.
Erbium	4007.97	10	Ac/N	PTDL	0.5	57
Er	3862.82	NI				
	3862.82/					
	4007.97	NI				
	3862.82/					
	4151.10	NI				
	3692.64 II/					
	3372.76 II	2			2.5	
	3692.64 II/					
	3264.79 II	NI				
	3692.64 II/					
	3499.11 II	NI				
	3692.64 II/					
	3616.58 II	NI				
	3692.64 II/					
	3692.64 II	NI				
	3692.64 II/					
	3896.25 II	NI				
	3692.64 II/					
	3903.34 II	NI				
Europium	4594.03/					
Eu	4627.22	10	Ac/N	PTDL	0.02	57
	4594.03/					
	4594.03	NI				
	4594.03/					
	4661.88	NI				
	4205.05 II/					
	3930.45 II	1			0.2	
	4205.05 II/					
	3819.67 II	NI				
	4205.05 II/					
	3907.10 II	NI				
	4205.05 II/					
	4129.70 II	NI				
	4205.05 II/					
	4205.05 II	NI				
Fermium Fm	No Information					

EXPERIMENTALLY OBSERVED
ATOMIC FLUORESCENCE LINES

Element	Line (Å)	Rel. Int.	Flame	Excit. Source	Limit of Detection (ppm)	Ref.
Fluorine F	No Information					
Francium Fr	No Information					
Gadolinium Gd	3768.39 II/ 3362.23 II	10	Ac/N	PTDL	0.8	57
	3768.39 II/ 3350.47 II	NI				
	3768.39 II/ 3422.47 II	NI				
	3768.39 II/ 3768.39 II	NI				
	3684.13	2			5.0	
	3684.13/ 3783.05	NI				
	3684.13/ 4225.85					
Gallium Ga	4172.06	10	H/A	EDT	0.007	1
	4172.06	10	H/O	EDT	1	22
	4172.06	10	H/N$_2$	EDT	0.3	2
	4032.98	NI	H/A	MVL	NI	25
	4172.06	NI				
	4172.06	10	H/A/Ar	EDT	NI	13
	4172.06	10	H/A/N$_2$			
	4172.06	10	PH/A	EDT	NI	6
	4172.06	10	PH/Ar			
	4172.06	10	PH/N			
	4172.06	10	PH/O			
	4172.06	10	Ac/A	HCL	40	43
	2943.64	0.4			1000	
	4172.06	10	SAc/A		20	
	2943.64	2			100	

EXPERIMENTALLY OBSERVED
ATOMIC FLUORESCENCE LINES

Element	Line (Å)	Rel. Int.	Flame	Excit. Source	Limit of Detection (ppm)	Ref.
Gallium						
Continued	4172.06	10	Ac/A	MVL	0.7	43
	2943.64	0.4			20	
	4172.06	10	SAc/A		0.5	
	2943.64	0.5			10	
	4032.98/					
	4172.05	10	H/A	PTDL	0.02	45
	4032.98	2			0.1	
Germanium	2651.18 }	10	Ac/O/N$_2$	EDT	15	26
Ge	2651.58					
	2651.18 }	10	Ac/N	EDT	3	1
	2651.58					
	2651.18 }	10	SAc/N	HIHC	2	44
	2651.58					
	2754.59	3.5				
	3039.06	3.5			17	
	2592.54	3.4				
	2709.63	3.1				
	2691.35	1.9				
	3269.49	1.2				
	2497.96	0.1				
	2651.18 }	10	H/A	EDT	10	14
	2651.58					
Gold	2675.95	10	H/A	DHC	0.05	27
Au	2675.95	10	H/O	EDT	0.2	21
	2675.95	10	H/Ar	Xe1.5	4	11
	2675.95	10	Ac/O	Xe9	NI	16
	2427.95	10	SH/O/Ar	HIHC	0.005	51
	2675.95	5.2			0.02	
	3029.21	0.1			NI	
	3122.82	0.5			2.5	
	2427.95	10	H/A	HIHC	0.015	

EXPERIMENTALLY OBSERVED
ATOMIC FLUORESCENCE LINES

Element	Line (Å)	Rel. Int.	Flame	Excit. Source	Limit of Detection (ppm)	Ref.
Gold Continued	2675.95	2	H/A	HIHC	0.07	51
	2427.95	10	Ac/A	HIHC	0.015	
	2675.95	2			0.07	
	2675.95	10	SH/O/Ar	HFHC	0.003	47
	2675.95	0.6		HIHC	0.05	
	2427.95	10	Ac/A	HCL	4.0	43
	2427.95	10		HIHC	0.15	
	2427.95	10	SAc/A	HCL	1.0	
	2427.95	10		HIHC	0.05	
	2427.95	10	SAc/A	HCL	7	55
	2427.95	10	Ac/A	HCL	14	
	2675.95	10	H/O	EDT	0.2	14
	2675.95	10	H/O	PDHC	0.2	60
	2427.95	4				
Hafnium Hf	3682.24/ 3777.64	10	Ac/N	PTDL	100	59
Helium He	No Information					
Holmium Ho	4103.84/ 4053.93	10	Ac/N	PTDL	0.15	57
	4103.84/ 4040.81	NI				
	4103.84/ 4103.84	NI				
	4103.84/ 4108.62	NI				
	4103.84/ 4127.16	NI				
	4103.84/ 4163.03	NI				
	4103.84/ 4173.23	NI				

EXPERIMENTALLY OBSERVED
ATOMIC FLUORESCENCE LINES

Element	Line (Å)	Rel. Int.	Flame	Excit. Source	Limit of Detection (ppm)	Ref.
Holmium Continued	4103.84/ 4227.04	NI	Ac/N	PTDL	0.15	57
Hydrogen H	No Information					
Indium In	4511.31	NI	H/A	MVL	NI	25
	4101.76	NI				
	4101.76	10	H/Ar	EDT	0.1	21
	4511.31	10	H/N$_2$	EDT	1	2
	4511.31	10	Ac/A	HCL	8	43
	3039.36	1			60	
	4511.31	10	SAc/A		4	
	3039.36	8			5	
	4511.31	10	Ac/A	MVL	0.9	
	3039.36	1			7	
	4511.31	10	SAc/A		0.4	
	3039.36	2			2	
	4101.76	10	H/Ar	EDT	0.1	14
	4101.76/ 4511.31	10	H/A	PTDL	0.002	45
	4101.76	2			0.01	
	3039.36	10	H/A	EDT	0.05	61
Iodine I	No Information					
Iridium Ir	2543.97	10	Ac/A	HCL	600	43
	2543.97	10	SAc/A		170	
Iron Fe	2483.27	10	H/A	HIHC	0.02	21
	2522.85	2				

EXPERIMENTALLY OBSERVED
ATOMIC FLUORESCENCE LINES

Element	Line (Å)	Rel. Int.	Flame	Excit. Source	Limit of Detection (ppm)	Ref.
Iron Continued	2719.02	0.5	H/A	HIHC	0.02	21
	3020.49 }	0.4				
	3020.64					
	2483.27	10	H/O/Ar		0.02	
	2483.27	10	Ac/A	HIHC	0.08	21
	2483.27	10	H/O/A		0.02	
	2483.27	10	H/Ar	EDT	0.3	24
	2483.27	10	H/N$_2$	EDT	0.008	2
	3734.87	0.2				
	2483.27	10	H/EA	Xe9	2	15
	2483.27	10	H/A/Ar	EDT	NI	13
	2483.27	10	H/A/N$_2$			
	2483.27	10	PH/Ar	EDT	NI	6
	2483.27	10	PH/N			
	2483.27	10	PH/O			
	2483.27	10	H/O	Xe9	NI	16
	2483.27	10	Ac/O			
	2483.27	10	Ac/A	HCL	3	43
	2483.27	10	SAc/A		0.8	
	2483.27	10		HIHC	0.05	
	2483.27	10	Ac/A	HCL	0.45	55
	2483.27	10	SAc/A		0.15	
	2483.27	10	H/A	Xe1.5	5.0	56
	3719.94	10	H/A	PTDL	0.03	45
	2483.27	10	H/A	EDT	0.02	61

| Krypton Kr | No Information | | | | | |

EXPERIMENTALLY OBSERVED
ATOMIC FLUORESCENCE LINES

Element	Line (Å)	Rel. Int.	Flame	Excit. Source	Limit of Detection (ppm)	Ref.
Lanthanum La	No Information					
Lead Pb	2833.06	10	H/N_2	EDT	0.06	2
	4057.83	2				
	4057.83	10	H/A	HIHC	0.02	23
	4057.83	10	H/O/Ar	Xe9	0.5	24
	4057.83	10	H/Ar		0.2	
	2833.06	10	P/A	Xe1.5	20	12
	2833.06	10	H/O	Xe9	NI	
	2833.06	10	Ac/O			
	4057.83	10	H/Ar	EDT	0.5	14
	4057.83	10	H/A	HIHC	0.10	48
	2833.06	8			0.12	
	3683.47	6.5			0.30	
	2802.00	5.5			0.80	
	3639.58	5.0			0.15	
	2613.65 } 2614.18	5.0			0.45	
	2169.99	1.5			0.20	
	4057.83	10	H/O/Ar	HIHC	0.02	48
	2833.06	7			0.05	
	3639.58	3.3			0.05	
	3683.47	2.5			0.07	
	2169.99	2.0			0.06	
	2613.65 } 2614.18	1.3			0.15	
	4057.83	10	SAc/A	HIHC	0.20	48
	2833.06	7			0.15	
	3639.58	3.5			0.40	
	2169.99	3			0.12	
	2613.65 } 2614.18	2			0.30	
	2802.00	1			1.0	

EXPERIMENTALLY OBSERVED
ATOMIC FLUORESCENCE LINES

Element	Line (Å)	Rel. Int.	Flame	Excit. Source	Limit of Detection (ppm)	Ref.
Lead Continued	4057.83	10	Ac/A	HIHC	NI	49
	2833.06	5.0				
	7229.11	3.9				
	3639.58	1.4				
	2169.99	0.8				
	2613.65} 2614.18	0.8				
	2802.00	0.6				
	2823.28	0.4				
	3683.48	0.2				
	2873.30	0.1				
	4019.64	weak				
	4062.14	weak				
	4057.83	10	SH/O/Ar	EDT	0.01	49
	7229.11	6.5				
	2833.06	5.2				
	3639.58	1.6				
	3683.48	0.8				
	2613.65} 2614.18	0.4				
	2169.99	0.3				
	2802.00	0.3				
	2823.28	0.2				
	2873.30	0.1				
	4019.64	0.06				
	4062.14	weak				
	4057.83	10	H/N	HIHC	NI	49
	2833.06	4.4				
	7229.11	2.3				
	3639.58	1.5				
	2169.99	0.9				
	3683.48	0.86				
	2613.65} 2614.18	0.8				
	2802.00	0.66				
	2823.28	0.5				
	2873.30	weak				
	4019.64	weak				
	4062.14	weak				
	4057.83	10	H/Ar	EDT	0.5	14

EXPERIMENTALLY OBSERVED
ATOMIC FLUORESCENCE LINES

Element	Line (Å)	Rel. Int.	Flame	Excit. Source	Limit of Detection (ppm)	Ref.
Lead Continued	4057.83	10	Ac/A	HCL	9	43
	2833.06	4			20	
	2833.06	10	SAc/A		3	
	4057.83	8			4	
	4057.83	10	Ac/A	HCL	2.3	55
	2833.06	4			6.0	
	4057.83	10	SAc/A		1.0	
	2833.06	3			3.0	
	4057.83	10	H/Ar	Xe1.5	20	56
Lithium Li	6707.80	10	H/O/Ar	MVL	NI	17
Lutetium Lu	4658.02/ 5135.09	10	Ac/N	PTDL	3	57
	4658.02	NI				
Magnesium Mg	2852.13	10	P/A	HIHC	0.001	29
	2852.13	10	Ac/N		0.005	
	2852.13	10	H/A	Xe9	0.002	1
	2852.13	10	H/Ar	Xe4.5	0.004	15
	2852.13	10	H/EA	Xe4.5	0.01	30
	2852.13	10	H/O	EDT	0.01	22
	2852.13	10	H/O/Ar	Xe9	0.2	24
	2852.13	10	PH/A	EDT	NI	6
	2852.13	10	PH/Ar			
	2852.13	10	PH/N			
	2852.13	10	PH/O			
	2852.13	10	Ac/O	Xe9	NI	16
	2852.13	10	Ac/A	HCL	0.0005	43
	2852.13	10		HIHC	0.0006	

EXPERIMENTALLY OBSERVED
ATOMIC FLUORESCENCE LINES

Element	Line (Å)	Rel. Int.	Flame	Excit. Source	Limit of Detection (ppm)	Ref.
Magnesium Continued	2852.13 2852.13	10 10	SAc/A	HCL HIHC	0.0008 0.00015	43
	2852.13	10	H/Ar	Xe1.5	0.04	56
	2852.13	10	H/Ar	EDT	0.008	14
Manganese Mn	2794.82 2798.27 } 2801.06	10	H/A	Xe4.5	0.003	1
	2794.82 2798.27 } 2801.06	10	H/Ar	EDT	0.006	21
	2794.82 2798.27 } 2801.06	10	H/N$_2$	EDT	0.01	2
	2794.82 2798.27 } 2801.06	10	H/EA	Xe9	0.1	15
	2794.82 2798.27 } 2801.06	10	P/A	Xe1.5	0.3	12
	2794.82 2798.27 } 2801.06	10	H/O	EDT	1	22
	2794.82 2798.27 } 2801.06	10	Ac/O	Xe9	NI	16
	2794.82 2798.27 } 2801.06	10	Ac/A	EDT	0.001	54
	4030.76 4033.07 } 4034.49	4.9			0.02	
	2576.10	1.2			1.5	
	3045.59	1.1			30	

EXPERIMENTALLY OBSERVED
ATOMIC FLUORESCENCE LINES

Element	Line (Å)	Rel. Int.	Flame	Excit. Source	Limit of Detection (ppm)	Ref.
Manganese Continued	3834.36	0.8	Ac/A	EDT	67	54
	2593.73	0.8			12	
	2605.69	0.8			12	
	3577.88	0.7			50	
	2933.06 } 2939.30	0.4			90	
	2949.21	0.3			90	
	2794.82	10	Ac/A	HCL	0.2	43
	2794.82	10	SAc/A		0.03	
	2794.82	10	Ac/A	HCL	0.13	55
	2794.82	10	SAc/A		0.05	
	2794.84	10	H/Ar	Xe1.5	0.2	56
	4030.76	10	H/A	PTDL	0.01	45
Mendelevium Md	No Information					
Mercury Hg	2536.52	10	H/A	MVL	0.1	31
	2536.52	10	H/O		0.1	
	2536.52	10	Ac/O		0.1	
	2536.52	10	NG/A		7	
	2536.52	10	H/N$_2$	EDT	0.1	2
	2536.52	10	Ac/A	HCL	400	43
	2536.52	10	SAc/A		50	
	2536.52	10	Ac/A	MVL	2	
	2536.52	10	SAc/A		0.5	
	2536.52	10	Ac/A	HCL	120	55
	2536.52	10	SAc/A		60	
	2536.52	10	H/A	Xe1.5	100	56
	2536.52	10	H/O	PDHC	2	60
	2536.52	10	H/A	EDT	0.02	61

EXPERIMENTALLY OBSERVED
ATOMIC FLUORESCENCE LINES

Element	Line (Å)	Rel. Int.	Flame	Excit. Source	Limit of Detection (ppm)	Ref.
Molybdenum Mo	3132.59	10	SAc/N	EDT	0.46	42
	3170.35	5.9				
	3193.97	4.6				
	3798.25	3.5				
	3864.11	2.5				
	3158.16	1.3				
	3902.96	0.7				
	3902.96	10	Ac/N	PTDL	0.3	45
	3798.25	0.3			10	
Neodymium Nd	4634.24/ 4896.93	10	Ac/N	PTDL	2	57
	4634.24	NI				
	4303.58 II/ 4247.38 II	0.5			40	
	4303.58 II	NI				
Neon Ne	No Information					
Neptunium Np	No Information					
Nickel Ni	2320.03	10	H/A	HIHC	0.005	21
	2310.96	0.5				
	3414.76	0.3				
	2289.98	0.2				
	2345.54	0.2				
	3002.49	0.2				
	3524.54	0.2				
	3050.82	0.1				
	3461.65	0.1				
	2312.34	weak				
	2313.98	weak				
	2325.79	weak				
	3003.63	weak				
	3101.55 } 3101.88	weak				

EXPERIMENTALLY OBSERVED
ATOMIC FLUORESCENCE LINES

Element	Line (Å)	Rel. Int.	Flame	Excit. Source	Limit of Detection (ppm)	Ref.
Nickel Continued	3134.11	weak	H/A	HIHC	0.005	21
	3515.05	weak				
	2320.03	10	H/O/Ar		0.003	
	2320.03	10	Ac/A		0.03	
	2320.03	10	H/O/Ar		0.006	
	2320.03	10	H/EA	Xe9	1	15
	2320.03	10	H/N$_2$	EDT	0.006	2
	3414.76	0.6				
	2320.03	10	Ac/A	HCL	0.6	43
	2320.03	10	SAc/A		0.3	
	2320.03	10	SAc/A	HCL	0.5	55
	2320.03	10	Ac/A		1.0	
	2320.03	10	H/A	Xe1.5	10	56
	3610.46/ 3524.54	10	Ac/A	PTDL	0.05	45
	3610.46	0.2			2	
	2320.03	10	H/A	EDT	0.04	14
	2320.03	10	H/A	PHC	1	60
	3524.54	7				
	3414.76	4				
	3461.65	3				
Niobium Nb	4079.73/ 4058.94	10	Ac/N	PTDL	1.5	59
	4079.73/ 4079.73	NI				
	4079.73/ 4100.92	NI				
	4079.73/ 4123.81	NI				
	4079.73/ 4137.10	NI				
	4079.73/ 4139.71	NI				

EXPERIMENTALLY OBSERVED
ATOMIC FLUORESCENCE LINES

Element	Line (Å)	Rel. Int.	Flame	Excit. Source	Limit of Detection (ppm)	Ref.
Niobium Continued	4079.73/ 4152.58	NI	Ac/N	PTDL	1.5	59
	4079.73/ 4163.66	NI				
	4079.73/ 4168.13	NI				
Nitrogen N	No Information					
Nobelium No	No Information					
Osmium Os	4420.47/ 4260.85	10	Ac/N	PTDL	150	59
Oxygen O	No Information					
Palladium Pd	3404.58	10	H/A	HCL	2	32
	3404.58	10	H/O/Ar	HIHC	0.04	52
	3634.70	9.8			0.05	
	3609.55	4.0			0.10	
	3242.70	3.2			0.20	
	3421.24	1.9			0.20	
	3516.94	1.7			0.30	
	2447.91	1.0			0.50	
	3302.13	0.7			0.50	
	3404.58	10	H/A	HIHC	0.06	52
	3634.70	8.0			0.12	
	3609.55	3.9			0.20	
	3242.70	2.7			0.35	
	3421.24	1.9			0.25	
	3516.94	1.9			0.45	
	2447.91	1.7			0.25	
	3302.13	1.0			0.50	

EXPERIMENTALLY OBSERVED
ATOMIC FLUORESCENCE LINES

Element	Line (Å)	Rel. Int.	Flame	Excit. Source	Limit of Detection (ppm)	Ref.
Palladium	3404.58	10	Ac/A	HIHC	0.08	52
Continued	3634.70	6.4			0.18	
	3609.55	2.9			0.30	
	3242.70	3.1			0.40	
	3421.24	1.8			0.35	
	3516.94	1.8			0.50	
	2447.91	1.5			0.25	
	3302.13	1.0			0.70	
	3404.58	10	SAc/A	HIHC	0.15	44
	3609.55	7.3				
	3634.70	6.3				
	3242.70	3.0				
	3421.24	2.0				
	3516.94	2.0				
	3433.45	1.2				
	3460.77	1.0				
	3302.13	1.0				
	3553.08	0.8				
	3481.15	0.6				
	3373.00	0.6				
	2447.91	0.5				
	2476.42	0.3			5.0	
	2447.91	10	Ac/A	HIHC	4	43
	2447.91	10	SAc/A		2	
	3404.58	10	Ac/A	HCL	3.5	55
	2476.42	NFO			ND	
	3404.58	10	SAc/A		1.0	
	2476.42	NFO			ND	
	3404.58	10	H/A	PDHC	3	60
	3609.55	6				
	3242.70	5				
	3516.94	2				

| Phosphorus P | No Information | | | | | |

EXPERIMENTALLY OBSERVED
ATOMIC FLUORESCENCE LINES

Element	Line (Å)	Rel. Int.	Flame	Excit. Source	Limit of Detection (ppm)	Ref.
Platinum Pt	2659.45	10	Ac/A	HCL	90	55
	2659.45	10	SAc/A		60	
	2659.45	10	SAc/A	HIHC	1.0	44
	3064.71	9.4				
	2702.40	5.2				
	2997.97	2.7				
	2734.50	2.5				
	2719.04	2.2				
	2650.86	1.7				
	2929.79	1.7				
	2830.30	1.6				
	2628.03	1.6				
	2174.67	1.4				
	3042.64	1.2				
	2487.17	0.9				
	2893.87	0.7				
	2440.06	0.6				
Plutonium Pu	No Information					
Polonium Po	No Information					
Potassium K	7664.91	NI	H/O/Ar	MVL	NI	28
	7698.98	NI				
Praseodymium Pr	4272.27 II/ 4305.76 II	10	Ac/N	PTDL	1	57
	4272.27 II/ 4118.48 II	NI				
	4272.27 II/ 4164.19 II	NI				
	4272.27 II/ 4179.42 II	NI				
	4272.27 II/ 4272.27 II	NI				

EXPERIMENTALLY OBSERVED
ATOMIC FLUORESCENCE LINES

Element	Line (Å)	Rel. Int.	Flame	Excit. Source	Limit of Detection (ppm)	Ref.
Promethium Pm	No Information					
Protactinium Pa	No Information					
Radium Ra	No Information					
Radon Rn	No Information					
Rhenium Re	No Information					
Rhodium Rh	3434.89	10	Ac/A	HCL	7	43
	3434.89	10	SAc/A		3	
	3692.36}/ 3700.91					
	3502.52	10	Ac/A	PTDL	0.15	59
	3692.36}/ 3700.91					
	3396.85	NI				
	3692.36}/ 3700.91					
	3434.89	NI				
	3692.36}/ 3700.91					
	3528.02	NI				
	3692.36}/ 3700.91					
	3597.15	NI				
	3692.36}/ 3700.91					
	3657.99	NI				
	3692.36}/ 3700.91					
	3692.36} 3700.91	NI				

EXPERIMENTALLY OBSERVED
ATOMIC FLUORESCENCE LINES

Element	Line (Å)	Rel. Int.	Flame	Excit. Source	Limit of Detection (ppm)	Ref.
Rubidium Rb	7947.60	10	H/O/Ar	MVL	NI	28
Ruthenium Ru	3728.03	10	Ac/A	HCL	15	43
	2874.98	NFO			ND	
	3728.03	10	SAc/A		5	
	2874.98	0.05			1000	
	3728.03/ 3498.94	10	Ac/N	PTDL	0.5	59
	3728.03/ 3428.31	NI				
	3728.03/ 3589.22	NI				
	3728.03/ 3593.02	NI				
	3728.03/ 3661.35	NI				
	3728.03/ 3726.93	NI				
	3728.03/ 3728.03	NI				
	3728.03/ 3730.43	NI				
	3728.03/ 3790.51	NI				
	3728.03/ 3798.05	NI				
	3728.03/ 3798.90	NI				
	3728.03/ 3799.35	NI				
	3728.03/ 3925.92	NI				
	3728.03/ 4199.90	NI				
	3728.03/ 4554.51	NI				
Samarium Sm	3661.36 II/ 3739.12 II	10	Ac/N	PTDL	0.15	57

EXPERIMENTALLY OBSERVED
ATOMIC FLUORESCENCE LINES

Element	Line (Å)	Rel. Int.	Flame	Excit. Source	Limit of Detection (ppm)	Ref.
Samarium Continued	3592.60 II/ 3592.60 II	NI	Ac/N	PTDL	0.15	57
	3592.60 II/ 3609.49 II	NI				
	3592.60 II/ 3634.29 II	NI				
	3592.60 II/ 3661.36 II	NI				
	3592.60 II/ 3739.12 II	NI				
	3592.60 II/ 4280.79 II	NI				
	3756.41/ 4296.74	2			0.6	
	3756.41/ 3756.41					
	3756.41/ 4271.86					
	3756.41/ 4470.89					
	3756.41/ 4716.10					
	3756.41/ 4728.42					
	3756.41/ 4760.27					
Scandium Sc	3907.48	NI	H/Ar	EDT	10	14
	3911.81/ 4020.40	10	Ac/N	PTDL	0.01	59
	3911.81/ 4023.69	NI				
	3911.81/ 3907.49	NI				
	3911.81/ 3911.81	NI				
Selenium Se	2039.85	10	P/A	EDT	0.2	33
	1960.26	7				
	2062.79	5				

EXPERIMENTALLY OBSERVED
ATOMIC FLUORESCENCE LINES

Element	Line (Å)	Rel. Int.	Flame	Excit. Source	Limit of Detection (ppm)	Ref.
Selenium Continued	2074.79	0.2	P/A	EDT	0.2	33
	1960.26	10	H/Ar	EDT	0.4	14
	1960.26	10	H/O		0.8	
	1960.26	10	H/N$_2$		0.2	
	1960.26	10	H/A	EDT	0.04	2
	2039.85	10	H/N$_2$		0.2	
	1960.26	NFO	Ac/A	HCL	NI	43
	1960.26	10	SAc/A		50	
	2039.85	10	Ac/A	HCL	45	55
	1960.26	2			180	
	2039.85	10	SAc/A		45	
	1960.26	2			180	
	1960.26	10	H/Ar	Xe1.5	1000	56
	1960.26	10	Ac/A	EDT	1.0	58
Silicon Si	2516.11	10	Ac/N	EDT	2	34
	2506.90	NI				
	2514.32	NI				
	2519.20	NI				
	2524.11	NI				
	2528.51	NI				
	2516.11	10	SAc/N	EDT	0.5	34
Silver Ag	3382.89	10	P/A	Xe1.5	0.2	12
	3280.68	6				
	3382.89	10	H/A		0.1	
	3280.68	10				
	3280.68	10	H/O	Xe4.5	0.003	30
	3280.68	10	H/EA		0.001	
	3280.68	10	H/Ar	EDT	0.0005	24
	3280.68	10	H/N$_2$	EDT	0.1	2

EXPERIMENTALLY OBSERVED
ATOMIC FLUORESCENCE LINES

Element	Line (Å)	Rel. Int.	Flame	Excit. Source	Limit of Detection (ppm)	Ref.
Silver Continued	3280.68	10	H/O/Ar	Xe9	0.5	24
	3280.68	10	Ac/A	HIHC	0.004	35
	3280.68	10	Ac/O	Xe9	NI	16
	3280.68	10	Ac/A	HCL	0.08	55
	3280.68	10	SAc/A		0.02	
	3280.68	10	H/A	Xe1.5	0.05	56
	3280.68	10	H/A	EDT	0.0001	14
	3382.89	10	H/O	PDHC	0.2	60
	3280.68	8				
Sodium Na	5895.92	10	H/A	Xe4.5	0.008	1
	5895.92	NI	H/O/Ar	MVL	NI	17
	5889.95	NI				
Strontium Sr	4607.33	10	H/Ar	EDT	0.03	14
	4607.33	10	H/A	PTDL	0.01	45
Sulfur S	No Information					
Tantalum Ta	No Information					
Technetium Tc	No Information					
Tellurium Te	2383.25 } 2385.76	10	P/A	EDT	0.05	33
	2142.75	7				

EXPERIMENTALLY OBSERVED
ATOMIC FLUORESCENCE LINES

Element	Line (Å)	Rel. Int.	Flame	Excit. Source	Limit of Detection (ppm)	Ref.
Tellurium Continued	2259.04	0.4	P/A	EDT	0.05	33
	2530.70	ND				
	2142.75	10	H/O	EDT	0.5	22
	2142.75	10	H/Ar	EDT	0.5	14
	2142.75	10	H/A	EDT	0.006	1
	2142.75	10	H/N$_2$	EDT	0.06	2
	2142.75	10	SAc/A	HCL	20	43
	2142.75	10	Ac/A	HCL	30	55
	2142.75	10	SAc/A		20	
	2142.75	10	H/A	Xe1.5	50	56
Terbium Tb	3702.85 II/ 3509.17 II	10	Ac/N	PTDL	0.5	57
	3702.85 II/ 3568.51 II	NI				
	3702.85 II/ 3600.44 II	NI				
	3702.85 II/ 3676.35 II	NI				
	3702.85 II/ 3702.85 II	NI				
	3702.85 II/ 3981.89 II	NI				
	3702.85 II/ 4005.57 II	NI				
	4326.47/ 4338.45	3			1.5	
	4326.47/ 3901.35	NI				
	4326.47/ 4318.85	NI				
	4326.47/ 4326.47	NI				

EXPERIMENTALLY OBSERVED
ATOMIC FLUORESCENCE LINES

Element	Line (Å)	Rel. Int.	Flame	Excit. Source	Limit of Detection (ppm)	Ref.
Thallium Tl	3775.72	10	H/A	EDT	0.1	36
	5350.46	4				
	2767.87	1				
	3775.72	10	H/Ar	EDT	0.008	14
	3775.72	10	H/A/Ar	EDT	NI	13
	3775.72	10	H/A/N$_2$			
	3775.72	10	PH/A	EDT	NI	6
	3775.72	10	PH/Ar			
	3775.72	10	PH/N			
	3775.72	10	PH/O			
	3775.72	10	H/O	MVL	0.04	37
	3775.72	10	H/N$_2$	EDT	0.2	2
	3775.72	10	Ac/O	MVL	5	31
	3775.72	10	NG/A		1	
	3775.72	10	H/EA	Xe4.5	0.07	15
	3775.72	10	H/O/Ar	MVL	NI	38
	3775.72	10	Ac/A	HCL	0.2	43
	2767.87	NFQ			ND	
	3775.72	10	SAc/A		2	
	2767.87	0.8			25	
	3775.72	10	Ac/A	MVL	0.4	
	2767.87	1			3	
	3775.72	10	SAc/A		0.1	
	2767.87	2			0.5	
	3775.72	10	Ac/A	HCL	3.0	55
	2767.87	2			14	
	3775.72	10	SAc/A		1.2	
	2767.87	2			7	
	3775.72	10	H/Ar	Xe1.5	1.0	56
	3775.72/ 5350.46	10	Ac/A	PTDL	0.02	45

EXPERIMENTALLY OBSERVED
ATOMIC FLUORESCENCE LINES

Element	Line (Å)	Rel. Int.	Flame	Excit. Source	Limit of Detection (ppm)	Ref.
Thallium Continued	3775.72	0.2	Ac/A	PTDL	1	45
Thorium Th	No Information					
Thulium Tm	3717.92/ 4094.19	10	Ac/N	PTDL	0.1	57
	3717.92/ 3717.92	NI				
	3717.92/ 3744.07	NI				
	3717.92/ 3883.13	NI				
	3717.92 3887.35	NI				
	3717.92/ 4105.84	NI				
	3717.92/ 4187.62	NI				
	3717.92/ 4203.73					
	3761.33 II}/ 3761.91 II					
	3761.33 II} 3761.91 II	0.2			5	
	3761.33 II}/ 3761.91 II					
	3462.20 II	NI				
	3761.33 II}/ 3761.91 II					
	3462.20 II	NI				
	3761.33 II}/ 3761.91 II					
	3700.26 II} 3701.36 II	NI				
	3761.33 II}/ 3761.91 II					
	3795.76 II	NI				
	3761.33 II}/ 3761.91 II					
	3848.02 II	NI				

EXPERIMENTALLY OBSERVED
ATOMIC FLUORESCENCE LINES

Element	Line (Å)	Rel. Int.	Flame	Excit. Source	Limit of Detection (ppm)	Ref.
Tin	3034.12	10	H/O/Ar	EDT	0.1	39
Sn	3175.05	9				
	2863.33	5				
	2839.99	4				
	3009.14	4				
	2706.51	2				
	3801.02	1				
	3330.62	0.7				
	3262.34	0.2				
	2546.55	0.1				
	2429.49	0.1				
	2246.05	0.07				
	2354.84	0.07				
	3175.05	10	H/Ar	EDT	0.3	39
	3034.12	6				
	2863.33	6				
	3009.14	4				
	2839.99	3				
	2706.51	3				
	3801.02	2				
	3262.34	2				
	2429.49	0.3				
	2546.55	0.2				
	3330.62	0.2				
	3034.12	10	Ac/A	EDT	0.6	39
	2839.99	8				
	3175.05	7				
	2863.33	4				
	3009.14	3				
	2706.51	3				
	3262.34	2				
	3801.02	1				
	3330.62	0.9				
	2429.49	0.4				
	2354.84	0.3				
	2546.55	0.3				
	2246.05	0.2				
	3034.12	10	SH/O/Ar	EDT	0.1	39
	3034.12	10	H/A	EDT	NI	13

EXPERIMENTALLY OBSERVED
ATOMIC FLUORESCENCE LINES

Element	Line (Å)	Rel. Int.	Flame	Excit. Source	Limit of Detection (ppm)	Ref.
Tin Continued	3034.12	10	H/A/Ar	EDT	NI	13
	3034.12	10	H/A/N$_2$			
	3034.12	10	H/N$_2$	EDT	0.6	2
	3034.12	10	H/A	Xe1.5	5	56
	3034.12	10	Ac/A	HCL	200	43
	2839.99	7			300	
	3034.12	10	SAc/A		70	
	2839.99	10			70	
	3034.12	10	Ac/A	HCL	100	55
	2863.33	3			300	
	3034.12	10	SAc/A		25	
	2863.33	3			75	
Titanium Ti	3635.46	NFO	SAc/N	EDT	—	42
	3642.68	NFO				
	3653.50	NFO				
	3752.86	NFO				
	3948.67	NFO				
	3956.33	NFO				
	3958.21	NFO				
	3981.76	NFO				
	3989.76	NFO				
	3998.64	NFO				
	3653.50	10	SAc/N	HCL	10	44
	3635.46	9.3				
	3642.68	8.6				
	3371.45	8.4			30	
	3998.64	8.2				
	3354.63	4.3				
	3199.92	4.0			4	
	3341.88	3.9				
	3191.99	3.3				
	3186.45	1.7				
	2948.24	1.0				
	2956.80	0.8			45	
	3646.20	0.8				

EXPERIMENTALLY OBSERVED
ATOMIC FLUORESCENCE LINES

Element	Line (Å)	Rel. Int.	Flame	Excit. Source	Limit of Detection (ppm)	Ref.
Titanium Continued	3998.64	10	Ac/N	PTDL	0.1	45
Tungsten W	4008.75	NFO	Ac/N	PTDL	ND	59
Uranium U	No Information					
Vanadium V	3183.98	10	SAc/N	EDT	0.07	42
	3066.38	0.8				
	4379.24	NFO				
	4384.72	NFO				
	4389.97	NFO				
	4395.23	NFO				
	4408.21	NFO				
	4408.51	NFO				
	3703.58/ 4111.78	10	Ac/N	PTDL	0.5	59
	3703.58/ 3703.58	NI				
	3703.58/ 3840.75	NI				
	3703.58/ 3855.37	NI				
	3703.58/ 3855.84	NI				
	3703.58/ 4099.80	NI				
	3703.58/ 4105.17	NI				
	3703.58/ 4115.18	NI				
	3703.58/ 4128.07	NI				
	3703.58/ 4132.02	NI				
	3703.58/ 4379.24	NI				

EXPERIMENTALLY OBSERVED
ATOMIC FLUORESCENCE LINES

Element	Line (Å)	Rel. Int.	Flame	Excit. Source	Limit of Detection (ppm)	Ref.
Vanadium Continued	3703.58/					
	4384.72	NI	Ac/N	PTDL	0.5	59
	3703.58/					
	4389.97	NI				
	3703.58/					
	4395.23	NI				
	3703.58/					
	4407.64	NI				
	3703.58/					
	4408.20	NI				
	3703.58/					
	4408.51	NI				
Xenon Xe	No Information					
Ytterbium Yb	3987.98/					
	3464.26	10	Ac/N	PTDL	0.01	57
	3987.98/					
	3987.98	NI				
	3987.98/					
	5556.48	NI				
	3694.19 II/					
	3289.37 II	3			0.03	
	3694.19 II/					
	2891.38 II	NI				
	3694.19 II/					
	2970.56 II	NI				
	3694.19 II/					
	3031.11 II	NI				
	3694.19 II/					
	3476.31 II	NI				
	3694.19 II/					
	3478.84 II	NI				
	3694.19 II/					
	3694.19 II	NI				
Yttrium Y	No Information					

EXPERIMENTALLY OBSERVED
ATOMIC FLUORESCENCE LINES

Element	Line (Å)	Rel. Int.	Flame	Excit. Source	Limit of Detection (ppm)	Ref.
Zinc Zn	2138.56	10	H/A	EDT	0.00004	40
	2138.56	10	H/O	MVL	0.0002	37
	2138.56	10	H/N$_2$	EDT	0.0002	2
	2138.56	10	Ac/O	MVL	0.04	31
	2138.56	10	NG/A		0.01	
	2138.56	10	H/EA	Xe4.5	0.01	15
	2138.56	10	P/A	MVL	0.003	12
	2138.56	10	SAc/A	MVL	0.0002	10
	2138.56	10	Ac/A	MVL	NI	41
	2138.56	10	PH/Ar	EDT	NI	15
	2138.56	10	PH/N			
	2138.56	10	PH/O			
	2138.56	10	PH/A			
	2138.56	10	Ac/A	HCL	0.3	43
	2138.56	10	SAc/A		0.15	
	2138.56	10	Ac/A	HIHC	0.005	
	2138.56	10	SAc/A		0.003	
	2138.56	10	Ac/A	HCL	0.13	55
	2138.56	10	SAc/A		0.10	
	2138.56	10	H/A	Xe1.5	7	56
	2138.56	10	H/O	PDHÇ	0.006	60
Zirconium Zr	3519.61	NFO	SAc/N	EDT	—	42
	3601.19	NFO				
	4072.70	NFO				
	4081.22	NFO				
	4687.80	NFO				
	4739.48	NFO				
	4772.31	NFO				

EXPERIMENTALLY OBSERVED
ATOMIC FLUORESCENCE LINES

Element	Line (Å)	Rel. Int.	Flame	Excit. Source	Limit of Detection (ppm)	Ref.
Zirconium Continued	3601.19	NFO	Ac/N	PTDL	ND	59

1. A. Hell and S. Ricchio, Talk given at the Pittsburgh Conference on Analytical Chemistry and Applied Spectroscopy, Cleveland, Ohio 1970. (Paper No. 23).

2. R.M. Dagnall, M.R.G. Taylor and T.S. West, Spectros. Letters, 1, 397 (1968).

3. R.M. Dagnall, K.C. Thompson and T.S. West, Talanta, 14, 1151 (1967).

4. R.M. Dagnall, K.C. Thompson and T.S. West, Talanta, 15, 677 (1967).

5. R.M. Dagnall and T.S. West, Appl. Optics, 7, 1287 (1968).

6. M.P. Bratzel, R.M. Dagnell and J.D. Winefordner, Anal. Chem. 41, 1527 (1969).

7. D.N. Hingle, G.F. Kirkbright and T.S. West, Analyst, 93, 522 (1968).

8. J.W. Robinson and C.J. Hsu, Anal. Chim. Acta, 43, 109 (1968).

9. R.M. Dagnall, K.C Thompson and T.S. West, Talanta, 14, 1467 (1967).

10. P.S. Hobbs, G.F. Kirkbright, M. Sargant, and T.S. West, Talanta, 15, 997 (1968).

11. C. Veillon, J.M. Mansfield, M.L. Parsons and J.D. Winefordner, Anal. Chem., 38, 204 (1966).

12. R.M. Dagnall, K.C. Thompson and T.S. West, Anal. Chim. Acta, 36, 269 (1966).

13. M.P. Bratzel, R.M. Dagnell and J.D. Winefordner, Anal. Chem., 41, 713 (1969).

14. K.E. Zacha, M.P. Bratzel, J.M. Mansfield and J.D. Winefordner, Anal. Chem., 40, 1733 (1968).

15. D.W. Ellis and D.R. Domers," Atomic Fluorescence Flame Spectrometry" in Trace Inorganics in Water, R.A. Bellar, ed. Advances in Chemistry Series, No. 73, Washington, D.C., 1968.

16. S.J. Pearce, L. deGalen and J.D. Winefordner, Spectrochim. Acta, 23B, 793 (1968).

17. D.R. Jenkins, Proc. Roy. Soc., London, A293, 493 (1966).

18. G. Rossi and N. Omenetto, Talanta, 16, 263 (1969).

19. N. Omenetto and G. Rossi, Anal. Chim. Acta, 40, 195 (1968).

20. B. Fleet, K.V. Liberty and T.S. West, Anal. Chim. Acta, 45, 205 (1969).

21. J. Matousek and V. Sychra, Anal. Chem., 41, 518 (1969).

22. J.M. Mansfield, M.P. Bratzel, H.O. Norgardon, D.N. Knapp,
     K.E. Zacha and J.D. Winefordner, Spectrochim. Acta, 23B, 389 (1968).

23. D.L. Manning and P. Heneage, At. Abs. Newsletter, 6, 124 (1967).

24. R. Smith, C.M. Stafford and J.D. Winefordner, Can. Spectros., 14,
     2 (1969).

25. N. Omenetto and G. Rossi, Spectrochim. Acta, 24B, 95 (1969).

26. R.M. Dagnall, K.C. Thompson and T.S. West, Anal. Chim. Acta, 41,
     551 (1968).

27. J.F. Dimmin, Anal. Chem., 39, 1491 (1967).

28. D.R. Jenkins, Proc. Roy. Soc., London, A 303, 453 (1968).

29. T.S West and X.K. Williams, Anal. Chim. Acta, 42, 29 (1968).

30. D.W. Ellis and D.R. Demors, Anal. Chem., 38, 1943 (1966).

31. J.D. Winefordner and R.A. Staab, Anal. Chem., 36, 1367 (1966).

32. D.L. Manning and P. Heneage, At. Abs. Newsletter, 7, 80 (1968).

33. R.M. Dagnall, K.C. Thompson and T.S. West, Talanta, 14, 557 (1967).

34. R.M. Dagnall, G.F. Kirkbright, T.S. West and R. Wood, Anal. Chim.
     Acta, 47, 407 (1969).

35. T.S. West and X.K. Williams, Anal. Chem., 40, 335 (1968).

36. R.F. Browner, R.M. Dagnall, and T.S. West, Talanta, 16, 75 (1969).

37. J.M. Mansfield, J.D. Winefordner and C. Veillon, Anal. Chem., 37,
     1049 (1965).

38. D.R. Jenkins, Proc. Roy. Soc., London, A 303, 467 (1968).

39. R.F. Browner, R.M. Dagnall and T.S. West, Anal. Chim. Acta, 46, 207
     (1969).

40. M.P. Bratzel and J.D. Winefordner, Anal. Letters, 1, 43 (1967).

41. G.J. Goodfellow, Anal. Chim. Acta, 36, 132 (1966).

42. R.M. Dagnall, G.F. Kirkbright, T.S. West and R. Wood, Anal. Chem.,
     42, 1029 (1970).

43. P.L. Larkins, Spectrochimica Acta B, 26B, 477 (1971).

44. P.L. Larkins and J.B. Willis, Spectrochimica Acta, 26B, 491 (1971).

45. L.M. Fraser and J.D. Winefordner, Anal. Chem., 44, 1444 (1972).

46. R.F. Browner and D.C. Manning, Anal. Chem., 44, 843 (1972).

47. H.G.C. Human, Spectrochimica Acta, 27B, 301 (1972).

48. V. Sychra and J. Matousek, Talanta, 17, 363 (1970).

49. R.F. Browner, R.M. Dagnall and T.S. West, Anal. Chim. Acta, 50, 375 (1970).

50. J.D Norris and T.S. West, Anal. Chim. Acta, 59, 355 (1972).

51. J. Matousek and V. Sychra, Anal. Chim. Acta, 49, 175 (1970).

52. V. Sychra, P.J. Slevin, J. Matousek and F. Bek, Anal. Chim. Acta, 52, 259 (1970).

53. D. Kolihova and V. Sychra, Anal. Chim. Acta, 63, 479 (1973).

54. L. Ebdon, G.F. Kirkbright and T.S. West, Talanta, 17, 965 (1970).

55. R.F. Browner and D.C. Manning, Anal. Chem., 44, 843 (1972).

56. M.P. Bratzel, R.M. Dagnall and J.D. Winefordner, Anal. Chim. Acta, 52, 157 (1970).

57. N. Omenetto, N.N Hatch, L.M. Fraser and J.D. Winefordner, Anal. Chem., 45, 195 (1973).

58. M.S. Cresser and T.S. West, Spec. Letters, 2, 9 (1969).

59. N. Omenetto, N.N. Hatch, L.M. Fraser and J.D. Winefordner, Spectrochim. Acta, 28B, 65 (1973).

60. J.O. Weide and M.L. Parsons, Anal. Chem., 45, 2417 (1973).

CHAPTER III

FUNDAMENTAL INFORMATION

A. FUNDAMENTAL EXPRESSIONS

In order to fully understand atomic spectroscopy it is necessary to know the parameters which determine the signal and the
parameters which control the precision, i.e. the noise, of the
technique. The equations presented are intended to acquaint the
reader with these fundamental expressions. Overall these expressions are not difficult, in fact, the mathematics consists of
simple algebra. One can obtain quite a bit of useful information
from a study of these expressions. For instance, information as
to whether the analytical signal is directly or inversely related
to a particular parameter or is more complex. This is readily
apparent from a study of these expressions. The same information
is obtainable from the noise expressions. Of course the affect
of the parameters on both the signal and the noise must be considered. It should also be remembered that the fundamental expressions do not help us with matrix problems except as they relate
to the noise expressions.

It is felt that sufficient information is presented in this
Handbook to estimate numerical values for most of the theoretical
quantities. The glaring exception to this lies in the noise expressions. To date, only gross estimates have been used for the
various "flicker factors", i.e. the noise due to "flicker", or
random fluctuations in the various excitation sources and sample
cells. The estimates usually are about 1 to 0.1% or factors of
0.01 to 0.001. It should be relatively easy to measure this value
by measuring the noise-to-signal ratio of an emission source by
a simple atomic emission experiment.

J.D. Winefordner and his research group are to be congratu-
lated on presenting the atomic spectroscopist with a unified approach to the theory of all three techniques both for theoretical
analytical signal expressions as well as noise expressions. These
are presented in Table III.A.1 for all three methods. Wherever
possible IUPAC definition and symbols have been used. A list of
symbols and definitions is given in Table III.A.2. These expressions are derived using ideal optical and experimental assumptions
for single spectrally isolated transitions.

Table III.A.1-a

FUNDAMENTAL EXPRESSIONS IN ATOMIC SPECTROSCOPY
[All terms defined in Table III.A.1-b]

AES

Signal (1)

$$B_E = \frac{h\nu_o n_t g_j A_{ji} \ell}{4 \times 10^7 \pi Z} \exp(-E_j/kT_f) \tag{1}$$

Noise (2)

$$\overline{\Delta i_t} = (\overline{\Delta i_p^2} + \overline{\Delta i_c^2})^{\frac{1}{2}} \tag{2}$$

where

$$\Delta i_p = (2eBM\Delta f i_t + 4kT_o \Delta f/R_o)^{\frac{1}{2}} \tag{3}$$

and

$$\Delta i_c = \gamma T_{opt} sh_m \Omega_E \Delta\lambda_m \overline{\Delta B_c} \, (\Delta f)^{\frac{1}{2}} \tag{4}$$

AAS (3,4)

Signal (Narrow line excitation source)

$$A = \log \frac{\phi_o}{\phi_t} = 0.43 k° \ell \rho \tag{5}$$

where

$$k° = \frac{2\sqrt{\ell n^2} \lambda_o^2 n_t g_j A_{ji} \delta}{8\pi\sqrt{\pi}\delta\lambda_D Z \exp(E_i/kT_f)} \tag{6}$$

Signal (Continuum excitation source)

$$A = 0.43 \, k°_c \ell \tag{7}$$

where

$$k°_c = \frac{2\sqrt{\ell n\pi}\lambda_o^2 n_t g_j A_{ji}}{8\pi\sqrt{\pi} \Delta\lambda_m Z \exp(E_i/kT_f)} \tag{8}$$

Noise (3)

$$\overline{\Delta i}_t = (\overline{\Delta i}_p^{\ 2} + \overline{\Delta i}^{\circ 2} + \overline{\Delta i}_c^{\ 2})^{\frac{1}{2}} \tag{9}$$

where $\overline{\Delta i}_p$ is given by Equation (3)

$\overline{\Delta i}_c$ is given by Equation (4)

and

$$\overline{\Delta i}^{\circ} = \gamma T_{opt} sh_m \Omega_E \overline{\Delta B^{\circ}} (\Delta f)^{\frac{1}{2}} \tag{10}$$

If a continuum source is used $\overline{\Delta i}^{\circ}$ becomes

$$\overline{\Delta i}_c^{\circ} = \gamma T_{opt} sh_m \Omega_E \Delta \lambda_m \overline{\Delta B_c^{\circ}} (\Delta f)^{\frac{1}{2}} \tag{11}$$

AFS (5)

Signal (Resonance Fluorescence, narrow excitation source)

$$B_F = \frac{2\sqrt{\ell n2}}{32\pi^{5/2}} B^{\circ} f_x \Omega_Z Y_p f_z \frac{\lambda_o n_t \ell g_j A_{ji}}{Z \delta \lambda_D} \tag{12}$$

(Resonance Fluorescence, continuum excitation source)

$$B_F = B_c^{\circ} \Omega_A Y_p \frac{\lambda_o^2 n_t \ell g_j A_{ji} f_x f_z}{32\pi^2 Z} \tag{13}$$

Noise (6)

$$\overline{\Delta i}_t = (\overline{\Delta i}_p^{\ 2} + \overline{\Delta i}_c^{\ 2} + \overline{\Delta i}_s^{\ 2})^{\frac{1}{2}} \tag{14}$$

where $\overline{\Delta i}_p$ is given by Equation (3)

$\overline{\Delta i}_c$ is given by Equation (4)

and $\overline{\Delta i}_s = \gamma T_{opt} sh_m \Omega_E \overline{\Delta B^{\circ}} (\Delta f)^{\frac{1}{2}}$ $\qquad$ (15)

for a narrow line excitation source and

$$\overline{\Delta i}_s = \gamma T_{opt} sh_m \Omega_E \overline{\Delta B_c^{\circ}} \Delta \lambda_m (\Delta f)^{\frac{1}{2}} \tag{16}$$

for a continuum excitation source.

The relationship between solution concentration and atomic

concentration if given by (7)

$$n_t = \frac{298E_n \beta_a F_\ell C_p}{10^3 N_A T_f A_r (1345F_\ell + F_u n_{T/n_{298}})}$$ (17)

The relationship between the atomic transition probability, A_{ji}, and oscillator strength is given by (8)

$$A_{ji} = \frac{6.670_2 \times 10^{15}}{\lambda_o^2} \frac{g_i}{g_j} f_{ij}$$ (18)

Table III.A.1-b

SYMBOLS, DEFINITIONS AND
PRACTICAL UNITS USED IN EQUATIONS (9)

Symbol	Definition	Practical Unit
A	Absorbance $= \log \dfrac{\phi_0}{\phi_t}$	(no units)
A_{ji}	Atomic transition probability	s^{-1}
A_r	Atomic weight of analyte	g/mol
BM	Photomultiplier surface factor x amplification factor of photomultiplier.	(no units)
B_E	Radiance of emission	$Wcm^{-2}sr^{-1}$
B_F	Radiance of fluorescence	$Wcm^{-2}sr^{-1}$
$B°$	Radiance of narrow line excitation source	$Wcm^{-2}sr^{-1}$
$B°_c$	Radiance of continuum excitation source	$Wcm^{-2}sr^{-1}nm^{-1}$
$\overline{\Delta B°}$	RMS fluctuation in the intensity of narrow line excitation source ($\overline{\Delta B°} = \chi B°$)	$Wcm^{-2}sr^{-1}s^{\frac{1}{2}}$
$\overline{\Delta B°_c}$	RMS fluctuation in the intensity of continuum excitation source ($\overline{\Delta B°_c} = \chi' B°_c$)	$Wcm^{-2}sr^{-1}nm^{-1}s^{\frac{1}{2}}$
$\overline{\Delta B_c}$	RMS fluctuation in the intensity of the flame background ($\overline{\Delta B_c} = \chi'' B_c$)	$Wcm^{-2}sr^{-1}nm^{-1}s^{\frac{1}{2}}$
C_p	Solution concentration of analyte	µg/l (ppm)
e	Charge on the electron	(Table III.B.9-a)
E_j	Energy associated with the jth level	J or eV
Δf	Frequency band pass of amplifier--read-out	s^{-1}
F_ℓ	Rate of analyte solution consumption	cm^3s^{-1}
F_u	Flow rate of unburnt gas mixture	cm^3s^{-1}
f_{ij}	Oscillator strength for absorptions	(no units)
g_j	Statistical weight of the j energy level ($g = 2J + 1$)	
k	Planck constant	(Table III.B.9-a)
h_m	Entrance slit height of monochromator	cm

Symbol	Definition	Practical Unit
i_t	Total current produced at the photomultiplier	A
$\overline{\Delta i_t}$	Total noise signal	A
$\overline{\Delta i_p}$	Phototube noise signal	A
$\overline{\Delta i_c}$	Flame background noise signal	A
$\overline{\Delta i^\circ}$	Excitation source noise signal	A
$\overline{\Delta i_s}$	Reflection and scatter noise signal in AFS	A
J_j	Inner quantum number associated with the jth level	(no units)
k	Boltzmann constant	(Table III.B.9-a)
k°	Atomic absorption coefficient at the line center	cm^{-1}
k_c°	Atomic absorption coefficient at the line center when a continuum excitation source is used	cm^{-1}
ℓ	Path length of flame cell	cm
N_A	Avogadro constant	(Table III.B.9-a)
n_t	Total (number) density of analyte species	cm^{-3}
n_{T}/n_{298}	Ratio of molar composition of flame gases at flame temperature with respect to 298°K	(no units)
R_d	Reciprocal linear dispersion	$nm\ cm^{-1}$
R_o	Resistance of load resistor in photomultiplier	ohms
s	Entrance slit width of the monochromator	cm
T_f	Temperature of the flame	K
T_o	Temperature of load resistor in photomultiplier	K
T_{opt}	Transmission factor including all of experimental optics	(no units)
Y_p	Power efficiency of fluorescence (radiant flux re-emitted per primary flux absorbed)	(no units)
β_a	Fraction of analyte atomized	(no units)

Symbol	Definition	Practical Unit
γ	Photosensitivity of photomultiplier	A watt^{-1}
δ	Broadening factor (other than doppler See Figure III.B.8)	(no units)
$\delta\lambda_D$	Doppler broadening (See III.B.7)	nm
ε_n	Efficiency of nebulization	(no units)
λ_o	Wavelength of the transition center $(\lambda_o = \frac{h}{\nu_o})$	nm
$\Delta\lambda_m$	Spectral bandividth of monochromator. (At large monochromator slit widths $\Delta\lambda_m = R_d s$)	nm
ν_o	Frequency of the transition center	s^{-1}
χ''	Fraction of total signal which can be attributed to noise in flame background emission	s$^{-\frac{1}{2}}$
π	pi (3.14159)	(no units)
ρ	Factor which accounts for hyperfine structure of excitation source and absorption line	(no units)
ϕ_o	Radiant flux incident on flame cell	W
ϕ_t	Radiant flux transmitted by flame cell	W
χ, χ', etc.	Fraction of total signal which can be attributed to noise in excitation source, etc.	s$^{-\frac{1}{2}}$
Ω_A	Solid angle over which radiation is absorbed by flame cell	sr
Ω_E	Solid angle over which emission is measured	sr

REFERENCES FOR TABLES III.A.1-a and b

1. J. D. Winefordner, W. W. McGee, J. M. Mansfield, M. L. Parsons, and K. E. Zecha, Anal. Chim. Acta, 36, 25 (1966).

2. J. D. Winefordner and T. J. Vickers, Anal. Chem, 36, 1939 (1964).

3. J. D. Winefordner and T. J. Vickers, Anal. Chem., 36, 1947 (1964).

4. L. DeGalan, W. W. McGee, and J. D. Winefordner, Anal. Chim. Acta, 37, 436 (1967).

5. J. D. Winefordner, M. L. Parsons, J. M. Mansfield, and W. J. McCarthy, Spectrochim. Acta, 23B, 37 (1967).

6. J. D. Winefordner, M. L. Parsons, J. M. Mansfield, and W. J. McCarthy, Anal. Chem., 39, 436 (1967).

7. L. DeGalan and G. F. Samaey, Spectrochim. Acta, 25B, 245 (1970).

8. W. L. Wiese, M. W. Smith, and B. M. Clennon, "Atomic Transition Probabilities Vol. 1," NSRDS-NBS 4, U. S. Government Printing Office, Washington, D.C., 1966).

9. C. Th. J. Alkemade, "Nomenclature, Symbols, Units, and their Usage in Spectrochemical Analysis III. Analytical Flame Spectroscopy and Associated Procedures," IUPAC, 1972. Plus personal communications, 1973.

B. FUNDAMENTAL CONSTANTS AND DATA

Spectral Data

Table III.B.1-a presents the wavelength, lower, E_i, and upper, E_j, energy levels, the statistical weight for the upper level, g_j, atomic transition probabilities, A_{ji}, a figure of merit and reference for the atomic transition probabilities, and a code to tell if the transition was observed in absorption; A; emission, E; fluorescence, F; or predicted to be observed in analytical flames, P. The atomic weight based on ^{12}C is also given.

This data is given for 53 elements, the most notable exclusion is that of the rare earth elements. Once again, the wavelength data conforms to NBS tabulations, (1,2,4). The atomic energy level data comes from Moore's Tables (6,7,8) as does the statistical weight of the upper level (g=2J+1). The figure of merit for the atomic transition probabilities follows that initiated by the NBS (1,4) as follows:

Figure of Merit	Uncertainty
AA	< 1%
A	< 3%
B	<10%
C	<25%
D	<50%
E	>50%

Data for the statistical weight of the lower level was omitted because this information is available in Table III.B.2.

Tables III.B.1-b, c, & d are wavelength locators for AAS, AES, and AFS, respectively. They include the observed and predicted transitions for the appropriate technique and are listed accordingly to wavelength.

References for Spectral Data Section:

1. W. L. Wiese, M. W. Smith, & B. M. Miles, "Atomic Transition Probabilities, Volume II, Sodium Through Calcium," NSRDS-NBS 22, U. S. Government Printing Office, Washington, D. C., 1969.

2. C. H. Corliss and William R. Bozman, "Experimental Transition Probabilities for Spectral Lines of Seventy Elements," NBS Monograph 53, U. S. Government Printing Office, Washington, D. C., 1961.

3.   B. M. Miles & W. L. Wiese, "Critically Evaluated Tran-
     sition Probabilities for BaI and II," NBS Technical Note
     474, U. S. Government Printing Office, Washington, D. C.,
     1969.

4.   W. L. Wiese, M. W. Smith, & B. M. Glennon, "Atomic Tran-
     sition Probabilities, Volume I Hydrogen Through Neon,"
     NSRDS-NBS 4, U. S. Government Printing Office, Washington,
     D. C., 1966.

5.   C. H. Corliss & J. L. Tech, "Oscillator Strengths and
     Transition Probabilities for 3288 Lines of FeI," NBS
     Monograph 108, U. S. Government Printing Office,
     Washington, D. C., 1968.

6.   C. E. Moore, "Atomic Energy Levels Volume I," NBS Circu-
     lar 467, U. S. Government Office, Washington, D. C., 1949.

7.   C. E. Moore, "Atomic Energy Levels Volume II," NBS Cir-
     cular 467, U. S. Government Printing Office, Washington,
     D. C., 1952.

8.   C. E. Moore, "Atomic Energy Levels Volume III," NBS Cir-
     cular 467, U. S. Government Printing Office, Washington,
     D. C., 1958.

9.   L. DeGalan & J. D. Winefordner, J. Quant. Spectr.
     Radiative Transfer, 7, 251 (1967).

Table III.B.1-a

Spectral Data

Wavelength (Å)	E_i (cm^{-1})	E_j (cm^{-1})	g_j	A_{ji} (10^8 sec^{-1})	Merit	Ref.	Exp. Obs.
ALUMINUM	ATOMIC	WEIGHT =		26.9815			
2263.46	0	44166	4	.64	C	1	P
2269.10	112	44169	6	.76	C	1	P
2269.22	112	44166	4	.13	D	1	P
2367.05	0	42234	4	.71	C	1	A,F
2372.07	0	42144	2	.0478	C	1	P
2373.12	112	42238	6	.85	C	1	A,F
2373.35	112	42234	4	.14	D	1	A,F
2373.35	112	42234	4	.14	D	1	A,F
2567.98	0	38929	4	.221	C	1	A,F
2575.10	112	38934	6	.264	C	1	A,F
2575.40	112	38929	4	.044	D	1	A,F
2652.48	0	37689	2	.133	C	1	P
2660.39	112	37689	2	.264	C	1	P
3082.15	0	32435	4	.61	C+	1	A,E,F
3092.71	112	32437	6	.73	C+	1	A,E,F
3092.84	112	32435	4	.12	D	1	A,E,F
3944.01	0	25348	2	.493	C+	1	A,E,F
3961.52	112	25348	2	.980	C+	1	A,E,F
ANTIMONY	ATOMIC	WEIGHT =		121.75			
2068.33	0	48332	6	42.0	E	2	A,E,F
2127.39	0	46991	2	8.60	E	2	A
2175.81	0	45945	4	13.8	E	2	A,E,F
2311.47	0	43249	2	3.75	E	2	A,E,F
2528.52	9854	49391	4	14.0	E	2	E
2598.05	8512	46991	2	32.0	E	2	E,F
2670.64	8512	45945	4	.75	E	2	F
2769.95	9854	45945	4	1.88	E	2	F
2877.92	8512	43249	2	2.85	E	2	F

Spectral Data

Wavelength (Å)	E_i (cm^{-1})	E_j (cm^{-1})	g_j	A_{ji} (10^{8}sec^{-1})	Merit	Ref.	Exp. Obs.
ARSENIC	ATOMIC	WEIGHT =		74.922			
1936.96	0	51610	4	44.0	E	2	A,E,F
1971.97	0	50694	2	100.	E	2	A,E,F
2288.12	10915	54605	4	26.0	E	2	E,F
2349.84	10592	60835	4	20.0	E	2	E,F
2381.18	10915	52898	6	1.8	E	2	F
2437.23	10592	51610	4	.58	E	2	F
2456.53	10915	51610	4	2.32	E	2	F
2492.91	10592	50694	2	4.95	E	2	E,F
2745.00	18186	54605	4	4.75	E	2	E,F
2780.22	18648	54605	4	15.0	E	2	E,F
2860.44	18186	53136	2	16.5	E	2	E,F
2898.71	18648	53136	2	3.75	E	2	F
3032.85	18648	51610	4	.62	E	2	F
BARIUM	ATOMIC	WEIGHT =		137.34			
3071.58	0	32547	3	.41	D	3	P
3501.11	0	28554	3	.29	D	3	A,E
4554.03 II	0	21952	4	1.19	B	3	E
4934.09 II	0	20262	2	.955	B	3	E
5535.48	0	18080	3	1.15	B	3	A,E
BERYLLIUM	ATOMIC	WEIGHT =		9.0122			
2348.61	0	42565	3	5.47	C+	4	A,E,F
BISMUTH	ATOMIC	WEIGHT =		208.980			
1953.89	0	51158	6	48.	E	2	A
1959.48	0	51059	4	14.5	E	2	A
2021.21	0	49456	4	10.0	E	2	A
2061.70	0	48489	6	40.0	E	2	A,F

Spectral Data

Wavelength (Å)	E_i (cm^{-1})	E_j (cm^{-1})	g_j	A_{ji} (10^{8}sec^{-1})	Merit	Ref.	Exp. Obs.
BISMUTH	CONTINUED						
2228.25	0	44865	4	.38	E	2	A,E,F
2230.61	0	44817	6	1.25	E	2	A,E,F
2276.58	0	43912	4	.31	E	2	A,E,F
2627.91	11418	49456	4	3.00	E	2	F
2696.76	11418	48489	6	.67	E	2	F
2780.52	11418	47373	2	2.05	E	2	F
2897.98	11418	45915	2	16.0	E	2	F
2938.30	15437	49456	4	15.2	E	2	F
2989.03	11418	44865	4	4.25	E	2	F
2993.34	11418	44817	6	.70	E	2	F
3024.64	15437	48490	6	6.33	E	2	F
3067.72	0	32588	2	3.50	E	2	A,E,F
3397.21	15437	44865	4	.95	E	2	F
3510.85	15437	43913	4	.68	E	2	F
4121.53	21660	45916	2	.80	E	2	F
4722.19	11418	32588	2	.090	E	2	E,F
BORON	ATOMIC WEIGHT = 10.811						
2088.84	0	47857	4	1.8	E	4	A
2089.57	15	47857	6	2.1	E	4	A
2496.77	0	40040	2	1.20	D	4	A,E
2497.72	15	40040	2	2.40	D	4	A,E
CADMIUM	ATOMIC WEIGHT = 112.40						
2288.02	0	43692	3	5.31	C	9	F
3261.06	0	30656	3	.003	E	2	F
CALCIUM	ATOMIC WEIGHT = 40.08						
2200.73	0	45425	3	.153	C	1	P
2275.46	0	43933	3	.301	C+	1	P

Spectral Data

Wavelength (Å)		E_i (cm^{-1})	E_j (cm^{-1})	g_j	A_{ji} (10^{8}sec^{-1})	Merit	Ref.	Exp. Obs.
CALCIUM	CONTINUED							
3933.66	II	0	25414	4	1.50	C+	1	A,E
3968.47	II	0	25192	2	1.46	C+	1	A,E
4226.73		0	23652	3	2.18	B+	1	A,E,F
CESIUM	ATOMIC WEIGHT =	132.905						
4555.36		0	21947	4	.35	E	2	A,E
4593.18		0	21766	2	.32	E	2	A,E
8521.10		0	11732	4	.32	E	2	A,E,F
8943.50		0	11178	2	.24	E	2	A,E
CHROMIUM	ATOMIC WEIGHT =	51.996						
2364.71		0	42275	9	.13	E	2	P
3578.69		0	27935	9	.92	E	2	A,E,F
3593.49		0	27820	7	1.00	E	2	A,E,F
3605.33		0	27650	9	.58	E	2	A,E,F
4254.35		0	23499	9	.22	E	2	A,E,F
4274.80		0	23386	7	.21	E	2	A,E,F
4289.72		0	23305	5	.19	E	2	A,E,F
5204.52		7593	26802	3	.53	E	2	A,F
5206.04		7593	26796	5	.50	E	2	E,F
5208.44		7593	26788	7	.47	E	2	A,F
COBALT	ATOMIC WEIGHT =	58.933						
2274.49		0	43952	12	.27	E	2	A
2295.23		0	43555	8	.46	E	2	A
2309.02		0	43295	10	3.40	E	2	A
2407.25		0	41529	12	3.08	E	2	A,F
2411.62		816	42269	10	3.70	E	2	A,F
2414.46		1407	42811	8	6.0	E	2	A,F
2415.30		1809	43200	6	8.00	E	2	A,F

Spectral Data

Wavelength (Å)	E_i (cm^{-1})	E_j (cm^{-1})	g_j	A_{ji} $(10^8 sec^{-1})$	Merit	Ref.	Exp. Obs.
COBALT	CONTINUED						
2432.21	816	41918	8	3.00	E	2	A,F
2435.83	0	4104	8	.50	E	2	A
2439.05	1809	42797	4	5.75	E	2	A
2521.36	0	39649	8	2.50	E	2	A,F
2528.97	816	40346	6	2.67	E	2	A,F
2535.96	1407	40828	4	4.75	E	2	A
2987.16	0	33467	8	.10	E	2	A
2989.59	0	33440	10	.082	E	2	A
3044.00	0	32842	10	.31	E	2	A
3405.12	3483	32842	10	1.50	E	2	A,E
3409.18	4143	33467	8	.91	E	2	P
3412.34	4143	33440	10	1.10	E	2	A,E
3412.63	0	29295	8	.14	E	2	A,E
3431.58	816	29949	6	.27	E	2	A
3433.04	5076	34196	4	2.28	E	2	E
3442.93	1407	30444	4	.28	E	2	A
3443.64	4143	33173	8	1.62	E	2	E
3445.23	1809	30743	2	.40	E	2	A
3449.44	3483	32465	10	.26	E	2	E
3453.50	3483	32431	10	2.60	E	2	A,E
3462.80	5076	33946	6	1.62	E	2	A
3465.80	0	28845	12	.19	E	2	A,E
3474.02	0	28777	8	.45	E	2	A,E
3502.28	3483	32028	8	1.38	E	2	A,E
3502.62	1407	29949	6	.060	E	2	A
3506.32	4143	32654	6	1.57	E	2	A,E
3512.64	4690	33151	4	1.85	E	2	A,E
3513.48	816	29270	10	.20	E	2	A,E
3526.85	0	28346	10	.26	E	2	A,E
3529.81	4143	32465	10	.94	E	2	A

Spectral Data

Wavelength (Å)	E_i (cm^{-1})	E_j (cm^{-1})	g_j	A_{ji} (10^8sec^{-1})	Merit	Ref.	Exp. Obs.
COBALT	CONTINUED						
3569.38	7442	35451	8	3.25	E	2	E
3575.36	816	28777	8	.15	E	2	A,E,F
3594.87	1407	29216	6	.17	E	2	E
3845.47	7442	33440	10	.88	E	2	E
3873.12	3843	29295	8	.28	E	2	E
3894.08	8461	34134	8	1.50	E	2	E
3909.93	0	25569	12	.002	E	2	A
3995.31	7442	32465	10	.60	E	2	E
4121.32	7442	31700	10	.37	E	2	E
4252.31	816	24326	8	.0008	E	2	E
COPPER	ATOMIC WEIGHT =	63.54					
2165.09	0	46173	4	1.30	E	2	A,F
2178.94	0	45879	4	1.55	E	2	A,F
2181.72	0	45821	2	2.30	E	2	A,F
2024.34	0	49383	2	4.25	E	2	A
2225.70	0	44916	2	.65	E	2	A,F
2441.64	0	40944	2	.030	E	2	A
2492.15	0	40114	4	.078	E	2	A,F
3247.54	0	30784	4	.948	B+	9	A,E,F
3273.96	0	30535	2	.95	E	2	A,E,F
5105.54	11203	30784	4	.013	E	2	E,F
5782.13	13245	30535	2	.027	E	2	F
GALLIUM	ATOMIC WEIGHT =	69.72					
2418.69	826	42158	2	.38	E	2	P
2450.07	0	40803	4	1.02	E	2	A
2500.17	826	40811	6	1.22	E	2	A
2500.70	826	40803	4	.17	E	2	A
2659.87	0	37585	2	.41	E	2	P

Spectral Data

Wavelength (Å)	E_i (cm^{-1})	E_j (cm^{-1})	g_j	A_{ji} $(10^8 sec^{-1})$	Merit	Ref.	Exp. Obs.
GALLIUM	CONTINUED						
2874.24	0	34782	4	1.48	E	2	A
2943.64	826	34788	6	1.83	E	2	A
2944.18	826	34782	4	.45	E	2	A,E,F
4032.98	0	24788	6	.122	B+	9	A,E,F
4172.06	826	24788	6	.33	E	2	A,E,F
GERMANIUM	ATOMIC WEIGHT =		72.59				
2041.69	0	48962	3	96.7	E	2	A
2043.76	1410	50323	7	35.7	E	2	A
2065.20	557	48962	3	40.0	E	2	A
2068.65	557	48882	5	82.0	E	2	A
2094.23	1410	49144	7	26.3	E	2	A
2198.70	7125	52592	7	1.36	E	2	A
2497.96	0	40020	3	.80	E	2	A,F
2533.23	557	40020	3	.63	E	2	P
2589.19	1410	40020	3	.25	E	2	P
2592.54	557	39118	5	2.20	E	2	A,E,F
2651.18	1410	39118	5	5.20	E	2	A,E,F
2651.58	0	37702	3	2.67	E	2	A,E,F
2691.34	557	37702	3	2.43	E	2	A,E,F
2709.63	557	37452	1	12.0	E	2	A,E,F
2754.59	1410	37702	3	3.27	E	2	A,E,F
3039.06	7125	40020	3	8.00	E	2	A,F
3269.49	7125	37702	3	.67	E	2	F
GOLD	ATOMIC WEIGHT =		196.967				
2012.00	9161	58845	4	44.0	E	2	P
2021.38	9161	58616	6	6.67	E	2	P
2352.65	9161	51654	6	.33	E	2	P
2387.75	9161	51029	8	.16	E	2	P

Spectral Data

Wavelength (Å)	E_i (cm^{-1})	E_j (cm^{-1})	g_j	A_{ji} (10^{8}sec^{-1})	Merit	Ref.	Exp. Obs.
GOLD	CONTINUED						
2641.49	9161	47007	4	.32	E	2	P
2675.95	0	37359	2	1.77	B+	9	A,E,F
2748.26	9161	45537	8	.48	E	2	P
3029.20	9161	42164	6	.078	E	2	F
3122.78	9161	41174	4	.48	E	2	A,F
6278.18	21435	37359	2	.024	E	2	A
HAFNIUM	ATOMIC WEIGHT =	178.49					
2779.37	2357	38325	5	.54	E	2	P
2866.37	0	34877	7	.83	E	2	A
2889.62	0	34596	3	.67	E	2	P
2898.26	2357	36850	9	.94	E	2	A
2904.41	4568	38988	9	1.22	E	2	A
2904.75	2357	36773	5	.82	E	2	A
2916.48	4568	38845	11	1.45	E	2	P
2918.58	2357	36610	7	.37	E	2	P
2940.77	0	33995	5	.84	E	2	A
2950.66	2357	36237	7	.73	E	2	A
2954.20	4568	38408	7	1.14	E	2	P
2964.88	2357	36075	9	.62	E	2	A
2980.81	0	33538	5	.42	E	2	P
3016.78	0	33138	3	.31	E	2	P
3018.31	0	33122	5	.34	E	2	P
3020.53	2357	35454	7	.56	E	2	A
3057.02	4568	37270	9	.67	E	2	P
3067.41	2357	34948	5	.50	E	2	P
3072.88	0	32533	7	.46	E	2	A
3332.73	0	29997	7	.10	E	2	P
3472.40	0	28790	3	.14	E	2	P
3497.49	0	28584	7	.084	E	2	P

Spectral Data

Wavelength (Å)	E_i (cm^{-1})	E_j (cm^{-1})	g_j	A_{ji} ($10^8 sec^{-1}$)	Merit	Ref.	Exp. Obs.
HAFNIUM	CONTINUED						
3682.24	0	27150	5	.15	E	2	A,E
3777.64	0	26464	3	.14	E	2	A,F
INDIUM	ATOMIC WEIGHT =	114.82					
2560.15	0	39048	4	1.18	E	2	A
2601.76	2213	40637	2	.70	E	2	P
2710.26	2213	39098	6	1.22	E	2	A
2713.94	2213	39048	4	.35	E	2	P
2753.88	0	36302	2	.75	E	2	A
2932.63	2213	36302	2	1.25	E	2	A
3039.36	0	32892	4	1.78	E	2	A,E,F
3256.09	2213	32916	6	2.00	E	2	A,E
3258.56	2213	32892	4	.72	E	2	A
4101.76	0	24373	2	.396	C	9	A,E,F
4511.31	2213	24373	2	1.10	E	2	A,E,F
IRIDIUM	ATOMIC WEIGHT =	192.2					
2033.57	0	49159	10	38.0	E	2	P
2088.82	0	47858	12	28.0	E	2	A,P
2092.63	2835	50606	12	48.0	E	2	P
2127.94	0	46979	8	10.5	E	2	P
2155.81	0	46372	10	4.50	E	2	P
2158.05	2835	49159	10	26.7	E	2	P
2175.24	0	45957	10	5.40	E	2	P
2372.77	0	42132	12	1.42	E	2	A
2455.61	0	40711	8	.56	E	2	P
2475.12	0	40390	10	1.00	E	2	A
2481.18	0	40291	8	.78	E	2	A
2502.98	0	39940	10	1.10	E	2	A
2543.97	2835	42132	10	4.00	E	2	A,F

Spectral Data

Wavelength (Å)	E_i (cm^{-1})	E_j (cm^{-1})	g_j	A_{ji} (10^{8}sec^{-1})	Merit	Ref.	Exp. Obs.
IRIDIUM	**CONTINUED**						
2639.71	0	37872	10	.56	E	2	A
2661.98	2835	40390	10	.89	E	2	A
2664.79	0	37515	8	.76	E	2	A
2694.23	2835	39940	12	1.08	E	2	P
2849.72	0	35081	10	.46	E	2	A
2882.64	2835	37515	8	.25	E	2	P
2924.79	0	34180	12	.35	E	2	A
3220.78	2835	33874	8	.82	E	2	P
3513.64	0	28452	12	.082	E	2	A,E
3800.12	0	26308	10	.059	E	2	A,E
IRON	**ATOMIC WEIGHT =**		**55.847**				
2166.77	0	46137	7	27.8	E	5	A
2462.65	0	40594	9	.67	E	5	A
2472.90	416	40842	7	1.71	E	5	A
2479.78	704	41018	5	2.55	E	5	A
2483.27	0	40257	11	3.04	E	5	A,F
2484.19	888	41131	3	4.04	E	5	P
2488.14	416	40594	9	3.88	E	5	A
2489.75	978	41131	3	6.23	E	5	A
2490.64	704	40842	7	3.69	E	5	A
2491.16	888	41018	5	3.57	E	5	A
2501.13	0	39970	7	1.33	E	5	A
2518.10	704	40405	3	2.92	E	5	A
2522.85	0	39626	9	3.68	E	5	A,F
2524.29	888	40491	1	6.60	E	5	A
2527.44	416	39970	3	2.53	E	5	P
2719.03	0	36767	7	1.82	E	5	A,F
2720.52	416	37163	5	1.43	E	5	A
2720.90	416	37158	5	1.43	E	5	A

Spectral Data

Wavelength (Å)	E_i (cm^{-1})	E_j (cm^{-1})	g_j	A_{ji} ($10^8 sec^{-1}$)	Merit	Ref.	Exp. Obs.
IRON	CONTINUED						
2947.88	416	34329	7	.41	E	5	A
2966.90	0	33695	11	.43	E	5	A
2973.13	704	34329	7	.31	E	5	A
2973.24	416	34040	7	.38	E	5	A
2983.57	0	33507	7	.47	E	5	A
2994.43	416	33802	5	.66	E	5	A
3000.95	704	34017	3	1.08	E	5	A
3020.49	704	33802	5	.28	E	5	A,E,F
3020.64	0	33096	9	.60	E	5	A,E,F
3021.07	416	33507	7	.67	E	5	A,E
3047.61	704	33507	7	.48	E	5	A
3059.09	416	33096	9	.29	E	5	A
3440.61	0	29056	7	.29	E	5	A,E
3440.99	418	29469	5	.17	E	5	A
3465.86	888	29733	3	.21	E	5	A
3570.10	7377	35379	11	1.17	E	5	A
3581.20	6928	34844	13	1.59	E	5	A,E
3608.86	8155	35856	5	1.82	E	5	A
3618.77	7986	35612	7	1.39	E	5	A
3631.46	7728	35257	9	.93	E	5	A
3679.92	0	27167	9	.032	E	5	A
3719.94	0	26875	11	.138	B	9	A,E,F
3723.68	704	37410	3	1.14	E	5	F
3734.87	6928	33695	11	1.62	E	5	A
3737.13	416	27167	9	.17	E	5	A,E
3745.56	704	27395	7	.16	E	5	A
3745.90	978	27666	3	.12	E	5	A,E
3758.24	7728	34329	7	1.38	E	5	A
3824.45	0	26140	7	.044	E	5	A
3825.88	7377	33507	7	1.18	E	5	A,E

Spectral Data

Wavelength (Å)	E_i (cm^{-1})	E_j (cm^{-1})	g_j	A_{ji} (10^8sec^{-1})	Merit	Ref.	Exp. Obs.
IRON	CONTINUED						
3859.91	0	25900	9	.13	E	5	A,E
3878.58	704	26479	3	.10	E	5	E
3886.28	416	26140	7	.072	E	5	A,E
3899.71	704	26340	5	.042	E	5	E
3920.26	978	26479	3	.047	E	5	A
3930.30	704	26140	7	.030	E	5	E
4385.55	11976	34782	11	1.02	E	5	E
5269.54	6928	25900	9	.012	E	5	E
5328.04	7377	26140	7	.012	E	5	E
LANTHANUM	ATOMIC WEIGHT =	138.91					
3574.43	0	27969	4	.65	E	2	E
3613.08	0	27669	6	.13	E	2	A
3641.53	1053	28506	6	.87	E	2	P
3649.53	0	27393	6	.15	E	2	A
3704.54	1053	28040	8	.18	E	2	P
3898.60	0	25643	4	.065	E	2	A
3927.56	0	25454	2	.40	E	2	A,E
4015.39	1053	25950	4	.28	E	2	E
4037.21	0	24763	4	.15	E	2	A,E
4060.32	4122	28743	12	.24	E	2	E
4064.79	3495	28089	10	.17	E	2	E
4079.18	0	24508	6	.080	E	2	A,E
4086.72 II	0	24463	5	.20	E	2	A
4089.61	3010	27455	8	.20	E	2	E
4104.87	2668	27023	6	.15	E	2	E
4137.04	1053	25218	6	.090	E	2	E
4160.26	1053	25083	8	.060	E	2	E
4187.32	0	23875	6	.16	E	2	A,E
4280.27	1053	24410	8	.15	E	2	E

Spectral Data

Wavelength (Å)		E_i (cm^{-1})	E_j (cm^{-1})	g_j	A_{ji} ($10^8 sec^{-1}$)	Merit	Ref.	Exp. Obs.
LANTHANUM		CONTINUED						
4567.91		3495	25380	8	.11	E	2	E
4570.02		4122	25997	10	.14	E	2	E
4662.51	II	0	21442	3	.009	E	2	A
4766.89		0	20972	6	.040	E	2	A,E
4850.82		1053	21663	8	.016	E	2	E
4949.77		0	20197	2	.20	E	2	A,E
5145.42		3010	22439	4	.24	E	2	E
5158.69		0	19379	6	.043	E	2	A,E
5177.31		3495	22804	6	.23	E	2	E
5211.86		4122	23303	8	.25	E	2	E
5234.27		4122	23221	8	.18	E	2	E
5253.46		1053	20083	4	.10	E	2	E
5271.19		1053	20019	4	.10	E	2	E
5455.15		1053	19379	6	.078	E	2	A,E
5501.34		0	18172	4	.080	E	2	A,E
5740.66		2668	20083	4	.055	E	2	E
5761.84		2668	20019	4	.048	E	2	E
5769.34		3010	20388	6	.080	E	2	E
5789.24		3495	20763	8	.059	E	2	E
5791.34		4122	21384	10	.079	E	2	E
5930.63		1053	17910	8	.028	E	2	E
6249.93		4122	20117	12	.081	E	2	E
6325.91		1053	16857	6	.009	E	2	E
6394.23		3495	19129	10	.046	E	2	E
6410.99		3010	18604	8	.024	E	2	E
6454.52		2668	18157	6	.012	E	2	E
6455.99		1053	16538	8	.015	E	2	E
6543.16		2668	17947	6	.017	E	2	E
6578.51		0	15197	2	.020	E	2	E
6616.59		3495	18604	8	.006	E	2	E

Spectral Data

Wavelength (Å)	E_i (cm^{-1})	E_j (cm^{-1})	g_j	A_{ji} $(10^8 sec^{-1})$	Merit	Ref.	Exp. Obs.
LANTHANUM	CONTINUED						
6650.81	0	15032	4	.005	E	2	E
6661.40	4122	19129	10	.010	E	2	E
LEAD	ATOMIC WEIGHT =	207.19					
2022.00	0	49440	3	4.8	E	2	A
2055.20	0	48687	3	5.1	E	2	A
2169.99	0	46069	3	4.5	E	2	A,E,F
2613.65	7819	46069	3	.63	E	2	A,E,F
2614.18	7819	46061	5	5.20	E	2	A,E,F
2801.99	10650	46329	7	6.1	E	2	F
2823.28	10650	46061	5	3.4	E	2	F
2833.06	0	35287	3	.609	B	9	A,E,F
2873.32	10650	45443	5	1.9	E	2	F
3639.58	7819	35287	3	.43	E	2	E,F
3683.48	7819	34960	1	3.10	E	2	E,F
4057.83	10650	35287	3	3.07	E	2	A,E,F
7229.00	21458	35287	3	.020	E	2	F
LITHIUM	ATOMIC WEIGHT =	6.939					
3232.63	0	39025	6	.117	B	4	A,E
4602.86	14904	36623	10	.230	B	4	E
6103.64	14904	31283	10	.716	B+	4	A,E
6707.80	0	14904	6	.372	A	4	A,E,F
MAGNESIUM	ATOMIC WEIGHT =	24.312					
2025.82	0	49347	3	1.20	D	1	A
2795.53	11	0 35761	4	2.68	B+	1	A
2802.70	11	0 35669	2	2.66	B+	1	A
2852.13	0	35051	3	4.95	B	1	A,E,F
5172.68	21870	41197	3	.346	B	1	E

Spectral Data

Wavelength (Å)		E_i (cm^{-1})	E_j (cm^{-1})	g_j	A_{ji} (10^{8}sec^{-1})	Merit	Ref.	Exp. Obs.
MANGANESE	ATOMIC	WEIGHT =	54.938					
2576.10	II	0	38807	9	8.89	E	2	F
2593.73	II	0	38543	9	8.9	E	2	F
2605.69	II	0	38366	5	6.6	E	2	F
2794.82		0	35770	8	1.04	E	2	A,E,F
2798.27		0	35726	6	1.12	E	2	A,E,F
2801.06		0	35690	4	1.22	E	2	A,E,F
2933.06	II	9473	43557	3	13.3	E	2	F
2939.30	II	9473	43485	5	10.8	E	2	F
2949.20	II	9473	43370	7	9.4	E	2	F
3045.59		25286	58110	10	2.0	E	2	F
3216.95		0	31076	4	.008	E	2	A
3577.88		17052	44994	8	4.9	E	2	F
3834.36		17452	43524	8	1.6	E	2	F
4030.76		0	24802	8	.176	B+	9	A,E,F
4033.07		0	24788	6	.16	E	2	E,F
4034.49		0	24779	4	.14	E	2	E,F
MERCURY	ATOMIC	WEIGHT =	200.59					
2536.52		0	39412	3	1.17	E	2	A,E,F
MOLYBDENUM	ATOMIC	WEIGHT =	95.94					
3112.12		0	32123	9	.11	E	2	A
3132.59		0	31913	9	1.09	E	2	A,E,F
3158.16		0	31655	7	.54	E	2	A,E,F
3170.35		0	31533	7	.77	E	2	A,E,F
3193.97		0	31300	5	.88	E	2	A,E,F
3208.83		0	31155	5	.34	E	2	A,E
3798.25		0	26321	9	.49	E	2	A,E,F

Spectral Data

Wavelength (Å)	E_i (cm^{-1})	E_j (cm^{-1})	g_j	A_{ji} (10^8sec^{-1})	Merit	Ref.	Exp. Obs.
MOLYBDENUM	CONTINUED						
3902.96	0	25624	5	.42	E	2	A,E,F
NICKEL	ATOMIC	WEIGHT =	58.71				
2289.98	0	43655	7	3.6	E	2	A,F
2310.96	0	43259	9	4.0	E	2	A,F
2312.34	1332	44565	7	5.5	E	2	F
2313.98	2217	45419	5	8.00	E	2	F
2320.03	0	43090	11	4.00	E	2	A,F
2325.79	1332	44315	9	3.89	E	2	F
2329.96	2217	45122	3	9.67	E	2	P
2337.49	0	42768	7	1.14	E	2	A
2337.82	1713	44475	5	.88	E	2	A
2345.54	0	42621	7	2.29	E	2	A,F
2346.63	1332	43933	5	.72	E	2	P
2347.52	0	42585	9	.63	E	2	A
2476.87	0	40361	7	.10	E	2	A
3002.49	205	33501	7	.96	E	2	A,E,F
3003.63	880	34163	5	.90	E	2	F
3012.00	3410	36601	5	3.0	E	2	A
3037.94	205	33112	7	.37	E	2	A
3050.82	205	32973	9	.57	E	2	A,F
3054.32	880	33611	5	.52	E	2	A
3057.64	1713	34409	3	1.4	E	2	A
3101.55	888	33112	7	.66	E	2	A,F
3101.88	3410	35639	7	.67	E	2	A,F
3134.11	1713	33611	5	1.16	E	2	F
3232.96	0	30923	11	.10	E	2	A
3369.57	0	29669	7	.30	E	2	A,E
3380.57	3410	32982	3	2.03	E	2	E
3391.05	0	29481	9	.10	E	2	A

Spectral Data

Wavelength (Å)		E_i (cm^{-1})	E_j (cm^{-1})	g_j	A_{ji} (10^8sec^{-1})	Merit	Ref.	Exp. Obs.
NICKEL		**CONTINUED**						
3413.94		880	29481	9	.031	E	2	P
3414.76		205	30913	3	1.90	E	2	A,E,F
3433.56		205	29321	7	.26	E	2	A,E
3437.28		0	29084	9	.069	E	2	A
3446.26		880	29888	5	.76	E	2	A,E
3458.47		1713	30619	5	.98	E	2	P
3461.65		205	29384	9	.022	E	2	A,E,F
3472.54		880	29669	7	.17	E	2	E
3492.96		880	29501	3	1.30	E	2	E
3510.34		1713	30192	1	2.30	E	2	E
3515.05		880	29321	7	.64	E	2	A,E,F
3524.54		205	28569	5	.92	E	2	A,E,F
3566.37		3410	31442	5	1.28	E	2	E
3610.46		880	28569	5	.15	E	2	E,F
3619.39		3410	31031	7	1.07	E	2	E
3624.73		0	27580	11	.005	E	2	A
3858.30		3410	29321	7	.16	E	2	E
NIOBIUM		**ATOMIC WEIGHT =**	**92.906**					
2697.06	II	1225	38291	9	1.5	E	2	A
2927.81	II	4146	38291	9	1.9	E	2	A
2950.68	II	4146	38024	11	1.4	E	2	A
3094.18	II	4146	36455	13	1.1	E	2	A
3130.79	II	3542	35474	11	.77	E	2	A
3163.40	II	3030	34632	9	.59	E	2	A
3349.10		2154	32005	10	.45	E	2	A
3535.30		0	28278	4	.50	E	2	E
3580.27		1050	28973	8	.78	E	2	A,E
3697.65		392	27427	8	.14	E	2	E
3713.01		1050	27975	10	.27	E	2	E

Spectral Data

Wavelength (Å)	E_i (cm^{-1})	E_j (cm^{-1})	g_j	A_{ji} (10^{8}sec^{-1})	Merit	Ref.	Exp. Obs.
NIOBIUM	CONTINUED						
3727.23	11044	37866	10	.18	E	2	A
3739.80	695	27427	8	.24	E	2	A,E
3741.78	0	26218	2	.080	E	2	A
3742.39	0	26713	4	.25	E	2	A,E
3759.55	392	26983	6	.015	E	2	A
3760.64	11428	37832	10	.053	E	2	A
3787.06	154	26552	2	.50	E	2	A,E
3787.48	2154	28549	6	.011	E	2	A
3790.15	1050	27427	8	.12	E	2	A,E
3791.21	1050	27420	10	.25	E	2	A,E
3798.12	392	26713	4	.40	E	2	A,E
3802.92	695	26983	6	.28	E	2	A,E
4058.94	1050	25680	12	.65	E	2	A,E,F
4079.73	695	25200	10	.49	E	2	A,E,F
4100.92	392	24770	8	.31	E	2	A,E,F
4110.40	392	24773	6	.028	E	2	A,E,F
4123.81	154	24397	6	.30	E	2	A,E,F
4137.10	0	24165	4	.18	E	2	A,E,F
4137.59	12018	36180	6	.12	E	2	A
4139.44	392	24543	6	.027	E	2	P
4139.71	1050	25200	6	.18	E	2	E,F
4152.58	695	24770	8	.21	E	2	A,E,F
4163.66	154	24165	4	.35	E	2	A,E,F
4164.66	392	24397	6	.23	E	2	A,E
4168.13	0	23985	2	.55	E	2	A,E,F
4606.77	2805	24507	10	.045	E	2	P
OSMIUM	ATOMIC WEIGHT =	190.2					
2424.56	0	41232	7	.33	E	2	P
2424.97	0	41225	11	.63	E	2	P

Spectral Data

Wavelength (Å)	E_i (cm^{-1})	E_j (cm^{-1})	g_j	A_{ji} (10^8sec^{-1})	Merit	Ref.	Exp. Obs.
OSMIUM	CONTINUED						
2590.76	4159	42747	9	1.22	E	2	P
2612.63	0	38264	7	.21	E	2	P
2621.82	0	38130	9	.18	E	2	P
2637.13	0	37909	9	1.22	E	2	A
2644.11	0	37809	7	.79	E	2	A
2714.64	0	36826	9	.73	E	2	A
2720.04	2740	39494	5	1.20	E	2	P
2806.91	0	35616	7	.64	E	2	A
2838.17	4159	39383	7	.31	E	2	A
2838.63	5144	40362	9	3.6	E	2	A
2850.76	2740	37809	7	.66	E	2	P
2909.06	0	34365	11	1.00	E	2	A,E
2912.33	4159	38486	7	1.16	E	2	P
3018.04	0	33124	7	.61	E	2	A,E
3030.70	5144	38130	9	.51	E	2	A
3040.90	2740	35616	7	.80	E	2	P
3058.66	0	32685	9	.82	E	2	A,E
3232.06	4159	35090	5	.32	E	2	A
3258.60	0	28332	9	.040	E	2	A
3262.29	4159	34804	9	.57	E	2	A
3262.75	8743	39383	7	.33	E	2	A
3267.94	0	30592	9	.18	E	2	A
3301.56	0	30280	11	.33	E	2	A,E
4260.85	0	23463	11	.034	E	2	A,F
4420.47	0	22616	9	.034	E	2	A,E
PALLADIUM	ATOMIC WEIGHT =	106.4					
2447.91	0	40839	3	.28	E	2	A,E,F
2476.42	0	40369	3	.37	E	2	A,F
2763.09	0	36181	3	.21	E	2	A

Spectral Data

Wavelength (Å)	E_i (cm^{-1})	E_j (cm^{-1})	g_j	A_{ji} (10^8sec^{-1})	Merit	Ref.	Exp. Obs.
PALLADIUM	CONTINUED						
3302.10	10094	40369	3	2.0	E	2	F
3373.00	7755	37394	7	.50	E	2	F
3404.58	6564	35928	9	1.33	E	2	E,F
3421.24	7755	36976	5	1.68	E	2	E,F
3433.45	11722	40839	3	3.3	E	2	F
3460.77	6564	35451	7	.47	E	2	E,F
3481.15	10094	38812	5	2.2	E	2	E,F
3516.94	7755	36181	3	2.13	E	2	E,F
3553.08	11722	39858	7	2.6	E	2	E,F
3609.55	7755	35451	7	1.29	E	2	A,E,F
3634.70	6564	34069	5	1.24	E	2	A,E,F
PLATINUM	ATOMIC	WEIGHT =	195.09				
2049.37	0	48779	7	20.7	E	2	A
2067.50	0	48352	9	4.89	E	2	A
2144.23	0	46622	7	5.14	E	2	A
2165.17	0	46170	5	4.00	E	2	A
2174.67	0	45984	0	0.	E	2	A,F
2202.22	776	46170	5	2.80	E	2	P
2274.38	776	44730	7	1.10	E	2	P
2357.10	776	43188	3	1.30	E	2	P
2440.06	0	40970	7	.60	E	2	A,F
2450.97	0	40788	5	.080	E	2	A
2467.44	0	40516	5	.54	E	2	A
2487.17	0	40194	9	.63	E	2	A,F
2490.12	0	40970	7	.20	E	2	P
2628.03	776	38816	5	.90	E	2	A,F
2646.89	0	37769	7	.44	E	2	A
2650.86	824	38536	11	.17	E	2	F
2659.45	0	37591	9	.91	E	2	A,E,F

Spectral Data

Wavelength (Å)	E_i (cm^{-1})	E_j (cm^{-1})	g_j	A_{ji} (10^8 sec^{-1})	Merit	Ref.	Exp. Obs.
PLATINUM	CONTINUED						
2702.40	776	37769	7	.89	E	2	A,F
2705.89	824	37769	7	.71	E	2	A
2719.04	824	37591	9	.43	E	2	A,F
2733.96	776	37342	5	1.00	E	2	A
2734.50	0	0	0	0.	E	2	F
2771.67	776	36845	3	.40	E	2	P
2830.30	0	35322	7	.33	E	2	A,F
2893.86	776	35322	7	.14	E	2	F
2929.79	0	34122	7	.30	E	2	A,F
2997.97	776	34122	7	.31	E	2	A,F
3064.71	0	32620	5	.52	E	2	A,E,F
3042.64	824	33681	11	.081	E	2	A,F
3315.05	0	30157	9	.003	E	2	A
3408.13	824	30157	9	.018	E	2	P
POTASSIUM	ATOMIC WEIGHT =	39.102					
4044.15	0	24720	4	.0142	C	1	A,E
4047.21	0	24701	2	.0142	C	1	P
7664.91	0	13043	4	.387	B+	1	A,E,F
7698.98	0	12985	2	.382	B+	1	A,E,F
RHENIUM	ATOMIC WEIGHT =	186.2					
2274.62	0	43950	6	1.63	E	2	A
2287.51	0	43702	4	3.15	E	2	A
2294.49	0	43569	6	1.75	E	2	A
2419.81	0	41313	6	.33	E	2	A
2428.58	0	41164	8	.55	E	2	A
2508.99	0	39845	8	.38	E	2	P
2674.34	0	37381	3	.30	E	2	P
3451.88	0	28962	4	.50	E	2	A,E

Spectral Data

Wavelength (Å)	E_i (cm^{-1})	E_j (cm^{-1})	g_j	A_{ji} (10^8sec^{-1})	Merit	Ref.	Exp. Obs.
RHENIUM	CONTINUED						
3464.73	0	28854	6	.82	E	2	A,E
4889.14	0	20448	8	.005	E	2	E
5275.56	0	18950	6	.003	E	2	E
5776.83	11584	28890	8	.003	E	2	E
RHODIUM	ATOMIC WEIGHT =	102.905					
3280.60	1530	32004	8	.31	E	2	A
3323.09	1530	31614	10	.41	E	2	E
3396.85	0	29431	10	.31	E	2	A,E,F
3434.89	0	29105	12	.34	E	2	A,E,F
3462.04	2598	31474	6	.97	E	2	P
3478.91	3310	32046	6	.42	E	2	P
3502.52	0	28543	10	.26	E	2	A,E,F
3507.32	2598	31102	8	.31	E	2	A
3583.10	1530	29431	10	.27	E	2	E
3596.19	2598	30397	4	.88	E	2	E
3597.15	3310	31102	8	.68	E	2	F
3657.99	1530	28860	6	.68	E	2	A,E,F
3690.70	3310	30397	4	.38	E	2	P
3692.36	0	27075	8	.35	E	2	A,E,F
3700.91	1530	28543	10	.35	E	2	A,E,F
3806.76	2598	28860	6	.11	E	2	P
3856.52	5691	31614	10	.67	E	2	E
4211.14	5691	29431	10	.22	E	2	E
4374.80	5691	28543	10	.23	E	2	E
RUBIDIUM	ATOMIC WEIGHT =	85.47					
4201.85	0	23793	4	.24	E	2	A,E
4215.56	0	23715	2	.24	E	2	A,E
7800.23	0	12817	4	.75	E	2	A,E

Spectral Data

Wavelength (Å)	E_i (cm^{-1})	E_j (cm^{-1})	g_j	A_{ji} (10^8sec^{-1})	Merit	Ref.	Exp. Obs.
RUTHENIUM	ATOMIC	WEIGHT =	101.07				
2874.98	0	34773	11	.61	E	2	E
2988.95	0	33447	9	.40	E	2	P
3428.31	0	30348	9	.46	E	2	E,F
3436.74	1191	30280	11	.67	E	2	E
3498.94	0	28572	13	.46	E	2	A,E,F
3589.22	3105	30959	5	1.9	E	2	F
3593.02	213	30537	7	1.27	E	2	E,F
3596.18	2092	29891	9	.77	E	2	P
3634.93	2092	29595	9	.33	E	2	P
3661.35	1191	28495	11	.40	E	2	E,F
3726.93	1191	28015	9	.58	E	2	F
3728.03	0	26816	11	.42	E	2	A,E,F
3730.43	2092	28891	7	.77	E	2	F
3790.51	2092	28466	5	.82	E	2	F
3798.05	3105	29427	5	.14	E	2	F
3798.90	1191	27507	7	.57	E	2	F
3799.35	0	26313	9	.32	E	2	E,F
3925.92	0	25465	9	.11	E	2	A,F
4199.90	6545	30348	9	1.10	E	2	F
4554.51	6545	28495	11	.4	E	2	F
5699.05	8771	26313	9	.009	E	2	E
SCANDIUM	ATOMIC	WEIGHT =	44.956				
2706.77	0	36934	4	.85	E	2	P
2707.95	168	37086	4	.50	E	2	P
2711.35	168	37040	6	.93	E	2	P
2974.01	0	33615	4	1.22	E	2	P
2980.75	168	33707	6	1.00	E	2	P

Spectral Data

Wavelength (Å)	E_i (cm^{-1})	E_j (cm^{-1})	g_j	A_{ji} (10^{8}sec^{-1})	Merit	Ref.	Exp. Obs.
SCANDIUM	CONTINUED						
3015.36	0	33154	6	1.30	E	2	P
3019.34	168	33279	8	1.24	E	2	E
3030.76	168	33154	6	.22	E	2	E
3255.69	0	30707	6	.55	E	2	A,E
3269.91	0	30573	2	4.55	E	2	A,E
3273.63	168	30707	6	2.00	E	2	A,E
3907.49	0	25585	6	2.00	E	2	A,E,F
3911.81	168	25725	8	1.88	E	2	A,E,F
3933.38	168	15585	6	.45	E	2	A,E
3996.61	0	24866	4	2.50	E	2	A,E
4020.40	0	24866	4	2.50	E	2	A,E,F
4023.69	168	25014	6	1.67	E	2	A,E,F
4047.79	168	24866	4	.35	E	2	E
4054.55	0	24657	4	.68	E	2	A,E
4082.40	168	24657	4	.72	E	2	A,E
4741.02	11610	32697	6	.80	E	2	E
4743.81	11677	32752	8	.91	E	2	E
4753.16	0	21033	6	.008	E	2	E
4779.35	168	21086	8	.006	E	2	E
5081.56	11677	31351	10	1.10	E	2	E
5083.72	11610	31275	8	.75	E	2	E
5342.96	0	18711	2	.007	E	2	E
5349.71	168	18856	4	.88	E	2	E
5671.81	11677	29304	12	.39	E	2	E
5686.84	11610	29190	10	.37	E	2	E
5700.21	11558	29096	8	.39	E	2	E
6210.68	0	16097	4	.012	E	2	E
6239.41	0	16023	6	.001	E	2	E
6239.78	0	16022	4	.006	E	2	E
6259.96	168	16141	6	.003	E	2	E

Spectral Data

Wavelength (Å)	E_i (cm^{-1})	E_j (cm^{-1})	g_j	A_{ji} (10^8 sec^{-1})	Merit	Ref.	Exp. Obs.
SCANDIUM	CONTINUED						
6305.67	168	16023	6	.009	E	2	E
6378.82	0	15673	4	.001	E	2	E
6413.35	168	15657	6	.001	E	2	E
SELENIUM	ATOMIC WEIGHT =	78.96					
1960.26	0	50997	2	100.	E	2	A,E,F
2039.85	1989	50997	2	65.0	E	2	A,E,F
2062.79	2534	50997	2	21.6	E	2	A,E,F
2074.79	0	48182	3	1.47	E	2	A,F
2164.00	0	0	0	.0	E	2	A
SILICON	ATOMIC WEIGHT =	28.086					
2207.98	0	45276	3	.311	C	1	P
2210.89	77	45294	5	.416	C	1	A
2211.74	77	42576	3	.232	C	1	P
2216.67	223	45322	7	.550	C	1	A
2218.06	223	45294	5	.138	C	1	P
2506.90	77	39955	5	.417	C	1	A,F
2514.32	0	39760	3	.550	C	1	A,F
2516.11	223	39955	5	1.21	C	1	A,E,F
2519.20	77	39760	3	.422	C	1	A,F
2524.11	77	39683	1	1.66	C	1	A,F
2528.51	223	39760	3	.690	C	1	A,F
2881.58	6299	40992	3	1.75	C	1	A
SILVER	ATOMIC WEIGHT =	107.870					
3280.68	0	50473	4	1.57	B+	9	A,E,F
3382.89	0	29552	4	.32	E	2	A,E,F

Spectral Data

Wavelength ($\overset{\circ}{A}$)	E_i (cm^{-1})		E_j (cm^{-1})	g_j	A_{ji} (10^8sec^{-1})	Merit	Ref.	Exp. Obs.
SODIUM	ATOMIC WEIGHT =		22.9898					
3302.37	0		30273	4	.0290	C+	1	A,E
3302.98	0		30267	2	.0293	C+	1	A,E
5889.95	0		16973	4	.628	A	1	A,E,F
5895.92	0		16956	2	.630	A	1	A,E,F
8194.82	16973		29173	6	.945	C	1	E
STRONTIUM	ATOMIC WEIGHT =		87.62					
2428.10	0		41172	3	.067	E	2	A
2569.47	0		38907	3	.037	E	2	A
2931.83	0		34098	3	.017	E	2	A
4077.71	II	0	24517	4	.16	E	2	A,E
4215.52	II	0	23715	2	.19	E	2	A,E
4607.33	0		21698	3	1.62	B	9	A,E,F
TANTALUM	ATOMIC WEIGHT =		180.948					
2507.45	2010		41879	6	.88	E	2	P
2526.35	2010		41581	6	1.65	E	2	P
2559.43	0		39060	6	.82	E	2	A
2608.20	2010		40339	4	.090	E	2	A
2608.63	2010		40333	4	2.40	E	2	A
2636.90	5621		43533	8	1.88	E	2	P
2647.47	0		37761	4	2.05	E	2	A
2653.27	2010		39688	6	2.50	E	2	P
2656.61	0		37630	6	1.05	E	2	P
2661.34	5621		43185	12	2.08	E	2	A
2661.89	5621		43177	8	.31	E	2	A
2691.31	0		37146	8	.12	E	2	P
2698.30	2010		39060	6	.87	E	2	P
2714.67	0		36826	6	1.17	E	2	A
2758.31	2010		38253	8	.52	E	2	A
2775.88	0		36014	6	.28	E	2	A
2850.98	5621		40686	10	1.60	E	2	P
2891.84	2010		36580	4	.52	E	2	P
2933.55	0		34078	4	.58	E	2	P

Spectral Data

Wavelength (Å)	E_i (com^{-1})	E_j (cm^{-1})	g_j	A_{ji} (10^8sec^{-1})	Merit	Ref.	Exp. Obs.
TANTALUM	CONTINUED						
2963.32	2010	35746	8	.42	E	2	P
2965.54	2010	35721	4	.42	E	2	P
3103.24	0	32215	4	.14	E	2	P
3124.97	2010	34001	6	.10	E	2	P
4740.16	9976	31066	4	.028	E	2	E
4812.75	0	20772	4	.002	E	2	E
5402.51	0	18505	2	.002	E	2	P
TELLERIUM	ATOMIC WEIGHT =	127.60					
1994.20	4751	54877	3	53.	E	2	P
2002.00	4751	54685	5	63.	E	2	P
2142.75	0	46653	3	38.	E	2	A,E,F
2259.04	0	44253	5	.66	E	2	A,F
2383.25	4707	46653	3	4.27	E	2	A,E,F
2385.76	4751	46653	3	5.47	E	2	E,F
2530.70	4751	44253	5	.15	E	2	A,F
THALLIUM	ATOMIC WEIGHT =	204.37					
2379.69	0	42011	4	.85	E	2	A
2580.14	0	38746	2	.65	E	2	A
2709.23	7793	44693	6	.72	E	2	A
2710.67	7793	44673	4	.13	E	2	A
2767.87	0	36118	4	1.02	E	2	A,E,F
2826.16	7793	43166	2	.95	E	2	A
2918.32	7793	42049	6	2.50	E	2	A
2921.52	7793	42011	4	.58	E	2	A
3229.75	7792	38746	2	1.35	E	2	P
3519.24	7793	36200	6	4.00	E	2	E
3529.43	7793	36118	4	1.48	E	2	P
3775.72	0	26478	2	.63	C	9	A,E,F

Spectral Data

Wavelength (Å)	E_i (cm^{-1})	E_j (cm^{-1})	g_j	A_{ji} (10^8 sec^{-1})	Merit	Ref.	Exp. Obs.
TIN	ATOMIC	WEIGHT =	118.69				
2073.08	0	48222	3	11.0	E	2	A
2151.43	3428	49894	7	5.43	E	2	P
2194.49	3428	48982	3	11.3	E	2	P
2199.34	1692	47146	5	6.80	E	2	A
2209.65	3428	48670	5	12.0	E	2	A
2246.05	0	44509	3	8.67	E	2	A,F
2268.91	3428	47488	7	7.71	E	2	A
2286.68	3428	47146	5	1.56	E	2	A
2334.80	1692	44509	3	3.20	E	2	A
2354.84	1692	44145	5	7.70	E	2	A,E,F
2421.70	8613	49894	7	9.4	E	2	A
2429.49	3428	44576	7	4.29	E	2	A,E,F
2483.39	3428	43683	5	1.20	E	2	A
2546.55	0	39257	3	1.27	E	2	A,F
2661.24	1692	39257	3	.77	E	2	A
2706.51	1692	38629	5	2.00	E	2	A,E,F
2839.99	3428	38629	5	4.20	E	2	A,E,F
2863.33	0	34914	3	.623	B	9	A,E,F
3009.14	1692	34914	3	1.30	E	2	A,E,F
3034.12	1692	34641	1	4.40	E	2	A,E,F
3175.05	3428	34914	3	1.07	E	2	A,E,F
3262.34	8613	39257	3	3.67	E	2	E,F
3330.62	8613	38629	5	.38	E	2	E,F
3801.02	8613	34914	3	.67	E	2	E,F
TITANIUM	ATOMIC	WEIGHT =	47.90				
2605.15	170	38544	7	1.29	E	2	A
2611.48	387	38671	9	1.33	E	2	A

Spectral Data

Wavelength (Å)	E_i (cm^{-1})	E_j (cm^{-1})	g_j	A_{ji} (10^8 sec^{-1})	Merit	Ref.	Exp. Obs.
TITANIUM	CONTINUED						
2644.26	170	37977	5	3.20	E	2	A
2646.64	387	38160	7	3.00	E	2	A
2942.00	0	33981	5	1.56	E	2	A
2948.24	170	34079	7	1.37	E	2	A,F
2956.13	387	34205	9	1.44	E	2	A
2956.80	170	33981	5	.26	E	2	F
3186.45	0	31374	7	1.16	E	2	A,E,F
3191.99	170	31489	9	1.22	E	2	A,E,F
3199.92	387	31629	11	1.27	E	2	A,E,F
3341.88	0	29915	7	1.86	E	2	A,E,F
3354.64	170	29971	9	1.08	E	2	A,E,F
3370.44	0	29661	3	.83	E	2	E
3371.45	0	30039	11	1.00	E	2	A,E,F
3377.48	387	29986	7	1.00	E	2	A,E
3385.95	387	29912	7	.49	E	2	E
3635.46	0	27888	7	.89	E	2	A,E,F
3642.68	0	27615	9	.98	E	2	A,E,F
3646.20	0	27418	5	.046	E	2	F
3653.50	0	27750	11	.91	E	2	A,E,F
3658.10	170	27499	7	.12	E	2	E
3668.97	170	27418	5	.10	E	2	E
3671.67	387	27615	9	.090	E	2	E
3729.82	387	26803	5	.62	E	2	A,E
3741.06	387	26893	7	.54	E	2	A,E
3752.86	170	27026	9	.68	E	2	A,E
3753.64	170	26803	5	.13	E	2	E
3771.66	387	26893	7	.10	E	2	E
3924.53	170	25644	7	.13	E	2	E
3929.88	0	25439	5	.14	E	2	E
3947.78	170	25494	5	.17	E	2	A,E

Spectral Data

Wavelength (Å)	E_i (cm^{-1})	E_j (cm^{-1})	g_j	A_{ji} (10^8sec^{-1})	Merit	Ref.	Exp. Obs.
TITANIUM	**CONTINUED**						
3956.34	170	25439	5	.72	E	2	A,E
3958.21	170	25644	7	.63	E	2	A,E
3962.85	0	25227	7	.10	E	2	E
3964.27	170	25388	9	.082	E	2	E
3981.76	170	25107	5	.68	E	2	A,E
3982.48	387	25103	5	.028	E	2	A,E
3989.76	0	25227	7	.61	E	2	A,E
3998.64	0	25388	9	.68	E	2	A,E,F
4008.93	170	25107	5	.14	E	2	E
4024.57	387	25227	7	.13	E	2	E
4300.56	6661	29907	5	1.74	E	2	E
4301.09	6743	29986	7	1.86	E	2	E
4305.92	6843	30060	9	2.11	E	2	E
4533.24	6843	28896	11	1.27	E	2	E
5039.95	170	20006	5	.056	E	2	E
5064.66	387	20126	7	.050	E	2	E
5173.75	0	19323	5	.044	E	2	E
5192.98	170	19422	7	.039	E	2	E
5210.39	387	19574	9	.034	E	2	E
TUNGSTEN	**ATOMIC WEIGHT =**	**183.85**					
2452.00	0	40771	3	1.33	E	2	A
2481.44	1670	41965	1	37.0	E	2	P
2550.37	1670	40868	5	.76	E	2	P
2551.35	0	42514	7	1.17	E	2	A
2606.39	0	38356	3	.57	E	2	A
2613.07	3326	41583	5	1.18	E	2	P
2613.82	2951	41198	9	.29	E	2	P
2656.54	2951	40583	9	1.01	E	2	A
2662.84	3326	40868	5	.98	E	2	P

Spectral Data

Wavelength (Å)	E_i (cm^{-1})	E_j (cm^{-1})	g_j	A_{ji} (10^8sec^{-1})	Merit	Ref.	Exp. Obs.
TUNGSTEN	CONTINUED						
2718.90	2951	39720	9	1.04	E	2	A
2724.35	2951	39646	7	1.57	E	2	A
2762.34	0	36190	3	.23	E	2	A
2831.38	2951	38259	9	.56	E	2	A
2879.11	2951	37674	7	.086	E	2	P
2879.39	0	34719	7	.10	E	2	A
2896.01	1670	36190	3	.26	E	2	A
2896.45	2951	37466	5	.78	E	2	A
2910.48	3326	37674	7	.087	E	2	P
2911.00	0	34342	3	.097	E	2	A
2944.40	2951	36904	5	1.06	E	2	A
2946.98	2951	36904	5	1.06	E	2	A
2947.38	4830	38748	9	.20	E	2	A
3191.57	0	31323	3	.063	E	2	A
3617.52	2951	30587	7	.13	E	2	A,E
3768.45	1970	28199	3	.077	E	2	A,E
3867.98	2951	28797	9	.052	E	2	E
4008.75	2951	27890	9	.20	E	2	A,E
4045.60	2951	27662	5	.036	E	2	E
4074.36	2951	27488	7	.14	E	2	A,E
4269.39	2951	26367	5	.040	E	2	A,E
4294.61	2951	26230	5	.11	E	2	A,E
4302.11	2951	26189	7	.043	E	2	A,E
VANADIUM	ATOMIC WEIGHT =	50.942					
2506.90	0	39878	2	3.90	E	2	P
2507.38	137	40001	8	1.62	E	2	A
2511.95	137	39935	4	2.38	E	2	A
2517.14	323	40039	10	.98	E	2	A
2519.62	323	40000	6	2.17	E	2	A

Spectral Data

Wavelength (Å)	E_i (cm^{-1})	E_j (cm^{-1})	g_j	A_{ji} (10^8 sec^{-1})	Merit	Ref.	Exp. Obs.
VANADIUM	CONTINUED						
2530.18	553	40064	12	1.08	E	2	A
2574.02	553	39391	10	1.00	E	2	A
2923.62	553	34747	8	1.25	E	2	A,E
3043.12	137	32989	8	.25	E	2	E
3043.56	0	32847	6	.33	E	2	E
3044.94	323	33155	10	.21	E	2	E
3050.89	0	32768	4	.35	E	2	E
3053.65	0	32738	4	.95	E	2	A,E
3056.33	137	32847	6	1.67	E	2	A,E
3060.46	323	32989	8	1.62	E	2	A,E
3066.38	553	33155	10	2.30	E	2	A,E,F
3183.41	137	31541	8	2.50	E	2	A,E
3183.98	323	31722	10	3.50	E	2	A,E,F
3185.40	553	31937	12	2.17	E	2	A,E
3198.01	137	31398	6	.53	E	2	E
3202.38	323	31541	8	.59	E	2	E
3207.41	553	31722	10	.30	E	2	E
3212.43	11101	42221	12	4.42	E	2	E
3263.24	0	30636	6	.11	E	2	E
3271.64	137	30694	8	.058	E	2	E
3283.31	323	30771	10	.047	E	2	E
3298.14	553	30864	12	.064	E	2	E
3675.70	2220	29418	8	.11	E	2	E
3703.58	2425	29418	8	1.50	E	2	A,E,F
3704.70	2311	29296	6	.92	E	2	A,E
3705.04	2220	29203	4	.42	E	2	E
3778.68	2311	28768	10	.072	E	2	E
3790.32	2220	28596	8	.16	E	2	E
3793.61	0	26353	6	.022	E	2	E
3794.96	2425	28768	10	.31	E	2	E

Spectral Data

Wavelength (Å)	E_i (cm^{-1})	E_j (cm^{-1})	g_j	A_{ji} (10^8 sec^{-1})	Merit	Ref.	Exp. Obs.
VANADIUM	CONTINUED						
3803.47	2311	28596	8	.19	E	2	E
3807.50	2112	28369	4	.18	E	2	E
3808.52	0	26249	4	.18	E	2	E
3809.60	2220	28462	6	.10	E	2	E
3813.49	137	26353	6	.23	E	2	E
3817.84	553	26738	10	.022	E	2	E
3818.24	0	26183	2	.85	E	2	E
3819.96	2425	28596	8	.074	E	2	E
3821.49	2153	28314	2	.28	E	2	E
3822.01	323	26480	8	.10	E	2	E
3822.89	2311	28462	6	.18	E	2	E
3823.21	2220	28369	4	.18	E	2	E
3828.56	137	26249	4	.58	E	2	A,E
3840.75	323	26353	6	.62	E	2	A,E,F
3844.44	0	26004	6	.080	E	2	E
3847.33	137	26122	8	.052	E	2	E
3855.37	0	25931	4	.38	E	2	A,E,F
3855.84	553	26480	8	.55	E	2	A,E,F
3864.86	137	26004	6	.28	E	2	E
3867.60	323	26172	10	.030	E	2	E
3875.08	323	26122	8	.25	E	2	E
3875.90	137	25931	4	.13	E	2	P
3876.09	553	26345	10	.080	E	2	E
3890.18	323	26022	8	.084	E	2	E
3892.86	323	26004	6	.073	E	2	E
3902.25	553	26172	10	.24	E	2	A,E
3909.89	553	26122	8	.086	E	2	E
4090.58	8716	33155	10	.86	E	2	E
4092.69	2311	26738	10	0.22	E	2	E
4095.49	8579	32989	8	0.80	E	2	E

Spectral Data

Wavelength (Å)	E_i (cm^{-1})	E_j (cm^{-1})	g_j	A_{ji} (10^8 sec^{-1})	Merit	Ref.	Exp. Obs.
VANADIUM	CONTINUED						
4102.16	8476	32847	6	.67	E	2	E
4105.17	2153	26506	6	.53	E	2	E,F
4109.79	2112	26438	4	.65	E	2	E
4111.78	2425	26738	10	1.10	E	2	A,E,F
4115.18	2311	26605	8	.65	E	2	A,E,F
4116.47	2220	26506	6	.35	E	2	E
4123.57	2153	26397	2	1.15	E	2	E
4128.07	2220	26438	4	.88	E	2	E,F
4132.02	2311	26506	6	.60	E	2	E,F
4134.49	2425	26605	8	.34	E	2	E
4179.42	2425	26345	10	.026	E	2	E
4182.59	2220	26122	8	.020	E	2	E
4189.84	2311	26172	10	.019	E	2	E
4209.86	2425	26172	10	.025	E	2	E
4306.21	137	23353	8	.010	E	2	E
4307.18	0	23211	6	.011	E	2	E
4309.80	323	23520	10	.009	E	2	E
4330.02	0	23088	4	.052	E	2	E
4332.82	137	23211	6	.042	E	2	E
4341.01	323	23353	8	.048	E	2	E
4355.94	137	23088	4	.018	E	2	E
4368.04	323	23211	6	.012	E	2	E
4379.24	2425	25254	12	.83	E	2	A,E,F
4384.72	2311	25112	10	.58	E	2	A,E,F
4389.97	2220	24993	8	.49	E	2	A,E,F
4395.23	2153	24899	6	.47	E	2	E,F
4400.58	2112	24830	4	.28	E	2	E
4406.64	2425	25112	10	.19	E	2	E
4407.64	2311	24993	8	.29	E	2	E,F
4408.20	2220	24899	6	.47	E	2	E,F

Spectral Data

Wavelength (Å)	E_i (cm^{-1})	E_j (cm^{-1})	g_j	A_{ji} (10^8 sec^{-1})	Merit	Ref.	Exp. Obs.
VANADIUM	CONTINUED						
4416.47	2153	24789	2	.24	E	2	E
4419.94	2220	24839	8	.011	E	2	E
4421.57	2220	24830	4	.12	E	2	E
4426.00	2311	24899	6	.060	E	2	E
4428.52	2153	24728	6	.038	E	2	E
4429.80	2425	24993	8	.024	E	2	E
4436.14	2112	24648	4	.080	E	2	E
4437.84	2311	24839	8	.062	E	2	E
4441.68	2220	24728	6	.10	E	2	E
4444.21	2153	24648	4	.12	E	2	E
4457.48	2220	24648	4	.075	E	2	E
4459.76	2311	24728	6	.13	E	2	E
4460.29	2425	24839	8	.20	E	2	E
4577.17	0	21841	6	.030	E	2	E
4580.40	137	21964	8	.029	E	2	E
4586.36	323	22121	10	.031	E	2	E
4594.11	553	22314	12	.042	E	2	E
4827.45	323	21033	8	.005	E	2	E
4831.64	137	20828	6	.006	E	2	E
4832.43	0	20686	4	.008	E	2	E
4851.48	0	20606	2	.040	E	2	E
4864.74	137	20688	4	.030	E	2	E
4875.48	323	20828	6	.027	E	2	E
4881.56	553	21033	8	.026	E	2	E
5627.64	8716	26480	8	.070	E	2	E
5670.85	8716	26345	10	.043	E	2	E
5727.03	8716	26172	10	.11	E	2	E
6039.73	8579	25131	6	.078	E	2	E
6081.44	8476	24915	4	.12	E	2	E
6090.22	8716	25131	6	.22	E	2	E

Spectral Data

Wavelength (Å)		E_i (cm^{-1})	E_j (cm^{-1})	g_j	A_{ji} (10^8 sec^{-1})	Merit	Ref.	Exp. Obs.
VANADIUM		CONTINUED						
6119.52		8579	24915	4	.15	E	2	E
6199.19		2311	18438	10	.007	E	2	E
6213.87		2425	18513	10	.002	E	2	E
6216.37		2220	18302	8	.009	E	2	E
6224.50		2311	18372	8	.002	E	2	E
6230.74		2153	18198	6	.011	E	2	E
6233.20		2220	18259	6	.002	E	2	E
6242.81		2112	18126	4	.006	E	2	E
6243.10		2425	18438	10	.011	E	2	E
6251.82		2311	18302	8	.006	E	2	E
6274.65		2153	18086	2	.012	E	2	E
6285.16		2220	18126	4	.008	E	2	E
6292.83		2311	18198	6	.005	E	2	E
6296.49		2425	18302	8	.003	E	2	E
YTTRIUM		ATOMIC WEIGHT =	88.905					
2760.10		530	36751	4	.40	E	2	P
2984.26		530	34030	8	.52	E	2	E
3552.69		0	28140	4	.19	E	2	E
3592.92		0	27824	2	1.85	E	2	E
3620.94		530	28140	4	1.55	E	2	E
3710.30	II	1450	28394	9	.56	E	2	E
3774.33	II	1045	27532	7	.46	E	2	E
4039.83		0	24747	6	.077	E	2	E
4047.63		0	24699	2	.55	E	2	E
4077.38		0	24519	6	.72	E	2	A,E
4083.71		0	24481	4	.23	E	2	E
4102.38		530	24900	8	.64	E	2	A,E
4128.31		530	24747	6	.73	E	2	A,E
4142.85		0	24131	4	.78	E	2	A,E

Spectral Data

Wavelength (Å)	E_i (cm^{-1})	E_j (cm^{-1})	g_j	A_{ji} (10^8 sec^{-1})	Merit	Ref.	Exp. Obs.
YTTRIUM	CONTINUED						
4174.13	530	24481	4	.23	E	2	E
4235.94	530	24131	4	.24	E	2	E
4527.25	11532	33614	8	.74	E	2	E
4643.70	0	21529	6	.063	E	2	E
4674.85	530	21915	8	.054	E	2	E
4760.98	530	21529	6	.013	E	2	E
5466.46	11532	29820	12	.11	E	2	E
5527.54	11278	29364	10	.12	E	2	E
6222.59	0	16066	6	.002	E	2	E
6402.01	530	16146	4	.0007	E	2	E
6435.00	530	16066	6	.007	E	2	E
6557.39	0	15246	6	.0004	E	2	E
6687.58	0	14949	4	.001	E	2	E
6793.71	530	15246	6	.001	E	2	E
ZINC	ATOMIC WEIGHT =	65.37					
2138.56	0	46745	3	6.3	C	9	A,E,F
2770.86	32501	68581	5	.24	E	2	E
2800.87	32890	68583	7	.26	E	2	E
3075.90	0	32502	3	.004	E	2	A,E
3345.02	32890	62777	7	4.0	E	2	E
4810.53	32890	53672	3	7.00	E	2	E
ZIRCONIUM	ATOMIC WEIGHT =	91.22					
2814.90	0	35515	5	1.30	E	2	P
2837.23	570	35806	7	1.16	E	2	P
2875.98	1241	36001	9	.67	E	2	P
2985.39	0	33487	3	2.40	E	2	A
3011.75	570	33764	5	1.58	E	2	A
3029.52	1241	34240	7	1.39	E	2	A

Spectral Data

Wavelength (Å)	E_i (cm^{-1})	E_j (cm^{-1})	g_j	A_{ji} (10^8 sec^{-1})	Merit	Ref.	Exp. Obs.
ZIRCONIUM	CONTINUED						
3519.60	0	28404	7	.71	E	2	A , E
3547.68	570	28750	9	.53	E	2	A
3550.46	0	28157	7	.21	E	2	P
3566.10	1241	29275	7	.49	E	2	P
3575.79	570	28528	9	.31	E	2	P
3586.29	0	27876	5	.38	E	2	P
3601.19	1241	29002	11	.91	E	2	A , E
3623.86	570	28157	7	.39	E	2	A
3663.65	1241	28528	9	.27	E	2	P
3835.96	0	26062	5	.50	E	2	P
3863.87	570	26444	7	.37	E	2	A
3885.42	0	25730	7	.16	E	2	P
3890.32	1241	26938	9	.33	E	2	A
3891.38	1241	26931	9	.22	E	2	P
3929.53	570	26012	9	.11	E	2	P
3968.26	1241	26434	11	.083	E	2	P
4633.98	570	22144	9	.023	E	2	P
4687.80	5899	27215	13	.23	E	2	A

Table III.B.1-b

Wavelength Locator for Atomic Absorption Lines

Wavelength (Å)	Element	Wavelength (Å)	Element
1936.96	ARSENIC	2138.56	ZINC
1953.89	BISMUTH	2142.75	TELLERIUM
1959.48	BISMUTH	2144.23	PLATINUM
1960.26	SELENIUM	2151.43	TIN
1971.97	ARSENIC	2155.81	IRIDIUM
1994.20	TELLERIUM	2158.05	IRIDIUM
2002.00	TELLERIUM	2164.00	SELENIUM
2012.00	GOLD	2165.09	COPPER
2021.21	BISMUTH	2165.17	PLATINUM
2021.38	GOLD	2166.77	IRON
2022.00	LEAD	2169.99	LEAD
2024.34	COPPER	2174.67	PLATINUM
2025.82	MAGNESIUM	2175.24	IRIDIUM
2033.57	IRIDIUM	2175.81	ANTIMONY
2039.85	SELENIUM	2178.94	COPPER
2041.69	GERMANIUM	2181.72	COPPER
2043.76	GERMANIUM	2194.49	TIN
2049.37	PLATINUM	2198.70	GERMANIUM
2053.20	LEAD	2199.34	TIN
2061.70	BISMUTH	2200.73	CALCIUM
2062.79	SELENIUM	2202.22	PLATINUM
2065.20	GERMANIUM	2207.98	SILICON
2067.50	PLATINUM	2209.65	TIN
2068.33	ANTIMONY	2210.89	SILICON
2068.65	GERMANIUM	2211.74	SILICON
2073.08	TIN	2216.67	SILICON
2074.79	SELENIUM	2218.06	SILICON
2088.82	IRIDIUM	2225.70	COPPER
2088.84	BORON	2228.25	BISMUTH
2089.57	BORON	2230.61	BISMUTH
2092.63	IRIDIUM	2246.05	TIN
2094.23	GERMANIUM	2259.04	TELLERIUM
2110.26	BISMUTH	2263.46	ALUMINUM
2127.39	ANTIMONY	2268.91	TIN
2127.94	IRIDIUM	2269.10	ALUMINUM

Wavelength Locator for Atomic Absorption Lines

Wavelength (Å)	Element	Wavelength (Å)	Element
2269.22	ALUMINUM	2387.75	GOLD
2274.38	PLATINUM	2398.56	CALCIUM
2274.49	COBALT	2407.25	COBALT
2274.62	RHENIUM	2411.62	COBALT
2275.46	CALCIUM	2414.46	COBALT
2276.58	BISMUTH	2415.30	COBALT
2286.68	TIN	2418.69	GALLIUM
2287.51	RHENIUM	2419.81	RHENIUM
2289.98	NICKEL	2421.70	TIN
2294.49	RHENIUM	2424.56	OSMIUM
2295.23	COBALT	2424.93	COBALT
2309.02	COBALT	2424.97	OSMIUM
2310.96	NICKEL	2427.95	GOLD
2311.47	ANTIMONY	2428.10	STRONTIUM
2320.03	NICKEL	2428.58	RHENIUM
2329.96	NICKEL	2429.49	TIN
2334.80	TIN	2432.21	COBALT
2337.49	NICKEL	2435.83	COBALT
2337.82	NICKEL	2439.05	COBALT
2345.54	NICKEL	2440.06	PLATINUM
2346.63	NICKEL	2441.64	COPPER
2347.52	NICKEL	2447.91	PALLADIUM
2348.61	BERYLLIUM	2450.07	GALLIUM
2352.65	GOLD	2450.97	PLATINUM
2354.84	TIN	2452.00	TUNGSTEN
2357.10	PLATINUM	2455.61	IRIDIUM
2364.71	CHROMIUM	2462.65	IRON
2367.05	ALUMINUM	2467.44	PLATINUM
2372.07	ALUMINUM	2472.90	IRON
2372.77	IRIDIUM	2475.12	IRIDIUM
2373.12	ALUMINUM	2476.42	PALLADIUM
2373.35	ALUMINUM	2476.84	OSMIUM
2373.35	ALUMINUM	2476.87	NICKEL
2379.69	THALLIUM	2479.78	IRON
2383.25	TELLERIUM	2481.18	IRIDIUM

Wavelength Locator for Atomic Absorption Lines

Wavelength (Å)	Element	Wavelength (Å)	Element
2481.44	TUNGSTEN	2526.35	TANTALUM
2483.27	IRON	2527.44	IRON
2483.39	TIN	2528.51	SILICON
2484.19	IRON	2528.97	COBALT
2487.17	PLATINUM	2530.18	VANADIUM
2488.14	IRON	2530.70	TELLERIUM
2489.75	IRON	2533.23	GERMANIUM
2490.12	PLATINUM	2535.96	COBALT
2490.64	IRON	2536.52	MERCURY
2491.16	IRON	2543.97	IRIDIUM
2492.15	COPPER	2546.55	TIN
2496.77	BORON	2550.37	TUNGSTEN
2497.72	BORON	2551.35	TUNGSTEN
2497.96	GERMANIUM	2559.43	TANTALUM
2500.17	GALLIUM	2560.15	INDIUM
2500.70	GALLIUM	2567.98	ALUMINUM
2501.13	IRON	2569.47	STRONTIUM
2502.98	IRIDIUM	2574.02	VANADIUM
2506.90	SILICON	2575.10	ALUMINUM
2506.90	VANADIUM	2575.40	ALUMINUM
2507.38	VANADIUM	2580.14	THALLIUM
2507.45	TANTALUM	2589.19	GERMANIUM
2508.99	RHENIUM	2590.76	OSMIUM
2511.95	VANADIUM	2592.54	GERMANIUM
2514.32	SILICON	2601.76	INDIUM
2516.11	SILICON	2605.15	TITANIUM
2517.14	VANADIUM	2606.39	TUNGSTEN
2518.10	IRON	2608.20	TANTALUM
2519.20	SILICON	2608.63	TANTALUM
2519.62	VANADIUM	2611.48	TITANIUM
2521.36	COBALT	2612.63	OSMIUM
2522.85	IRON	2613.07	TUNGSTEN
2524.11	SILICON	2613.65	LEAD
2524.29	IRON	2613.82	TUNGSTEN
2526.22	VANADIUM	2614.18	LEAD

Wavelength Locator for Atomic Absorption Lines

Wavelength (Å)	Element	Wavelength (Å)	Element
2621.82	OSMIUM	2697.06	NIOBIUM
2628.03	PLATINUM	2698.30	TANTALUM
2636.90	TANTALUM	2702.40	PLATINUM
2637.13	OSMIUM	2705.89	PLATINUM
2639.42	IRIDIUM	2706.51	TIN
2639.71	IRIDIUM	2706.77	SCANDIUM
2641.10	TITANIUM	2707.95	SCANDIUM
2641.49	GOLD	2709.23	THALLIUM
2644.11	OSMIUM	2709.63	GERMANIUM
2644.26	TITANIUM	2710.26	INDIUM
2646.64	TITANIUM	2710.67	THALLIUM
2646.89	PLATINUM	2711.35	SCANDIUM
2647.47	TANTALUM	2713.94	INDIUM
2651.18	GERMANIUM	2714.64	OSMIUM
2651.58	GERMANIUM	2714.67	TANTALUM
2652.48	ALUMINUM	2718.90	TUNGSTEN
2653.27	TANTALUM	2719.03	IRON
2656.54	TUNGSTEN	2719.04	PLATINUM
2656.61	TANTALUM	2719.65	GALLIUM
2659.45	PLATINUM	2720.04	OSMIUM
2659.87	GALLIUM	2720.52	IRON
2660.39	ALUMINUM	2720.90	IRON
2661.24	TIN	2724.35	TUNGSTEN
2661.34	TANTALUM	2733.96	PLATINUM
2661.89	TANTALUM	2748.26	GOLD
2661.98	IRIDIUM	2753.88	INDIUM
2662.84	TUNGSTEN	2754.59	GERMANIUM
2664.79	IRIDIUM	2758.31	TANTALUM
2674.34	RHENIUM	2760.10	YTTRIUM
2675.95	GOLD	2762.34	TUNGSTEN
2677.15	PLATINUM	2763.09	PALLADIUM
2681.41	TUNGSTEN	2767.87	THALLIUM
2691.31	TANTALUM	2771.67	PLATINUM
2691.34	GERMANIUM	2775.88	TANTALUM
2694.23	IRIDIUM	2779.37	HAFNIUM

Wavelength Locator for Atomic Absorption Lines

Wavelength (Å)	Element	Wavelength (Å)	Element
2794.82	MANGANESE	2910.48	TUNGSTEN
2795.53	MAGNESIUM	2911.00	TUNGSTEN
2798.27	MANGANESE	2912.33	OSMIUM
2801.06	MANGANESE	2916.48	HAFNIUM
2802.70	MAGNESIUM	2918.32	THALLIUM
2806.91	OSMIUM	2918.58	HAFNIUM
2814.90	ZIRCONIUM	2921.52	THALLIUM
2826.16	THALLIUM	2923.62	VANADIUM
2830.30	PLATINUM	2924.79	IRIDIUM
2831.38	TUNGSTEN	2927.81	NIOBIUM
2833.06	LEAD	2929.79	PLATINUM
2837.23	ZIRCONIUM	2931.83	STRONTIUM
2838.17	OSMIUM	2932.63	INDIUM
2838.63	OSMIUM	2933.55	TANTALUM
2839.99	TIN	2936.90	IRON
2849.72	IRIDIUM	2940.22	TANTALUM
2850.76	OSMIUM	2940.77	HAFNIUM
2850.98	TANTALUM	2942.00	TITANIUM
2852.13	MAGNESIUM	2943.64	GALLIUM
2863.33	TIN	2944.18	GALLIUM
2866.37	HAFNIUM	2944.40	TUNGSTEN
2874.24	GALLIUM	2946.98	TUNGSTEN
2875.98	ZIRCONIUM	2947.38	TUNGSTEN
2879.11	TUNGSTEN	2947.88	IRON
2879.39	TUNGSTEN	2948.24	TITANIUM
2881.58	SILICON	2950.68	HAFNIUM
2882.64	IRIDIUM	2950.88	NIOBIUM
2889.62	HAFNIUM	2954.20	HAFNIUM
2891.84	TANTALUM	2956.13	TITANIUM
2896.01	TUNGSTEN	2963.32	TANTALUM
2896.45	TUNGSTEN	2964.88	HAFNIUM
2898.26	HAFNIUM	2965.54	TANTALUM
2904.41	HAFNIUM	2966.90	IRON
2904.75	HAFNIUM	2973.13	IRON
2909.06	OSMIUM	2973.24	IRON

Wavelength Locator for Atomic Absorption Lines

Wavelength (Å)	Element	Wavelength (Å)	Element
2974.01	SCANDIUM	3053.65	VANADIUM
2980.75	SCANDIUM	3054.32	NICKEL
2980.81	HAFNIUM	3056.33	VANADIUM
2983.57	IRON	3057.02	HAFNIUM
2985.39	ZIRCONIUM	3057.64	NICKEL
2987.16	COBALT	3058.66	OSMIUM
2988.95	SCANDIUM	3059.09	IRON
2988.95	RUTHENIUM	3060.46	VANADIUM
2989.59	COBALT	3064.71	PLATINUM
2994.43	IRON	3066.38	VANADIUM
2997.97	PLATINUM	3067.41	HAFNIUM
3000.95	IRON	3067.72	BISMUTH
3002.49	NICKEL	3071.58	BARIUM
3009.14	TIN	3072.88	HAFNIUM
3011.75	ZIRCONIUM	3075.90	ZINC
3012.00	NICKEL	3082.15	ALUMINUM
3015.36	SCANDIUM	3092.71	ALUMINUM
3016.78	HAFNIUM	3092.84	ALUMINUM
3018.04	OSMIUM	3094.18	NIOBIUM
3018.31	HAFNIUM	3101.55	NICKEL
3020.49	IRON	3101.88	NICKEL
3020.53	HAFNIUM	3103.24	TANTALUM
3020.64	IRON	3112.12	MOLYBDENUM
3021.07	IRON	3122.78	GOLD
3029.52	ZIRCONIUM	3124.97	TANTALUM
3030.70	OSMIUM	3130.79	NIOBIUM
3034.12	TIN	3132.59	MOLYBDENUM
3037.94	NICKEL	3158.16	MOLYBDENUM
3039.06	GERMANIUM	3163.40	NIOBIUM
3039.36	INDIUM	3170.35	MOLYBDENUM
3040.90	OSMIUM	3175.05	TIN
3042.64	PLATINUM	3183.41	VANADIUM
3044.00	COBALT	3183.98	VANADIUM
3047.61	IRON	3185.40	VANADIUM
3050.82	NICKEL	3186.45	TITANIUM

Wavelength Locator for Atomic Absorption Lines

Wavelength (Å)	Element	Wavelength (Å)	Element
3191.57	TUNGSTEN	3377.48	TITANIUM
3191.99	TITANIUM	3382.89	SILVER
3193.97	MOLYBDENUM	3391.05	NICKEL
3199.92	TITANIUM	3392.99	NICKEL
3208.83	MOLYBDENUM	3396.85	RHODIUM
3216.95	MANGANESE	3405.12	COBALT
3220.78	IRIDIUM	3408.13	PLATINUM
3229.75	THALLIUM	3409.18	COBALT
3232.06	OSMIUM	3412.34	COBALT
3232.63	LITHIUM	3412.63	COBALT
3232.96	NICKEL	3413.94	NICKEL
3242.70	PALLADIUM	3414.76	NICKEL
3247.54	COPPER	3431.58	COBALT
3255.69	SCANDIUM	3433.56	NICKEL
3256.09	INDIUM	3434.89	RHODIUM
3258.56	INDIUM	3437.28	NICKEL
3258.60	OSMIUM	3440.61	IRON
3262.29	OSMIUM	3440.99	IRON
3262.75	OSMIUM	3442.93	COBALT
3267.94	OSMIUM	3445.23	COBALT
3269.91	SCANDIUM	3446.26	NICKEL
3273.63	SCANDIUM	3451.88	RHENIUM
3273.96	COPPER	3453.50	COBALT
3280.60	RHODIUM	3458.47	NICKEL
3280.68	SILVER	3460.46	RHENIUM
3301.56	OSMIUM	3461.65	NICKEL
3302.37	SODIUM	3462.04	RHODIUM
3302.98	SODIUM	3462.80	COBALT
3315.05	PLATINUM	3464.73	RHENIUM
3332.73	HAFNIUM	3465.80	COBALT
3341.88	TITANIUM	3465.86	IRON
3349.10	NIOBIUM	3472.40	HAFNIUM
3354.64	TITANIUM	3474.02	COBALT
3369.57	NICKEL	3478.91	RHODIUM
3371.45	TITANIUM	3497.49	HAFNIUM

Wavelength Locator for Atomic Absorption Lines

Wavelength (Å)	Element	Wavelength (Å)	Element
3498.94	RUTHENIUM	3613.08	LANTHANUM
3501.11	BARIUM	3617.52	TUNGSTEN
3502.28	COBALT	3618.77	IRON
3502.52	RHODIUM	3623.86	ZIRCONIUM
3502.62	COBALT	3624.73	NICKEL
3506.32	COBALT	3631.46	IRON
3507.32	RHODIUM	3634.70	PALLADIUM
3509.32	ZIRCONIUM	3634.93	RUTHENIUM
3512.64	COBALT	3635.46	TITANIUM
3513.48	COBALT	3641.53	LANTHANUM
3513.64	IRIDIUM	3642.68	TITANIUM
3515.05	NICKEL	3649.53	LANTHANUM
3519.60	ZIRCONIUM	3653.50	TITANIUM
3524.54	NICKEL	3657.99	RHODIUM
3526.85	COBALT	3663.65	ZIRCONIUM
3529.43	THALLIUM	3679.92	IRON
3529.81	COBALT	3682.24	HAFNIUM
3536.62	HAFNIUM	3690.70	RHODIUM
3547.68	ZIRCONIUM	3692.36	RHODIUM
3550.46	ZIRCONIUM	3700.91	RHODIUM
3564.95	COBALT	3703.58	VANADIUM
3566.10	ZIRCONIUM	3704.54	LANTHANUM
3570.10	IRON	3704.70	VANADIUM
3575.36	COBALT	3719.94	IRON
3575.79	ZIRCONIUM	3726.24	NIOBIUM
3578.69	CHROMIUM	3727.23	NIOBIUM
3580.27	NIOBIUM	3728.03	RUTHENIUM
3581.20	IRON	3729.82	TITANIUM
3586.29	ZIRCONIUM	3734.87	IRON
3593.49	CHROMIUM	3737.13	IRON
3596.18	RUTHENIUM	3739.80	NIOBIUM
3601.19	ZIRCONIUM	3741.06	TITANIUM
3605.33	CHROMIUM	3741.78	NIOBIUM
3608.86	IRON	3742.39	NIOBIUM
3609.55	PALLADIUM	3745.56	IRON

Wavelength Locator for Atomic Absorption Lines

Wavelength (Å)	Element	Wavelength (Å)	Element
3745.90	IRON	3902.96	MOLYBDENUM
3752.86	TITANIUM	3907.49	SCANDIUM
3758.24	IRON	3909.93	COBALT
3759.55	NIOBIUM	3911.81	SCANDIUM
3760.64	NIOBIUM	3920.26	IRON
3768.45	TUNGSTEN	3925.92	RUTHENIUM
3775.72	THALLIUM	3927.56	LANTHANUM
3777.64	HAFNIUM	3929.53	ZIRCONIUM
3787.06	NIOBIUM	3933.38	SCANDIUM
3787.48	NIOBIUM	3933.66	CALCIUM
3790.15	NIOBIUM	3944.01	ALUMINUM
3791.21	NIOBIUM	3947.78	TITANIUM
3798.12	NIOBIUM	3948.67	TITANIUM
3798.25	MOLYBDENUM	3956.34	TITANIUM
3800.12	IRIDIUM	3958.21	TITANIUM
3802.92	NIOBIUM	3961.52	ALUMINUM
3806.76	RHODIUM	3968.26	ZIRCONIUM
3824.45	IRON	3968.47	CALCIUM
3825.88	IRON	3981.76	TITANIUM
3827.83	IRON	3982.48	TITANIUM
3828.56	VANADIUM	3989.76	TITANIUM
3835.96	ZIRCONIUM	3996.61	SCANDIUM
3840.75	VANADIUM	3998.64	TITANIUM
3855.37	VANADIUM	4008.75	TUNGSTEN
3855.84	VANADIUM	4020.40	SCANDIUM
3859.91	IRON	4023.69	SCANDIUM
3863.87	ZIRCONIUM	4030.76	MANGANESE
3864.11	MOLYBDENUM	4032.98	GALLIUM
3875.90	VANADIUM	4037.21	LANTHANUM
3885.42	ZIRCONIUM	4044.15	POTASSIUM
3886.28	IRON	4047.21	POTASSIUM
3890.32	ZIRCONIUM	4054.55	SCANDIUM
3891.38	ZIRCONIUM	4057.83	LEAD
3898.60	LANTHANUM	4058.94	NIOBIUM
3902.25	VANADIUM	4074.36	TUNGSTEN

Wavelength Locator for Atomic Absorption Lines

Wavelength (Å)	Element	Wavelength (Å)	Element
4077.38	YTTRIUM	4379.24	VANADIUM
4077.71	STRONTIUM	4384.72	VANADIUM
4079.18	LANTHANUM	4389.97	VANADIUM
4079.73	NIOBIUM	4420.47	OSMIUM
4082.40	SCANDIUM	4511.31	INDIUM
4086.72	LANTHANUM	4555.36	CESIUM
4100.92	NIOBIUM	4593.18	CESIUM
4101.76	INDIUM	4606.77	NIOBIUM
4102.38	YTTRIUM	4607.33	STRONTIUM
4110.40	NIOBIUM	4633.98	ZIRCONIUM
4111.78	VANADIUM	4662.51	LANTHANUM
4115.18	VANADIUM	4687.80	ZIRCONIUM
4123.81	NIOBIUM	4766.89	LANTHANUM
4128.31	YTTRIUM	4949.77	LANTHANUM
4137.10	NIOBIUM	5158.69	LANTHANUM
4137.59	NIOBIUM	5204.52	CHROMIUM
4139.44	NIOBIUM	5208.44	CHROMIUM
4142.85	YTTRIUM	5350.46	THALLIUM
4152.58	NIOBIUM	5402.51	TANTALUM
4163.66	NIOBIUM	5455.15	LANTHANUM
4164.66	NIOBIUM	5501.34	LANTHANUM
4168.13	NIOBIUM	5535.48	BARIUM
4172.06	GALLIUM	5889.95	SODIUM
4187.32	LANTHANUM	5895.92	SODIUM
4201.85	RUBIDIUM	6103.64	LITHIUM
4215.52	STRONTIUM	6278.18	GOLD
4215.56	RUBIDIUM	6707.80	LITHIUM
4226.73	CALCIUM	7664.91	POTASSIUM
4254.35	CHROMIUM	7698.98	POTASSIUM
4260.85	OSMIUM	7800.23	RUBIDIUM
4269.39	TUNGSTEN	7947.60	RUBIDIUM
4274.80	CHROMIUM	8521.10	CESIUM
4289.72	CHROMIUM	8943.50	CESIUM
4294.61	TUNGSTEN		
4302.11	TUNGSTEN		

Table III.B.1-c

Wavelength Locator for Atomic Emission Lines

Wavelength (Å)	Element	Wavelength (Å)	Element
1936.96	ARSENIC	2288.12	ARSENIC
1960.26	SELENIUM	2311.47	ANTIMONY
1971.97	ARSENIC	2329.96	NICKEL
1994.20	TELLERIUM	2346.63	NICKEL
2002.00	TELLERIUM	2348.61	BERYLLIUM
2012.00	GOLD	2349.84	ARSENIC
2021.38	GOLD	2352.65	GOLD
2033.57	IRIDIUM	2354.84	TIN
2039.85	SELENIUM	2357.10	PLATINUM
2062.79	SELENIUM	2364.71	CHROMIUM
2068.33	ANTIMONY	2372.07	ALUMINUM
2092.63	IRIDIUM	2383.25	TELLERIUM
2127.94	IRIDIUM	2385.76	TELLERIUM
2138.56	ZINC	2387.75	GOLD
2142.75	TELLERIUM	2418.69	GALLIUM
2151.43	TIN	2424.56	OSMIUM
2155.81	IRIDIUM	2424.97	OSMIUM
2158.05	IRIDIUM	2427.95	GOLD
2169.99	LEAD	2429.49	TIN
2175.24	IRIDIUM	2447.91	PALLADIUM
2175.81	ANTIMONY	2455.61	IRIDIUM
2194.49	TIN	2476.84	OSMIUM
2200.73	CALCIUM	2481.44	TUNGSTEN
2202.22	PLATINUM	2484.19	IRON
2207.98	SILICON	2490.12	PLATINUM
2211.74	SILICON	2492.91	ARSENIC
2218.06	SILICON	2496.77	BORON
2228.25	BISMUTH	2497.72	BORON
2230.61	BISMUTH	2506.90	VANADIUM
2263.46	ALUMINUM	2507.45	TANTALUM
2269.10	ALUMINUM	2508.99	RHENIUM
2269.22	ALUMINUM	2516.11	SILICON
2274.38	PLATINUM	2526.35	TANTALUM
2275.46	CALCIUM	2527.44	IRON
2276.58	BISMUTH	2528.52	ANTIMONY

Wavelength Locator for Atomic Emission Lines

Wavelength (Å)	Element	Wavelength (Å)	Element
2533.23	GERMANIUM	2711.35	SCANDIUM
2536.52	MERCURY	2713.94	INDIUM
2550.37	TUNGSTEN	2720.04	OSMIUM
2589.19	GERMANIUM	2745.00	ARSENIC
2590.76	OSMIUM	2748.26	GOLD
2592.54	GERMANIUM	2754.59	GERMANIUM
2598.05	ANTIMONY	2760.10	YTTRIUM
2601.76	INDIUM	2767.87	THALLIUM
2612.63	OSMIUM	2770.86	ZINC
2613.07	TUNGSTEN	2771.67	PLATINUM
2613.65	LEAD	2779.37	HAFNIUM
2613.82	TUNGSTEN	2780.22	ARSENIC
2614.18	LEAD	2794.82	MANGANESE
2621.82	OSMIUM	2798.27	MANGANESE
2636.90	TANTALUM	2800.87	ZINC
2641.49	GOLD	2801.06	MANGANESE
2651.18	GERMANIUM	2814.90	ZIRCONIUM
2651.58	GERMANIUM	2833.06	LEAD
2652.48	ALUMINUM	2837.23	ZIRCONIUM
2653.27	TANTALUM	2839.99	TIN
2656.61	TANTALUM	2850.76	OSMIUM
2659.45	PLATINUM	2850.98	TANTALUM
2659.87	GALLIUM	2852.13	MAGNESIUM
2660.39	ALUMINUM	2860.44	ARSENIC
2662.84	TUNGSTEN	2863.33	TIN
2674.34	RHENIUM	2874.98	RUTHENIUM
2675.95	GOLD	2875.98	ZIRCONIUM
2691.31	TANTALUM	2879.11	TUNGSTEN
2691.34	GERMANIUM	2882.64	IRIDIUM
2694.23	IRIDIUM	2889.62	HAFNIUM
2698.30	TANTALUM	2891.84	TANTALUM
2706.51	TIN	2909.06	OSMIUM
2706.77	SCANDIUM	2910.48	TUNGSTEN
2707.95	SCANDIUM	2912.33	OSMIUM
2709.63	GERMANIUM	2916.48	HAFNIUM

Wavelength Locator for Atomic Emission Lines

Wavelength (Å)	Element	Wavelength (Å)	Element
2918.58	HAFNIUM	3058.66	OSMIUM
2923.62	VANADIUM	3060.46	VANADIUM
2933.55	TANTALUM	3064.71	PLATINUM
2940.22	TANTALUM	3066.38	VANADIUM
2944.18	GALLIUM	3067.41	HAFNIUM
2954.20	HAFNIUM	3067.72	BISMUTH
2963.32	TANTALUM	3071.58	BARIUM
2965.54	TANTALUM	3075.90	ZINC
2974.01	SCANDIUM	3082.15	ALUMINUM
2980.75	SCANDIUM	3092.71	ALUMINUM
2980.81	HAFNIUM	3092.84	ALUMINUM
2984.26	YTTRIUM	3103.24	TANTALUM
2988.95	SCANDIUM	3124.97	TANTALUM
2988.95	RUTHENIUM	3132.59	MOLYBDENUM
3002.49	NICKEL	3158.16	MOLYBDENUM
3009.14	TIN	3170.35	MOLYBDENUM
3015.36	SCANDIUM	3175.05	TIN
3016.78	HAFNIUM	3183.41	VANADIUM
3018.04	OSMIUM	3183.98	VANADIUM
3018.31	HAFNIUM	3185.40	VANADIUM
3019.34	SCANDIUM	3186.45	TITANIUM
3020.49	IRON	3191.99	TITANIUM
3020.64	IRON	3193.97	MOLYBDENUM
3021.07	IRON	3198.01	VANADIUM
3030.76	SCANDIUM	3199.92	TITANIUM
3034.12	TIN	3202.38	VANADIUM
3039.36	INDIUM	3207.41	VANADIUM
3040.90	OSMIUM	3208.83	MOLYBDENUM
3043.12	VANADIUM	3212.43	VANADIUM
3043.56	VANADIUM	3220.78	IRIDIUM
3044.94	VANADIUM	3229.75	THALLIUM
3050.89	VANADIUM	3232.63	LITHIUM
3053.65	VANADIUM	3242.70	PALLADIUM
3056.33	VANADIUM	3247.54	COPPER
3057.02	HAFNIUM	3255.69	SCANDIUM

Wavelength Locator for Atomic Emission Lines

Wavelength (Å)	Element	Wavelength (Å)	Element
3256.09	INDIUM	3414.76	NICKEL
3262.34	TIN	3421.24	PALLADIUM
3263.24	VANADIUM	3428.31	RUTHENIUM
3269.91	SCANDIUM	3433.04	COBALT
3271.64	VANADIUM	3433.56	NICKEL
3273.63	SCANDIUM	3434.89	RHODIUM
3273.96	COPPER	3436.74	RUTHENIUM
3280.68	SILVER	3440.61	IRON
3283.31	VANADIUM	3443.64	COBALT
3298.14	VANADIUM	3446.26	NICKEL
3301.56	OSMIUM	3449.44	COBALT
3302.37	SODIUM	3451.88	RHENIUM
3302.98	SODIUM	3453.50	COBALT
3323.09	RHODIUM	3458.47	NICKEL
3330.62	TIN	3460.46	RHENIUM
3332.73	HAFNIUM	3460.77	PALLADIUM
3341.88	TITANIUM	3461.65	NICKEL
3345.02	ZINC	3462.04	RHODIUM
3354.64	TITANIUM	3464.73	RHENIUM
3369.57	NICKEL	3465.80	COBALT
3370.44	TITANIUM	3472.40	HAFNIUM
3371.45	TITANIUM	3472.54	NICKEL
3377.48	TITANIUM	3474.02	COBALT
3380.57	NICKEL	3478.91	RHODIUM
3382.89	SILVER	3481.15	PALLADIUM
3385.95	TITANIUM	3492.96	NICKEL
3392.99	NICKEL	3497.49	HAFNIUM
3396.85	RHODIUM	3498.94	RUTHENIUM
3404.58	PALLADIUM	3501.11	BARIUM
3405.12	COBALT	3502.28	COBALT
3408.13	PLATINUM	3502.52	RHODIUM
3409.18	COBALT	3506.32	COBALT
3412.34	COBALT	3510.34	NICKEL
3412.63	COBALT	3512.64	COBALT
3413.94	NICKEL	3513.48	COBALT

Wavelength Locator for Atomic Emission Lines

Wavelength (Å)	Element	Wavelength (Å)	Element
3513.64	IRIDIUM	3617.52	TUNGSTEN
3515.05	NICKEL	3619.39	NICKEL
3516.94	PALLADIUM	3620.94	YTTRIUM
3519.24	THALLIUM	3634.70	PALLADIUM
3519.60	ZIRCONIUM	3634.93	RUTHENIUM
3524.54	NICKEL	3635.46	TITANIUM
3526.85	COBALT	3639.58	LEAD
3529.43	THALLIUM	3641.53	LANTHANUM
3535.30	NIOBIUM	3642.68	TITANIUM
3536.62	HAFNIUM	3653.50	TITANIUM
3550.46	ZIRCONIUM	3657.99	RHODIUM
3552.69	YTTRIUM	3658.10	TITANIUM
3553.08	PALLADIUM	3661.35	RUTHENIUM
3564.95	COBALT	3663.65	ZIRCONIUM
3566.10	ZIRCONIUM	3668.97	TITANIUM
3566.37	NICKEL	3671.67	TITANIUM
3569.38	COBALT	3675.70	VANADIUM
3574.43	LANTHANUM	3682.24	HAFNIUM
3575.36	COBALT	3683.48	LEAD
3575.79	ZIRCONIUM	3690.70	RHODIUM
3578.69	CHROMIUM	3692.36	RHODIUM
3580.27	NIOBIUM	3697.85	NIOBIUM
3581.20	IRON	3700.91	RHODIUM
3583.10	RHODIUM	3703.58	VANADIUM
3586.29	ZIRCONIUM	3704.54	LANTHANUM
3592.92	YTTRIUM	3704.70	VANADIUM
3593.02	RUTHENIUM	3705.04	VANADIUM
3593.49	CHROMIUM	3710.30	YTTRIUM
3594.87	COBALT	3713.01	NIOBIUM
3596.18	RUTHENIUM	3719.94	IRON
3596.19	RHODIUM	3726.24	NIOBIUM
3601.19	ZIRCONIUM	3728.03	RUTHENIUM
3605.33	CHROMIUM	3729.82	TITANIUM
3609.55	PALLADIUM	3737.13	IRON
3610.46	NICKEL	3739.80	NIOBIUM

Wavelength Locator for Atomic Emission Lines

Wavelength (Å)	Element	Wavelength (Å)	Element
3741.06	TITANIUM	3823.21	VANADIUM
3742.39	NIOBIUM	3825.88	IRON
3745.90	IRON	3828.56	VANADIUM
3752.86	TITANIUM	3835.96	ZIRCONIUM
3753.64	TITANIUM	3840.75	VANADIUM
3768.45	TUNGSTEN	3844.44	VANADIUM
3771.66	TITANIUM	3845.47	COBALT
3774.33	YTTRIUM	3847.33	VANADIUM
3775.72	THALLIUM	3855.37	VANADIUM
3778.68	VANADIUM	3855.84	VANADIUM
3787.06	NIOBIUM	3856.52	RHODIUM
3790.15	NIOBIUM	3858.30	NICKEL
3790.32	VANADIUM	3859.91	IRON
3791.21	NIOBIUM	3864.11	MOLYBDENUM
3793.61	VANADIUM	3864.86	VANADIUM
3794.96	VANADIUM	3867.60	VANADIUM
3798.12	NIOBIUM	3867.98	TUNGSTEN
3798.25	MOLYBDENUM	3873.12	COBALT
3799.35	RUTHENIUM	3875.08	VANADIUM
3799.91	VANADIUM	3875.90	VANADIUM
3800.12	IRIDIUM	3876.09	VANADIUM
3801.02	TIN	3878.58	IRON
3802.92	NIOBIUM	3885.42	ZIRCONIUM
3803.47	VANADIUM	3886.28	IRON
3806.76	RHODIUM	3890.18	VANADIUM
3807.50	VANADIUM	3891.38	ZIRCONIUM
3808.52	VANADIUM	3892.86	VANADIUM
3809.60	VANADIUM	3894.08	COBALT
3813.49	VANADIUM	3899.71	IRON
3817.84	VANADIUM	3902.25	VANADIUM
3818.24	VANADIUM	3902.96	MOLYBDENUM
3819.96	VANADIUM	3907.49	SCANDIUM
3821.49	VANADIUM	3909.89	VANADIUM
3822.01	VANADIUM	3911.81	SCANDIUM
3822.89	VANADIUM	3924.53	TITANIUM

Wavelength Locator for Atomic Emission Lines

Wavelength (Å)	Element	Wavelength (Å)	Element
3927.56	LANTHANUM	4045.60	TUNGSTEN
3929.53	ZIRCONIUM	4047.21	POTASSIUM
3929.88	TITANIUM	4047.63	YTTRIUM
3930.30	IRON	4047.79	SCANDIUM
3933.38	SCANDIUM	4054.55	SCANDIUM
3933.66	CALCIUM	4057.83	LEAD
3944.01	ALUMINUM	4058.94	NIOBIUM
3947.78	TITANIUM	4060.32	LANTHANUM
3948.67	TITANIUM	4064.79	LANTHANUM
3956.34	TITANIUM	4074.36	TUNGSTEN
3958.21	TITANIUM	4077.38	YTTRIUM
3961.52	ALUMINUM	4077.71	STRONTIUM
3962.85	TITANIUM	4079.18	LANTHANUM
3964.27	TITANIUM	4079.73	NIOBIUM
3968.26	ZIRCONIUM	4082.40	SCANDIUM
3968.47	CALCIUM	4083.71	YTTRIUM
3981.76	TITANIUM	4089.61	LANTHANUM
3982.48	TITANIUM	4090.58	VANADIUM
3989.76	TITANIUM	4092.69	VANADIUM
3995.31	COBALT	4095.49	VANADIUM
3996.61	SCANDIUM	4099.80	VANADIUM
3998.64	TITANIUM	4100.92	NIOBIUM
4008.75	TUNGSTEN	4101.76	INDIUM
4008.93	TITANIUM	4102.16	VANADIUM
4015.39	LANTHANUM	4102.38	YTTRIUM
4020.40	SCANDIUM	4104.87	LANTHANUM
4023.69	SCANDIUM	4105.17	VANADIUM
4024.57	TITANIUM	4109.79	VANADIUM
4030.76	MANGANESE	4110.40	NIOBIUM
4032.98	GALLIUM	4111.78	VANADIUM
4033.07	MANGANESE	4115.18	VANADIUM
4034.49	MANGANESE	4116.47	VANADIUM
4037.21	LANTHANUM	4121.32	COBALT
4039.83	YTTRIUM	4123.57	VANADIUM
4044.15	POTASSIUM	4123.81	NIOBIUM

Wavelength Locator for Atomic Emission Lines

Wavelength (Å)	Element	Wavelength (Å)	Element
4128.07	VANADIUM	4300.56	TITANIUM
4128.31	YTTRIUM	4301.09	TITANIUM
4132.02	VANADIUM	4302.11	TUNGSTEN
4134.49	VANADIUM	4305.92	TITANIUM
4137.04	LANTHANUM	4306.21	VANADIUM
4137.10	NIOBIUM	4307.18	VANADIUM
4139.44	NIOBIUM	4309.80	VANADIUM
4139.71	NIOBIUM	4330.02	VANADIUM
4142.85	YTTRIUM	4332.82	VANADIUM
4152.58	NIOBIUM	4341.01	VANADIUM
4160.26	LANTHANUM	4355.94	VANADIUM
4163.66	NIOBIUM	4368.04	VANADIUM
4164.66	NIOBIUM	4374.80	RHODIUM
4167.52	YTTRIUM	4379.24	VANADIUM
4168.13	NIOBIUM	4383.55	IRON
4172.06	GALLIUM	4384.72	VANADIUM
4174.13	YTTRIUM	4389.97	VANADIUM
4179.42	VANADIUM	4395.23	VANADIUM
4182.59	VANADIUM	4400.58	VANADIUM
4187.32	LANTHANUM	4406.64	VANADIUM
4189.84	VANADIUM	4407.64	VANADIUM
4201.85	RUBIDIUM	4408.20	VANADIUM
4209.86	VANADIUM	4408.51	VANADIUM
4211.14	RHODIUM	4416.47	VANADIUM
4215.52	STRONTIUM	4419.94	VANADIUM
4215.56	RUBIDIUM	4420.47	OSMIUM
4226.73	CALCIUM	4421.57	VANADIUM
4235.94	YTTRIUM	4426.00	VANADIUM
4252.31	COBALT	4428.52	VANADIUM
4254.35	CHROMIUM	4429.80	VANADIUM
4269.39	TUNGSTEN	4436.14	VANADIUM
4274.80	CHROMIUM	4437.84	VANADIUM
4280.27	LANTHANUM	4441.68	VANADIUM
4289.72	CHROMIUM	4444.21	VANADIUM
4294.61	TUNGSTEN	4457.48	VANADIUM

Wavelength Locator for Atomic Emission Lines

Wavelength (Å)	Element	Wavelength (Å)	Element
4459.76	VANADIUM	4851.48	VANADIUM
4460.29	VANADIUM	4864.74	VANADIUM
4511.31	INDIUM	4875.48	VANADIUM
4527.25	YTTRIUM	4881.56	VANADIUM
4533.24	TITANIUM	4889.14	RHENIUM
4549.50	LANTHANUM	4934.09	BARIUM
4554.03	BARIUM	4949.77	LANTHANUM
4555.36	CESIUM	5039.95	TITANIUM
4567.91	LANTHANUM	5064.66	TITANIUM
4570.02	LANTHANUM	5081.56	SCANDIUM
4577.17	VANADIUM	5083.72	SCANDIUM
4580.40	VANADIUM	5105.54	COPPER
4586.36	VANADIUM	5145.42	LANTHANUM
4593.18	CESIUM	5158.69	LANTHANUM
4594.11	VANADIUM	5172.68	MAGNESIUM
4602.86	LITHIUM	5173.75	TITANIUM
4606.77	NIOBIUM	5177.31	LANTHANUM
4607.33	STRONTIUM	5183.60	MAGNESIUM
4633.98	ZIRCONIUM	5192.98	TITANIUM
4643.70	YTTRIUM	5206.04	CHROMIUM
4674.85	YTTRIUM	5210.39	TITANIUM
4722.19	BISMUTH	5211.86	LANTHANUM
4740.16	TANTALUM	5234.27	LANTHANUM
4741.02	SCANDIUM	5253.46	LANTHANUM
4743.81	SCANDIUM	5269.54	IRON
4753.16	SCANDIUM	5271.19	LANTHANUM
4760.98	YTTRIUM	5275.56	RHENIUM
4766.89	LANTHANUM	5328.04	IRON
4779.35	SCANDIUM	5342.96	SCANDIUM
4810.53	ZINC	5349.71	SCANDIUM
4812.75	TANTALUM	5350.46	THALLIUM
4827.45	VANADIUM	5402.51	TANTALUM
4831.64	VANADIUM	5455.15	LANTHANUM
4832.43	VANADIUM	5466.46	YTTRIUM
4850.82	LANTHANUM	5501.34	LANTHANUM

Wavelength Locator for Atomic Emission Lines

Wavelength (Å)	Element	Wavelength (Å)	Element
5527.54	YTTRIUM	6243.10	VANADIUM
5535.48	BARIUM	6249.93	LANTHANUM
5627.64	VANADIUM	6251.82	VANADIUM
5670.85	VANADIUM	6259.96	SCANDIUM
5671.81	SCANDIUM	6274.65	VANADIUM
5686.84	SCANDIUM	6276.31	SCANDIUM
5699.05	RUTHENIUM	6285.16	VANADIUM
5700.21	SCANDIUM	6292.83	VANADIUM
5727.03	VANADIUM	6296.49	VANADIUM
5740.66	LANTHANUM	6305.67	SCANDIUM
5761.84	LANTHANUM	6325.91	LANTHANUM
5769.34	LANTHANUM	6378.82	SCANDIUM
5776.83	RHENIUM	6394.23	LANTHANUM
5789.24	LANTHANUM	6402.01	YTTRIUM
5791.34	LANTHANUM	6410.99	LANTHANUM
5889.95	SODIUM	6413.35	SCANDIUM
5895.92	SODIUM	6435.00	YTTRIUM
5930.63	LANTHANUM	6454.52	LANTHANUM
6039.73	VANADIUM	6455.99	LANTHANUM
6081.44	VANADIUM	6543.16	LANTHANUM
6090.22	VANADIUM	6557.39	YTTRIUM
6103.64	LITHIUM	6578.51	LANTHANUM
6111.67	VANADIUM	6616.59	LANTHANUM
6119.52	VANADIUM	6644.41	LANTHANUM
6199.19	VANADIUM	6650.81	LANTHANUM
6210.68	SCANDIUM	6661.40	LANTHANUM
6213.87	VANADIUM	6687.58	YTTRIUM
6216.37	VANADIUM	6707.80	LITHIUM
6222.59	YTTRIUM	6793.71	YTTRIUM
6224.50	VANADIUM	7664.91	POTASSIUM
6230.74	VANADIUM	7698.98	POTASSIUM
6233.20	VANADIUM	7800.23	RUBIDIUM
6239.41	SCANDIUM	7947.60	RUBIDIUM
6239.78	SCANDIUM	8194.82	SODIUM
6242.81	VANADIUM	8521.10	CESIUM
		8943.50	CESIUM

Table III.B.1-d

Wavelength Locator for Atomic Fluorescence Lines

Wavelength (Å)	Element	Wavelength (Å)	Element
1936.96	ARSENIC	2230.61	BISMUTH
1960.26	SELENIUM	2246.05	TIN
1971.97	ARSENIC	2259.04	TELLERIUM
1994.20	TELLERIUM	2263.46	ALUMINUM
2002.00	TELLERIUM	2269.10	ALUMINUM
2012.00	GOLD	2269.22	ALUMINUM
2021.38	GOLD	2274.38	PLATINUM
2033.57	IRIDIUM	2275.46	CALCIUM
2039.85	SELENIUM	2276.58	BISMUTH
2061.70	BISMUTH	2288.02	CADMIUM
2062.79	SELENIUM	2288.12	ARSENIC
2068.33	ANTIMONY	2289.98	NICKEL
2074.79	SELENIUM	2310.96	NICKEL
2092.63	IRIDIUM	2311.47	ANTIMONY
2127.94	IRIDIUM	2312.34	NICKEL
2138.56	ZINC	2313.98	NICKEL
2142.75	TELLERIUM	2320.03	NICKEL
2151.43	TIN	2325.79	NICKEL
2155.81	IRIDIUM	2329.96	NICKEL
2158.05	IRIDIUM	2345.54	NICKEL
2165.09	COPPER	2346.63	NICKEL
2169.99	LEAD	2348.61	BERYLLIUM
2174.67	PLATINUM	2349.84	ARSENIC
2175.24	IRIDIUM	2352.65	GOLD
2175.81	ANTIMONY	2354.84	TIN
2178.94	COPPER	2357.10	PLATINUM
2181.72	COPPER	2364.71	CHROMIUM
2194.49	TIN	2367.05	ALUMINUM
2200.73	CALCIUM	2372.07	ALUMINUM
2202.22	PLATINUM	2373.12	ALUMINUM
2207.98	SILICON	2373.35	ALUMINUM
2211.74	SILICON	2373.35	ALUMINUM
2218.06	SILICON	2381.18	ARSENIC
2225.70	COPPER	2383.25	TELLERIUM
2228.25	BISMUTH	2385.76	TELLERIUM

Wavelength Locator for Atomic Fluorescence Lines

Wavelength (Å)	Element	Wavelength (Å)	Element
2387.75	GOLD	2521.36	COBALT
2398.56	CALCIUM	2522.85	IRON
2407.25	COBALT	2524.11	SILICON
2411.62	COBALT	2526.35	TANTALUM
2414.46	COBALT	2527.44	IRON
2415.30	COBALT	2528.51	SILICON
2418.69	GALLIUM	2528.97	COBALT
2424.56	OSMIUM	2530.70	TELLERIUM
2424.93	COBALT	2533.23	GERMANIUM
2424.97	OSMIUM	2536.52	MERCURY
2427.95	GOLD	2543.97	IRIDIUM
2429.49	TIN	2546.55	TIN
2432.21	COBALT	2550.37	TUNGSTEN
2437.23	ARSENIC	2567.98	ALUMINUM
2440.06	PLATINUM	2575.10	ALUMINUM
2447.91	PALLADIUM	2575.40	ALUMINUM
2455.61	IRIDIUM	2576.10	MANGANESE
2456.53	ARSENIC	2589.19	GERMANIUM
2476.42	PALLADIUM	2590.76	OSMIUM
2476.84	OSMIUM	2592.54	GERMANIUM
2481.44	TUNGSTEN	2593.73	MANGANESE
2483.27	IRON	2598.05	ANTIMONY
2484.19	IRON	2601.76	INDIUM
2487.17	PLATINUM	2605.69	MANGANESE
2490.12	PLATINUM	2612.63	OSMIUM
2492.15	COPPER	2613.07	TUNGSTEN
2492.91	ARSENIC	2613.65	LEAD
2497.96	GERMANIUM	2613.82	TUNGSTEN
2506.90	SILICON	2614.18	LEAD
2506.90	VANADIUM	2621.82	OSMIUM
2507.45	TANTALUM	2627.91	BISMUTH
2508.99	RHENIUM	2628.03	PLATINUM
2514.32	SILICON	2636.90	TANTALUM
2516.11	SILICON	2641.49	GOLD
2519.20	SILICON	2650.86	PLATINUM

Wavelength Locator for Atomic Fluorescence Lines

Wavelength (Å)	Element	Wavelength (Å)	Element
2651.18	GERMANIUM	2779.37	HAFNIUM
2651.58	GERMANIUM	2780.22	ARSENIC
2652.48	ALUMINUM	2780.52	BISMUTH
2653.27	TANTALUM	2794.82	MANGANESE
2656.61	TANTALUM	2798.27	MANGANESE
2659.45	PLATINUM	2801.06	MANGANESE
2659.87	GALLIUM	2801.99	LEAD
2660.39	ALUMINUM	2814.90	ZIRCONIUM
2662.84	TUNGSTEN	2823.28	LEAD
2670.64	ANTIMONY	2830.30	PLATINUM
2674.34	RHENIUM	2833.06	LEAD
2675.95	GOLD	2837.23	ZIRCONIUM
2691.31	TANTALUM	2839.99	TIN
2691.34	GERMANIUM	2850.76	OSMIUM
2694.23	IRIDIUM	2850.98	TANTALUM
2696.76	BISMUTH	2852.13	MAGNESIUM
2698.30	TANTALUM	2860.44	ARSENIC
2702.40	PLATINUM	2863.33	TIN
2706.51	TIN	2873.32	LEAD
2706.77	SCANDIUM	2875.98	ZIRCONIUM
2707.95	SCANDIUM	2877.92	ANTIMONY
2709.63	GERMANIUM	2879.11	TUNGSTEN
2711.35	SCANDIUM	2882.64	IRIDIUM
2713.94	INDIUM	2889.62	HAFNIUM
2719.03	IRON	2891.84	TANTALUM
2719.04	PLATINUM	2893.86	PLATINUM
2720.04	OSMIUM	2897.98	BISMUTH
2734.50	PLATINUM	2898.71	ARSENIC
2745.00	ARSENIC	2910.48	TUNGSTEN
2748.26	GOLD	2912.33	OSMIUM
2754.59	GERMANIUM	2916.48	HAFNIUM
2760.10	YTTRIUM	2918.58	HAFNIUM
2767.87	THALLIUM	2929.79	PLATINUM
2769.95	ANTIMONY	2933.06	MANGANESE
2771.67	PLATINUM	2933.55	TANTALUM

Wavelength Locator for Atomic Fluorescence Lines

Wavelength (Å)	Element	Wavelength (Å)	Element
2938.30	BISMUTH	3050.82	NICKEL
2939.30	MANGANESE	3057.02	HAFNIUM
2940.22	TANTALUM	3064.71	PLATINUM
2944.18	GALLIUM	3066.38	VANADIUM
2948.24	TITANIUM	3067.41	HAFNIUM
2949.20	MANGANESE	3067.72	BISMUTH
2954.20	HAFNIUM	3071.58	BARIUM
2956.80	TITANIUM	3082.15	ALUMINUM
2963.32	TANTALUM	3092.71	ALUMINUM
2965.54	TANTALUM	3092.84	ALUMINUM
2974.01	SCANDIUM	3101.55	NICKEL
2980.75	SCANDIUM	3101.88	NICKEL
2980.81	HAFNIUM	3103.24	TANTALUM
2988.95	SCANDIUM	3122.78	GOLD
2988.95	RUTHENIUM	3124.97	TANTALUM
2989.03	BISMUTH	3132.59	MOLYBDENUM
2993.34	BISMUTH	3134.11	NICKEL
2997.97	PLATINUM	3158.16	MOLYBDENUM
3002.49	NICKEL	3170.35	MOLYBDENUM
3003.63	NICKEL	3175.05	TIN
3009.14	TIN	3183.98	VANADIUM
3015.36	SCANDIUM	3186.45	TITANIUM
3016.78	HAFNIUM	3191.99	TITANIUM
3018.31	HAFNIUM	3193.97	MOLYBDENUM
3020.49	IRON	3199.92	TITANIUM
3020.64	IRON	3220.78	IRIDIUM
3024.64	BISMUTH	3229.75	THALLIUM
3029.20	GOLD	3242.70	PALLADIUM
3032.85	ARSENIC	3247.54	COPPER
3034.12	TIN	3261.06	CADMIUM
3039.06	GERMANIUM	3262.34	TIN
3039.36	INDIUM	3269.49	GERMANIUM
3040.90	OSMIUM	3273.96	COPPER
3042.64	PLATINUM	3280.68	SILVER
3045.59	MANGANESE	3302.10	PALLADIUM

Wavelength Locator for Atomic Fluorescence Lines

Wavelength (Å)	Element	Wavelength (Å)	Element
3330.62	TIN	3553.08	PALLADIUM
3332.73	HAFNIUM	3564.95	COBALT
3341.88	TITANIUM	3566.10	ZIRCONIUM
3354.64	TITANIUM	3575.36	COBALT
3371.45	TITANIUM	3575.79	ZIRCONIUM
3373.00	PALLADIUM	3577.88	MANGANESE
3382.89	SILVER	3578.69	CHROMIUM
3396.85	RHODIUM	3586.29	ZIRCONIUM
3397.21	BISMUTH	3589.22	RUTHENIUM
3404.58	PALLADIUM	3593.02	RUTHENIUM
3408.13	PLATINUM	3593.49	CHROMIUM
3409.18	COBALT	3596.18	RUTHENIUM
3413.94	NICKEL	3597.15	RHODIUM
3414.76	NICKEL	3605.33	CHROMIUM
3421.24	PALLADIUM	3609.55	PALLADIUM
3428.31	RUTHENIUM	3610.46	NICKEL
3433.45	PALLADIUM	3634.70	PALLADIUM
3434.89	RHODIUM	3634.93	RUTHENIUM
3458.47	NICKEL	3635.46	TITANIUM
3460.77	PALLADIUM	3639.58	LEAD
3461.65	NICKEL	3641.53	LANTHANUM
3462.04	RHODIUM	3642.68	TITANIUM
3472.40	HAFNIUM	3646.20	TITANIUM
3478.91	RHODIUM	3653.50	TITANIUM
3481.15	PALLADIUM	3657.99	RHODIUM
3497.49	HAFNIUM	3661.35	RUTHENIUM
3498.94	RUTHENIUM	3663.65	ZIRCONIUM
3502.52	RHODIUM	3683.48	LEAD
3510.85	BISMUTH	3690.70	RHODIUM
3515.05	NICKEL	3692.36	RHODIUM
3516.94	PALLADIUM	3700.91	RHODIUM
3524.54	NICKEL	3703.58	VANADIUM
3529.43	THALLIUM	3704.54	LANTHANUM
3536.62	HAFNIUM	3719.94	IRON
3550.46	ZIRCONIUM	3723.68	IRON

Wavelength Locator for Atomic Fluorescence Lines

Wavelength (Å)	Element	Wavelength (Å)	Element
3726.93	RUTHENIUM	4034.49	MANGANESE
3728.03	RUTHENIUM	4047.21	POTASSIUM
3730.43	RUTHENIUM	4057.83	LEAD
3775.72	THALLIUM	4058.94	NIOBIUM
3777.64	HAFNIUM	4079.73	NIOBIUM
3790.51	RUTHENIUM	4099.80	VANADIUM
3798.05	RUTHENIUM	4100.92	NIOBIUM
3798.25	MOLYBDENUM	4101.76	INDIUM
3798.90	RUTHENIUM	4105.17	VANADIUM
3799.35	RUTHENIUM	4110.40	NIOBIUM
3801.02	TIN	4111.78	VANADIUM
3806.76	RHODIUM	4115.18	VANADIUM
3834.36	MANGANESE	4121.53	BISMUTH
3835.96	ZIRCONIUM	4123.81	NIOBIUM
3840.75	VANADIUM	4128.07	VANADIUM
3855.37	VANADIUM	4132.02	VANADIUM
3855.84	VANADIUM	4137.10	NIOBIUM
3864.11	MOLYBDENUM	4139.44	NIOBIUM
3875.90	VANADIUM	4139.71	NIOBIUM
3885.42	ZIRCONIUM	4152.58	NIOBIUM
3891.38	ZIRCONIUM	4163.66	NIOBIUM
3902.96	MOLYBDENUM	4168.13	NIOBIUM
3907.49	SCANDIUM	4172.06	GALLIUM
3911.81	SCANDIUM	4199.90	RUTHENIUM
3925.92	RUTHENIUM	4226.73	CALCIUM
3929.53	ZIRCONIUM	4254.35	CHROMIUM
3944.01	ALUMINUM	4260.85	OSMIUM
3961.52	ALUMINUM	4274.80	CHROMIUM
3968.26	ZIRCONIUM	4289.72	CHROMIUM
3998.64	TITANIUM	4379.24	VANADIUM
4020.40	SCANDIUM	4384.72	VANADIUM
4023.69	SCANDIUM	4389.97	VANADIUM
4030.76	MANGANESE	4395.23	VANADIUM
4032.98	GALLIUM	4407.64	VANADIUM
4033.07	MANGANESE	4408.20	VANADIUM

Wavelength Locator for Atomic Fluorescence Lines

Wavelength (Å)	Element	Wavelength (Å)	Element
4408.51	VANADIUM		
4511.31	INDIUM		
4554.51	RUTHENIUM		
4606.77	NIOBIUM		
4607.33	STRONTIUM		

Electronic Partition Functions, Z

    L. DeGalan, R. Smith, and J. D. Winefordner (1) have presented
the means to calculate the electronic partition functions for the
atoms and ions of 73 elements. The general equation is given below:

$$Z = \sum_{i=o}^{m} g_i \exp \left({}^{-E_i/kT} \right) \tag{1}$$

where g_i is the statistical weight for the ith level, and E_i is
the energy of the ith level, k is the Boltzmann constant, T is
the absolute temperature.

    The Z functions have been calculated for the common analytical
flame temperature region and the results are presented in
Table III.B.2.

Reference:

    1. L. DeGalan, R. Smith, and J. D. Winefordner, Spectrochim.
        Acta, 23 B, 521, (1968).

Table III.B.2

PARTITION FUNCTIONS OF ATOMS AND IONS FOR COMMON FLAME TEMPERATURES

Element	Temperature °K				
	2000	2300	2500	2800	3000
Ac	5.23	5.54	5.75	6.07	6.29
Ac II	1.25	1.43	1.57	1.81	1.99
Ag	2.00	2.00	2.00	2.00	2.00
Ag II	1.00	1.00	1.00	1.00	1.00
Al	5.69	5.73	5.75	5.78	5.79
Al II	1.00	1.00	1.00	1.00	1.00
As	4.00	4.01	4.02	4.04	4.06
As II	3.21	3.58	3.80	4.12	4.32
Au	2.01	2.02	2.03	2.05	2.07
Au II	1.00	1.00	1.00	1.00	1.01
B	5.95	5.96	5.96	5.97	5.97
B II	1.00	1.00	1.00	1.00	1.00
Ba	1.02	1.05	1.08	1.15	1.21
Ba II	2.22	2.36	2.47	2.65	2.78
Be	1.00	1.00	1.00	1.00	1.00
Be II	2.00	2.00	2.00	2.00	2.00
Bi	4.00	4.00	4.01	4.01	4.02
Bi II	1.00	1.00	1.00	1.00	1.01
Br	4.14	4.20	4.24	4.30	4.34
Br II	5.38	5.52	5.61	5.75	5.85
C	8.81	8.84	8.86	8.89	8.91
C II	5.82	5.84	5.86	5.87	5.88
Ca	1.00	1.00	1.00	1.00	1.00
Ca II	2.00	2.00	2.00	2.01	2.01
Cd	1.00	1.00	1.00	1.00	1.00
Cd II	2.00	2.00	2.00	2.00	2.00
Cl	5.06	5.15	5.20	5.25	5.31
Cl II	7.31	7.48	7.58	7.67	7.79
Co	19.30	20.91	21.95	22.97	24.45
Co II	16.09	17.56	18.54	19.50	20.93
Cr	7.10	7.20	7.30	7.43	7.65
Cr II	6.00	6.01	6.03	6.04	6.08
Cs	2.00	2.01	2.01	2.02	2.03
Cs II	1.00	1.00	1.00	1.00	1.00
Cu	2.00	2.01	2.01	2.02	2.03
Cu II	1.00	1.00	1.00	1.00	1.00

PARTITION FUNCTIONS OF ATOMS AND IONS FOR COMMON FLAME TEMPERATURES

Element	Temperature °K				
	2000	2300	2500	2800	3000
F	5.50	5.55	5.59	5.63	5.65
F II	8.05	8.16	8.22	8.29	8.34
Fe	19.46	20.23	20.72	21.46	21.94
Fe II	28.17	30.21	31.46	33.21	34.30
Ga	4.21	4.39	4.49	4.62	4.69
Ga II	1.00	1.00	1.00	1.00	1.00
Ge	4.85	5.25	5.48	5.81	6.00
Ge II	3.12	3.32	3.45	3.61	3.71
Hf	6.77	7.38	7.81	8.48	8.94
Hf II	5.25	5.79	6.19	6.82	7.27
Hg	1.00	1.00	1.00	1.00	1.00
Hg II	2.00	2.00	2.00	2.00	2.00
I	4.01	4.02	4.03	4.04	4.05
I II	5.03	5.05	5.08	5.12	5.15
In	2.81	3.00	3.12	3.28	3.38
In II	1.00	1.00	1.00	1.00	1.00
Ir	11.75	12.45	12.95	13.75	14.31
K	2.00	2.00	2.00	2.01	2.01
K II	1.00	1.00	1.00	1.00	1.00
La	9.41	10.76	11.70	13.14	14.12
La II	14.78	16.57	17.70	19.32	20.34
Li	2.00	2.00	2.00	2.00	2.00
Li II	1.00	1.00	1.00	1.00	1.00
Mg	1.00	1.00	1.00	1.00	1.00
Mg II	2.00	2.00	2.00	2.00	2.00
Mn	6.00	6.00	6.00	6.00	6.00
Mn II	7.01	7.02	7.03	7.05	7.08
Mo	7.01	7.02	7.04	7.08	7.12
Mo II	6.00	6.01	6.02	6.04	6.07
N	4.00	4.00	4.00	4.00	4.00
N II	8.45	8.52	8.55	8.60	8.63
Na	2.00	2.00	2.00	2.00	2.00
Na II	1.00	1.00	1.00	1.00	1.00
Nb	26.68	29.30	30.96	33.38	34.96
Nb II	18.55	20.88	22.41	24.71	26.24

PARTITION FUNCTIONS OF ATOMS AND IONS FOR COMMON FLAME TEMPERATURES

Element	Temperature °K				
	2000	2300	2500	2800	3000
Ni	22.70	23.96	24.71	25.73	26.35
Ni II	7.39	7.65	7.83	8.10	8.30
O	8.53	8.58	8.62	8.66	8.68
O II	4.00	4.00	4.00	4.00	4.00
Os	10.41	11.03	11.50	12.25	12.78
Os II	11.07	11.57	11.93	12.51	12.92
P	4.00	4.01	4.01	4.03	4.04
P II	7.23	7.45	7.57	7.73	7.83
Pb	1.01	1.03	1.04	1.08	1.10
Pb II	2.00	2.00	2.00	2.00	2.00
Pd	1.08	1.16	1.23	1.36	1.46
Pd II	6.31	6.44	6.52	6.65	6.73
Po	5.00	5.01	5.01	5.02	5.03
Pt	14.90	15.58	15.98	16.53	16.86
Pt II	6.34	6.55	6.71	6.98	7.18
Ra	1.00	1.00	1.01	1.01	1.02
Ra II	2.00	2.00	2.01	2.01	2.02
Rb	2.00	2.00	2.00	2.01	2.01
Rb II	1.00	1.00	1.00	1.00	1.00
Re	6.00	6.01	6.02	6.05	6.07
Re II	7.00	7.00	7.01	7.01	7.02
Rh	14.71	15.89	16.68	17.87	18.65
Rh II	10.64	11.13	11.46	11.94	12.26
Ru	17.64	19.00	19.92	21.34	22.31
Ru II	14.16	15.06	15.65	16.53	17.12
S	7.92	8.05	8.13	8.23	8.30
S II	4.00	4.00	4.00	4.00	4.01
Sb	4.01	4.03	4.05	4.09	4.12
Sb II	1.42	1.59	1.71	1.90	2.04
Sc	9.32	9.42	9.48	9.58	9.65
Sc II	15.35	16.06	16.56	17.31	17.81
Se	5.88	6.08	6.21	6.39	6.50
Se II	4.00	4.00	4.00	4.01	4.02
Si	8.15	8.30	8.40	8.54	8.63
Si II	5.25	5.34	5.39	5.45	5.49

PARTITION FUNCTIONS OF ATOMS AND
IONS FOR COMMON FLAME TEMPERATURES

Element	Temperature °K				
	2000	2300	2500	2800	3000
Sn	2.32	2.65	2.86	3.18	3.38
Sn II	2.19	2.28	2.35	2.45	2.52
Sr	1.00	1.00	1.00	1.00	1.01
Sr II	2.00	2.00	2.00	2.01	2.01
Ta	6.16	6.90	7.43	8.28	8.88
Ta II	7.71	8.93	9.78	11.11	12.02
Tc	9.10	10.14	10.84	11.90	12.60
Tc II	8.37	8.98	9.42	10.09	10.55
Te	5.13	5.21	5.27	5.37	5.44
Te II	4.00	4.01	4.02	4.03	4.05
Ti	18.35	19.02	19.49	20.26	20.82
Ti II	37.34	39.55	40.91	42.81	44.02
Tl	2.01	2.03	2.05	2.07	2.10
Tl II	1.00	1.00	1.00	1.00	1.00
V	28.32	30.41	31.71	33.54	34.71
V II	25.98	27.80	28.97	30.69	31.81
W	3.52	4.32	4.87	5.71	6.28
W II	4.39	5.10	5.61	6.40	6.96
Y	8.11	8.34	8.48	8.68	8.81
Y II	7.98	8.95	9.55	10.40	10.94
Zn	1.00	1.00	1.00	1.00	1.00
Zn II	2.00	2.00	2.00	2.00	2.00
Zr	14.64	16.16	17.20	18.81	19.91
Zr II	20.85	23.35	25.01	27.50	29.14

Percent of Atoms in the Lower Energy Levels

Utilizing the data in electronic partition functions Z listed in the last section, and Moore's energy level data (ref. 6,7,8, first section of this chapter), it is possible to calculate the relative number of atoms in the low lying energy levels. The equation is given (1) below:

$$\%\text{atoms in the ith level} = \frac{n_i}{n_t} \times 100 = \frac{g_i}{Z} \exp\left(-E_i/kT\right) \qquad (1)$$

where n_i is the number of atoms in the ith level, n_t is the total number of atoms, g_i is the statistical weight of the ith level, Z is the electronic partition function, E_i is the energy of the ith level, k is the Boltzmann constant, and T is the absolute temperature. Of course all units must be consistant. This data is presented in Table III.B.3. It should be noted that given approximately the same flame, the same atomization, and the same transition probability, an element with 100% of its atoms in the lower level of a given transition will be twice as sensitive as one with only 50% of its atoms in the lower level.

Reference:

M. L. Parsons, B. W. Smith, P. M. McElfresh, Appl. Spectrosc. 27, 471 (1973).

Table III.B.3

Lower Energy Level Populations (in Percent)

at Various Flame Temperatures (°K)

Element	Level[a]	2000	2300	2500	2800	3000	3200
Aluminum	0.0	35.1	34.9	34.8	34.6	34.5	34.5
	112.040	64.9	65.1	65.2	65.4	65.5	65.5
Total		100.0	100.0	100.0	100.0	100.0	100.0
Antimony	0.0	99.7	99.2	98.7	97.8	97.0	96.1
	8512.100	0.2	0.5	0.7	1.2	1.6	2.1
	9854.100	0.1	0.3	0.5	0.9	1.3	1.7
Total		100.0	100.0	99.9	99.9	99.9	99.9
Arsenic	0.0	99.9	99.7	99.5	99.0	98.5	98.0
	10592.500	0.0	0.1	0.2	0.4	0.6	0.8
	10914.600	0.1	0.2	0.3	0.5	0.8	1.1
Total		100.0	100.0	100.0	99.9	99.9	99.9
Barium	0.0	98.0	95.2	92.6	87.0	82.6	78.1
	9033.985	0.4	1.0	1.5	2.5	3.3	4.0
	9215.518	0.6	1.5	2.3	3.8	5.0	6.2
	9596.551	0.7	1.7	2.6	4.4	5.8	7.3
	11395.382	0.1	0.4	0.7	1.2	1.8	2.3
Total		99.8	99.8	99.7	98.9	98.5	97.9
Beryllium	0.0	100.0	100.0	100.0	100.0	100.0	99.9
Bismuth	0.0	100.0	99.9	99.8	99.7	99.5	99.3
Boron	0.0	33.5	33.5	33.5	33.5	33.5	33.5
	16.0	66.5	66.5	66.5	66.5	66.5	66.5
Total		100.0	100.0	100.0	100.0	100.0	100.0
Cadmium	0.0	100.0	100.0	100.0	100.0	100.0	100.0
Calcium	0.0	100.0	99.9	99.8	99.6	99.3	98.9
Cesium	0.0	99.9	99.5	99.5	98.9	98.3	97.6
	11178.240	0.0	0.1	0.2	0.3	0.5	0.6
	11732.350	0.0	0.1	0.2	0.5	0.7	1.0
Total		99.9	99.7	99.9	99.7	99.5	99.2

Element	Level[a]	2000	2300	2500	2800	3000	3200
Chromium	0.0	98.6	97.2	95.9	93.5	91.5	89.4
	7593.160	0.3	0.6	0.9	1.4	1.7	2.1
	7750.780	0.1	0.1	0.2	0.2	0.3	0.4
	7810.820	0.1	0.1	0.2	0.2	0.3	0.4
	7927.470	0.1	0.3	0.4	0.7	0.9	1.1
	8095.210	0.2	0.4	0.7	1.0	1.3	1.7
	8307.570	0.3	0.5	0.8	1.3	1.7	2.1
Total		99.7	99.2	99.1	98.3	97.7	97.2
Cobalt	0.0	51.8	47.8	45.6	42.6	40.9	39.3
	816.000	23.1	23.0	22.8	22.4	22.1	21.8
	1406.840	11.3	11.9	12.2	12.4	12.5	12.5
	1809.330	5.6	6.2	6.4	6.7	6.9	7.0
	3482.820	4.2	5.4	6.1	7.1	7.7	8.2
	4142.660	2.1	2.9	3.4	4.1	4.5	4.9
	4690.180	1.1	1.5	1.8	2.3	2.6	2.9
	5075.830	0.5	0.8	1.0	1.3	1.4	1.6
	7442.410	0.2	0.4	0.5	0.7	0.9	1.1
Total		99.9	99.9	99.8	99.6	99.5	99.3
Copper	0.0	99.9	99.7	99.4	98.8	98.3	97.6
	11202.565	0.1	0.3	0.5	0.9	1.4	1.9
Total		100.0	100.0	99.9	99.7	98.7	99.5
Gallium	0.0	47.5	45.6	44.6	43.3	42.6	42.0
	826.240	52.5	54.4	55.4	56.7	57.4	58.0
Total		100.0	100.0	100.0	100.0	100.0	100.0
Germanium	0.0	20.6	19.0	18.2	17.2	16.6	16.1
	557.100	41.4	40.3	39.7	38.8	38.3	37.7
	1409.900	37.4	39.4	40.5	41.7	42.4	42.9
	7125.260	0.6	1.1	1.5	2.2	2.7	3.3
Total		100.0	99.8	99.9	99.9	100.0	100.0
Gold	0.0	99.5	99.0	98.5	97.4	96.4	95.2
	9161.300	0.4	1.0	1.5	2.6	3.5	4.7
Total		99.9	100.0	100.0	100.0	99.9	99.9
Hafnium	0.0	73.9	67.8	64.0	59.0	55.9	53.1
	2356.680	19.0	21.7	23.1	24.6	25.3	25.8
	4567.640	5.0	7.0	8.3	10.2	11.3	12.3
	5521.780	0.3	0.4	0.5	0.7	0.8	0.9
	5638.620	1.3	2.0	2.5	3.3	3.7	4.2
	6572.550	0.4	0.7	0.9	1.2	1.4	1.7
Total		99.9	99.6	99.3	99.0	98.4	98.0
Indium	0.0	71.0	66.6	64.1	60.9	59.1	57.5
	2212.560	29.0	33.4	35.9	39.1	40.9	42.5
Total		100.0	100.0	100.0	100.0	100.0.	100.0

Element	Level[a]	2000	2300	2500	2800	3000	3200
Iridium	0.0	85.1	80.3	77.2	72.7	69.9	66.6
	2834.980	11.1	13.6	15.1	17.0	18.0	18.6
	4078.940	1.8	2.5	3.0	3.6	4.0	4.3
	5784.620	0.8	1.3	1.7	2.2	2.6	3.0
	6323.910	0.7	1.2	1.6	2.3	2.7	3.1
	7106.610	0.4	0.8	1.0	1.5	1.9	2.2
Total		99.9	99.7	99.6	99.3	99.1	97.8
Iron	0.0	46.2	44.5	43.4	41.9	41.0	40.1
	415.933	26.7	26.7	26.6	26.3	26.1	25.9
	704.003	15.5	15.9	16.1	16.2	16.3	16.2
	888.132	8.1	8.5	8.7	8.9	8.9	9.0
	978.074	2.5	2.7	2.7	2.8	2.9	2.9
	6928.280	0.4	0.7	1.0	1.5	1.8	2.2
	7376.775	0.2	0.4	0.6	0.9	1.2	1.5
	7728.071	0.1	0.3	0.4	0.6	0.8	1.0
	7985.795	0.1	0.2	0.2	0.4	0.5	0.6
Total		99.8	99.9	99.7	99.5	99.5	99.4
Lanthanum	0.0	42.5	37.2	34.2	30.4	28.3	26.3
	1053.200	29.9	28.9	28.0	26.6	25.6	24.6
	2668.200	6.2	7.0	7.4	7.7	7.9	7.9
	3010.010	7.3	8.5	9.1	9.7	10.0	10.2
	3494.580	6.9	8.4	9.2	10.1	10.6	11.0
	4121.610	5.5	7.1	8.0	9.2	9.8	10.3
	7011.900	0.4	0.7	0.9	1.2	1.5	1.7
	8052.150	0.3	0.5	0.7	1.0	1.2	1.4
	7231.360	0.1	0.2	0.3	0.4	0.4	0.5
	7490.460	0.2	0.3	0.5	0.6	0.8	0.9
	7679.940	0.3	0.5	0.6	0.9	1.1	1.3
Total		99.6	99.3	98.9	97.8	97.2	96.1
Lead	0.0	98.7	97.0	95.7	92.6	90.8	88.4
	7819.350	1.1	2.2	3.2	5.0	6.4	7.9
	10650.470	0.2	0.6	1.1	2.0	2.8	3.7
Total		100.0	99.8	100.0	99.6	100.0	100.0
Lithium	0.0	100.0	100.0	99.9	99.9	99.8	99.5
Magnesium	0.0	100.0	100.0	100.0	100.0	100.0	99.9
Manganese	0.0	100.0	100.0	100.0	99.9	99.8	99.5
Mercury	0.0	100.0	100.0	100.0	100.0	100.0	100.0
Molybdenum	0.0	99.9	99.7	99.4	98.7	98.1	97.6

Element	Level[a]	2000	2300	2500	2800	3000	3200
Nickel	0.0	39.5	37.6	36.4	35.0	34.2	33.4
	204.786	26.6	25.7	25.2	24.5	24.1	23.7
	879.813	11.7	12.0	12.2	12.4	12.4	12.5
	1332.153	11.8	12.7	13.2	13.7	14.0	14.3
	1713.080	3.9	4.3	4.5	4.8	5.0	5.2
	2216.519	4.5	5.2	5.7	6.2	6.6	6.9
	3409.925	1.9	2.5	2.8	3.4	3.7	4.0
Total		100.0	100.0	100.0	100.0	100.0	100.0
Niobium	0.0	7.5	6.8	6.5	6.0	5.7	5.5
	154.190	13.4	12.4	11.8	11.1	10.6	10.2
	391.990	17.0	16.0	15.4	14.7	14.2	13.7
	695.250	18.2	17.7	17.3	16.8	16.4	16.0
	1050.260	17.6	17.7	17.7	17.5	17.3	17.0
	1142.790	6.6	6.7	6.7	6.7	6.6	6.5
	1586.900	7.2	7.6	7.8	8.0	8.0	8.0
	2154.110	6.4	7.1	7.5	7.9	8.1	8.3
	2805.360	5.0	5.9	6.4	7.1	7.5	7.7
	4998.170	0.2	0.3	0.4	0.5	0.5	0.6
	5297.920	0.3	0.5	0.6	0.8	0.9	1.0
	5965.450	0.3	0.5	0.6	0.8	1.0	1.1
Total		99.7	99.2	98.7	97.9	96.8	95.6
Osmium	0.0	86.5	81.6	78.3	73.5	70.4	67.3
	2740.490	6.7	8.2	9.0	10.0	10.5	10.9
	4159.320	3.4	4.7	5.6	6.7	7.5	8.1
	5143.920	2.6	4.0	5.0	6.4	7.3	8.1
	5766.140	0.5	0.7	0.9	1.3	1.5	1.7
	6092.790	0.1	0.2	0.3	0.4	0.4	0.5
	8742.830	0.2	0.3	0.5	0.8	1.1	1.3
Total		100.0	99.7	99.6	99.1	98.7	97.9
Palladium	0.0	91.7	85.3	80.3	72.4	67.2	62.4
	6464.110	6.2	10.5	13.8	18.6	21.6	24.2
	7754.990	1.8	3.4	4.7	6.9	8.3	9.7
	10093.940	0.2	0.5	0.7	1.2	1.6	2.0
	11721.770	0.1	0.3	0.5	0.9	1.2	1.6
Total		100.0	100.0	100.0	100.0	99.9	99.9
Platinum	0.0	47.0	44.9	43.8	42.3	41.5	40.7
	775.900	19.2	19.8	20.0	20.3	20.4	20.5
	823.700	33.4	34.5	35.1	35.7	36.0	36.2
	6140.000	0.1	0.1	0.2	0.3	0.3	0.4
	6567.500	0.3	0.5	0.7	1.0	0.3	1.5
Total		100.0	99.8	99.8	99.6	99.5	99.3
Potassium	0.0	100.0	99.9	99.8	99.5	99.4	99.0
Rhenium	0.0	99.9	99.7	99.5	98.9	98.3	97.4

Element	Level[a]	2000	2300	2500	2800	3000	3200
Rhodium	0.0	68.0	62.9	60.0	56.0	53.6	51.4
	1529.970	18.1	19.3	19.9	20.4	20.6	20.7
	2598.030	6.3	7.4	8.1	8.8	9.3	9.6
	3309.860	3.8	4.8	5.4	6.1	6.6	7.0
	3472.680	2.2	2.9	3.3	3.8	4.1	4.3
	5657.970	0.5	0.8	0.9	1.2	1.4	1.6
	5690.970	0.9	1.4	1.8	2.4	2.8	3.2
Total		99.8	99.4	99.4	98.7	98.4	97.8
Rubidium	0.0	100.0	99.9	99.8	99.5	99.3	98.9
Ruthenium	0.0	62.4	57.9	55.2	51.5	49.3	47.1
	1190.640	21.7	22.5	22.8	22.9	22.8	22.6
	2091.540	8.8	10.0	10.6	11.2	11.5	11.7
	2713.240	4.0	4.8	5.3	5.8	6.1	6.3
	3105.490	1.8	2.3	2.5	2.9	3.0	3.2
	6545.030	0.5	0.8	1.0	1.5	1.8	2.0
	7483.070	0.2	0.4	0.6	0.9	1.1	1.3
	8084.120	0.1	0.2	0.3	0.5	0.7	0.8
	9183.660	0.0	0.1	0.1	0.2	0.3	0.3
Total		99.4	99.0	98.4	97.4	96.6	95.3
Scandium	0.0	42.9	42.5	42.2	41.8	41.5	40.9
	168.340	57.0	57.3	57.4	57.4	57.4	56.8
Total		99.9	99.8	99.6	99.2	98.9	97.7
Selenium	0.0	85.0	82.2	80.5	78.2	76.8	75.6
	1989.490	12.2	14.2	15.4	16.9	17.8	18.6
	2534.350	2.7	3.4	3.7	4.3	4.6	4.8
	9576.080	0.1	0.2	0.3	0.6	0.8	1.0
Total		100.0	100.0	99.9	100.0	100.0	100.0
Silicon	0.0	12.3	12.0	11.9	11.7	11.6	11.5
	77.150	34.8	34.4	34.2	33.8	33.5	33.2
	223.310	52.2	52.4	52.3	52.2	52.1	51.9
	6298.810	0.7	1.2	1.6	2.3	2.8	3.4
Total		100.0	100.0	100.0	100.0	100.0	100.0
Silver	0.0	100.0	100.0	100.0	100.0	100.0	100.0
Sodium	0.0	100.0	100.0	100.0	100.0	99.9	99.8
Strontium	0.0	100.0	99.9	99.8	99.4	98.9	98.0

Element	Level[a]	2000	2300	2500	2800	3000	3200
Tantalum	0.0	64.9	58.0	53.8	48.3	45.0	41.8
	2010.100	23.0	24.7	25.4	25.8	25.8	25.4
	3963.920	7.5	9.7	11.0	12.6	13.5	14.1
	5621.040	2.9	4.3	5.3	6.7	7.6	8.4
	6049.420	0.4	0.7	0.8	1.1	1.2	1.4
	6068.910	0.8	1.3	1.6	2.1	2.5	2.7
	9253.430	0.1	0.3	0.4	0.6	0.8	1.0
	9705.380	0.1	0.3	0.4	0.7	0.9	1.1
Total		99.7	99.3	98.7	97.9	97.3	95.9
Technetium	0.0	65.9	59.2	55.4	50.4	47.6	45.0
	2572.890	17.3	19.7	21.0	22.4	23.1	23.6
	3250.910	8.5	10.3	11.4	12.7	13.4	13.9
	3700.550	4.6	5.9	6.6	7.5	8.1	8.5
	4002.580	2.5	3.2	3.7	4.3	4.7	5.0
	4178.720	1.1	1.4	1.7	2.0	2.1	2.3
Total		99.9	99.7	99.8	99.3	99.0	98.3
Tellurium	0.0	97.4	96.0	94.9	93.1	91.9	90.6
	4751.000	1.9	3.0	3.7	4.9	5.7	6.4
	4707.000	0.7	1.0	1.3	1.7	1.9	2.2
Total		100.0	100.0	99.9	99.7	99.5	99.2
Thallium	0.0	99.3	98.5	97.6	96.5	95.2	94.3
	7792.700	0.7	1.5	2.2	3.5	4.5	5.7
Total		100.0	100.0	99.8	100.0	99.7	100.0
Tin	0.0	43.0	37.7	34.9	31.4	29.6	27.9
	1691.800	38.3	39.3	39.6	39.6	39.4	39.2
	3427.700	18.3	22.1	24.3	27.0	28.6	29.9
	8613.000	0.4	0.9	1.2	1.9	2.4	2.9
		100.0	100.0	100.0	99.9	100.0	99.9
Titanium	0.0	27.2	26.3	25.7	24.7	24.0	23.2
	170.132	33.8	33.1	32.6	31.7	31.0	30.1
	386.873	37.1	37.2	37.0	36.4	35.9	35.1
	6556.860	0.1	0.3	0.4	0.5	0.6	0.7
	6598.830	0.2	0.4	0.6	0.8	1.0	1.2
	6661.000	0.3	0.6	0.8	1.1	1.4	1.6
	6742.790	0.4	0.7	1.0	1.4	1.7	2.0
	6843.000	0.4	0.8	1.1	1.6	2.0	2.4
	7255.290	0.1	0.3	0.4	0.6	0.7	0.9
Total		99.6	99.7	99.6	98.8	98.3	97.2

Element	Level[a]	2000	2300	2500	2800	3000	3200
Tungsten	0.0	28.4	23.1	20.5	17.5	15.9	14.5
	1670.300	25.6	24.4	23.6	22.3	21.5	20.6
	2951.290	23.8	25.6	26.3	26.9	27.1	27.0
	3325.530	13.0	14.5	15.2	15.9	16.2	16.3
	4830.000	6.2	7.9	8.9	10.3	11.0	11.6
	6219.330	2.9	4.3	5.2	6.5	7.3	8.0
Total		99.9	99.8	99.7	99.4	99.0	98.0
Vanadium	0.0	14.1	13.2	12.6	11.9	11.5	11.1
	137.380	19.2	18.1	17.5	16.7	16.2	15.7
	323.420	22.4	21.5	20.9	20.2	19.7	19.2
	553.020	23.7	23.3	22.9	22.4	22.1	21.7
	2112.320	1.5	1.8	1.9	2.0	2.1	2.2
	2153.200	3.0	3.4	3.7	3.9	4.1	4.2
	2220.130	4.3	4.9	5.3	5.7	6.0	6.2
	2311.370	5.4	6.2	6.7	7.3	7.6	7.9
	2424.890	6.2	7.2	7.8	8.6	9.0	9.3
Total		99.8	99.6	99.3	98.7	98.3	97.5
Yttrium	0.0	49.3	48.0	47.2	46.1	45.4	44.3
	530.360	50.5	51.6	52.2	52.6	52.8	52.4
Total		99.8	99.6	99.4	98.7	98.2	97.7
Zinc	0.0	100.0	100.0	100.0	100.0	100.0	100.0
Zirconium	0.0	34.2	30.9	29.1	26.6	25.1	23.6
	570.410	31.7	30.3	29.3	27.8	26.7	25.6
	1240.840	25.2	25.6	25.6	25.3	24.9	24.4
	4186.110	1.7	2.3	2.6	3.1	3.4	3.6
	4196.850	0.3	0.4	0.5	0.6	0.7	0.7
	4376.280	0.9	1.2	1.4	1.7	1.8	2.0
	4870.530	0.6	0.9	1.1	1.3	1.5	1.6
	5023.410	0.9	1.3	1.6	2.0	2.3	2.5
	5101.680	0.9	1.3	1.5	1.9	2.2	2.4
	5249.070	1.1	1.6	2.0	2.5	2.8	3.1
	5540.540	1.1	1.7	2.2	2.8	3.2	3.5
	5888.930	1.1	1.7	2.2	2.8	3.3	3.7
	8057.300	0.2	0.4	0.5	0.8	1.0	1.1
Total		99.9	99.6	99.6	99.2	98.9	97.8

Ionization Potentials of the Elements

Ionization of the analytes in analytical flames will remove the atomic species which are required for analysis, thus rendering the analysis less sensitive. (See Chapter I.E, equation 3 for the reaction involed.) The ionization potential gives an indication of a specific element's tendency to ionize (See discussion in Chapter I.E under Chemical Interferences). Table III.B.4 gives the ionization potential for nearly all of the elements.

Table III.B.4

IONIZATION POTENTIALS

Element	eV	Kcal/mole	KJ/mole
Actinium, Ac	6.9	159	666
Aluminum, Al	5.986	138.0	577.6
Americium, Am	6.0	138	579
Antimony, Sb	8.641	199.3	833.7
Argon, Ar	15.759	363.42	1520.5
Arsenic, As	9.81	226	947
Astatine, At	no information		
Barium, Ba	5.212	120.2	502.9
Berkelium, Bk	no information		
Beryllium, Be	9.322	215.0	899.5
Bismuth, Bi	7.289	168.1	703.3
Boron, B	8.298	191.4	799.8
Bromine, Br	11.814	272.44	1139.9
Cadmium, Cd	8.993	207.4	867.7
Calcium, Ca	6.113	141.0	589.8
Californium, Cf	no information		
Carbon, C	11.260	259.67	1086.4
Cerium, Ce	5.47	126	528
Cesium, Cs	3.894	89.80	375.7
Chlorine, Cl	12.967	299.03	1251.1
Chromium, Cr	6.766	156.0	652.8
Cobalt, Co	7.86	181	758
Copper, Cu	7.726	178.2	745.5
Curium, Cm	no information		
Dysprosium, Dy	5.93	137	572
Einsteinium, Es	no information		
Erbium, Er	6.10	141	589
Europium, Eu	5.67	131	547
Fermium, Fm	no information		
Fluorine, F	17.422	401.77	1681.0
Francium, Fr	no information		
Gadolinium, Gd	6.14	142	592
Gallium, Ga	5.999	138.3	578.8
Germanium, Ge	7.899	182.2	762.2
Gold, Au	9.225	212.7	890.1
Hafnium, Hf	7.0	161	675

Element	eV	Kcal/mole	KJ/mole
Helium, He	24.587	567.00	2372.3
Holmium, Ho	6.02	139	581
Hydrogen, H	13.598	313.58	1312.0
Indium, In	5.786	133.4	558.3
Iodine, I	10.451	241.01	1008.4
Iridium, Ir	9.1	210	878
Iron, Fe	7.870	181.5	759.4
Krypton, Kr	13.999	322.83	1350.7
Lanthanum, La	5.577	128.6	538.1
Lead, Pb	7.416	171.0	715.5
Lithium, Li	5.392	124.3	520.3
Lutetium, Lu	5.426	125.1	523.5
Magnesium, Mg	7.646	176.3	737.7
Manganese, Mn	7.435	171.5	717.4
Mendelevium, Md	no information		
Mercury, Hg	10.437	240.69	1007.0
Molybdenum, Mo	7.099	163.7	685.0
Neodymium, Nd	5.49	127	530
Neon, Ne	21.564	497.29	2080.6
Neptunium, Np	no information		
Nickel, Ni	7.635	176.1	736.7
Niobium, Nb	6.88	159	664
Nitrogen, N	14.534	335.16	1402.3
Nobelium, No	no information		
Osmium, Os	8.7	201	839
Oxygen, O	13.618	314.04	131.40
Palladium, Pd	8.34	192	805
Phosphorus, P	10.486	241.82	1011.8
Platinium, Pt	9.0	208	868
Plutonium, Pu	5.8	134	560
Polonium, Po	8.42	194	812
Potassium, K	4.341	100.1	418.8
Praseodymium, Pr	5.42	125	523
Promethium, Pm	5.55	128	536
Protactinium, Pa	no information		
Radium, Ra	5.279	121.7	509.4
Radon, Rn	10.748	247.86	1037.0
Rhenium, Re	7.88	182	760

Element	eV	Kcal/mole	KJ/mole
Rhodium, Rh	7.46	172	720
Rubidium, Rb	4.177	96.33	403.0
Ruthenium, Ru	7.37	170	711
Samarium, Sm	5.63	130	543
Scandium, Sc	6.54	151	631
Selenium, Se	9.752	224.9	940.9
Silver, Ag	7.576	174.3	731.0
Sodium, Na	5.139	118.3	495.8
Strontium, Sr	5.695	131.3	549.5
Sulfur, S	10.360	238.91	999.61
Tantalum, Ta	7.89	182	761
Technetium, Tc	7.28	168	702
Terbium, Tb	5.85	135	564
Thallium, Tl	6.108	140.9	589.3
Thorium, Th *	7.5	173	724
Thulium, Tm	6.18	143	596
Tin, Sn	7.344	169.4	708.6
Titanium, Ti	6.82	157	658
Tungsten, W	7.98	184	770
Uranium, U *	6.3	145	608
Vanadium, V	6.74	155	650
Xenon, Xe	12.130	279.73	1170.4
Ytterbium, Yb	6.254	144.2	603.4
Yttrium, Y	6.38	147	616
Zinc, Zn	9.394	216.6	906.4
Zirconium, Zr	6.84	158	660

C.E. Moore, "Ionization Potentials and Ionization Limits Derived From the Analyses of Optical Spectra", NSRDS-NBS 34, U.S. Government Printing Office, Washington, D.C., 1970.

* R. Avni and F.S. Klein, Spectrochim. Acta, 28B, 331 (1973).

Molecular Dissociation Energies

Similarly to ionization, the formation of stable molecules in the flame gases remove atoms from the total sample that would otherwise be available for excitation. The formation of stable oxides in some flames makes the flame unsuitable for certain analyses. The dissociation energy of the bonds is an indication of the stability of the particular molecule in the flame. The greater the bond dissociation energy, the more stable the molecular species. A general discussion of the types of molecular forming reactions and the extent of the problem is given in Chapter I.E. Values of the bond dissociation energy are given in Table III.B.5 with their references.

Table III.B.5

SELECTED MOLECULAR BOND
DISSOCIATION ENERGIES

Element	Bond	Dissociation Energy			Ref.
		Kcal/mole	KJ/mole	eV	
Actinium	no information				
Aluminum	Al-Al	39	163	1.7	1
	Al-Br	105	439	4.6	
	Al-Cl	117	490	5.1	
	ClAl-Cl	95.5	400	4.1	
	Cl_2Al-Cl	89	372	3.9	
	Al-F	157.5	659.0	6.8	
	FAl-F	130.5	546.0	5.7	
	F_2Al-F	130	544	5.6	
	Al-H	67.0	280.3	2.9	
	Al-I	87	364	3.8	
	Al-O	115	481	5.0	
	Al-P	51	213	2.2	
	Al-S	96	402	4.2	
	Al-N	70	293	3	7
Americium	no information				
Antimony	Sb-Sb	71	295	3.1	1
	Sb-O	102	427	4.4	2
		74	310	3.2	3
		88	368	3.8	4
		74	310	3.2	6
		92	385	4	
	Sb-Br	74	310	3.2	7
	Sb-Cl	85	356	3.7	
	Sb-F	104	435	4.5	
	Sb-N	71	297	3.1	
Arsenic	As-As	91	380	3.9	1
	As-Cl	106	444	4.6	
	As-N	138	577	6.0	
	As-O	114	477	4.9	
Astatine	no information				

Element	Bond	Dissociation Energy			Ref.
		Kcal/mole	KJ/mole	eV	
Barium	Ba-Cl	118	494	5.1	1
	ClBa-Cl	109	456	4.7	
	Ba-F	136	569	5.9	
	FBa-F	140	586	6.1	
	Ba-H	47	196	2.0	
	Ba-O	134	561	5.8	
	Ba-OH	113	473	4.9	
	(OH)Ba-OH	110	460	4.8	
	Ba-S	95	396	4.1	
Berkelium	no information				
Beryllium	Be-Cl	92	358	4.0	1
	ClBe-Cl	128	536	5.6	
	Be-F	137	573	5.9	
	FBe-F	165	690	7.2	
	Be-H	53	222	2.3	
	Be-O	106	444	4.6	
	Be-S	88	368	3.8	7
Bismuth	Bi-Bi	46	192	2.0	1
	Bi-Br	63	264	2.7	7
	Bi-Cl	72	301	3.1	
	Bi-F	61	255	2.6	
	Bi-H	58	242	2.5	
	Bi-I	58	242	2.5	
	Bi-S	69	289	3.0	
	Bi-O	85	356	3.7	2
		92	385	4.0	3
		67	280	2.9	4
		71	297	3.1	7
Boron	B-B	70	293	3.0	1
	B-Br	104	433	4.5	
	B-C	106	444	4.6	
	B-Cl	127	531	5.5	
	B-F	182	759	7.9	
	FB-F	125	523	5.4	
	F_2B-F	133	557	5.8	
	B-H	78	326	3.4	
	B-N	92	385	4.0	
	B-O	187	782	8.1	
	B-S	118	494	5.1	

Element	Bond	Dissociation Energy			Ref.
		Kcal/mole	KJ/mole	eV	
Bromine	Br-H	87	305	3.8	6
	Br-O	55	230	2.4	7
	(See also specific elements)				
Cadmium	Cd-O	88	368	3.8	2
		88	368	3.8	3
Calcium	Ca-Cl	102	427	4.4	1
	ClCa-Cl	103	431	4.5	
	Ca-F	132	552	5.7	
	FCa-F	131	548	5.7	
	Ca-O	110	460	4.8	
	Ca-OH	100	418	4.3	
	Ca-H	39	163	1.7	6
	Ca-S	89	372	3.9	
	(OH)Ca-OH	117	490	5.1	
	Ca-S	74	310	3.2	
	Ca-I	69	289	3	7
Californium	no information				
Carbon	C-C	144	602	6.2	1
	C-H	80	335	3.5	
	HC-H	99	415	4.3	
	H_2C-H	112.3	469.9	4.9	
	H_3C-H	102	425	4.4	
	H-CN	127	531	5.5	
	H-CO	28.6	119.7	1.2	
	C-N	174.5	730.1	7.6	
	C-O	256.2	1071.9	11.1	
	OC-O	125.7	525.9	5.5	
	C-S	181	757	7.8	
	(See also specific elements)				
Cerium	Ce-O	185	774	8.0	2
		192	803	8.3	5
		190	795	8.2	6
	Ce-S	135	565	5.8	6
	Ce-N	124	519	5.4	
	Ce-C	108	452	4.7	

Element	Bond	Dissociation Energy			Ref.
		Kcal/mole	KJ/mole	eV	
Cesium	Cs-Br	99.5	416.3	4.3	1
	Cs-Cl	104	435	4.5	
	Cs-F	120	502	5.2	
	Cs-I	80	335	3.5	
	Cs-OH	91	381	3.9	
	Cs-Cs	10	42	0.4	7
	Cs-H	41	172	1.8	
Chlorine	Cl-Cl	57.3	239.7	2.5	1
	Cl-H	102.3	428	4.4	
	Cl-O	64	268	2.8	
	(See also specific elements)				
Chromium	Cr-Cr	<40	<167	<1.7	1
	Cr-O	101	423	4.4	
	OCr-O	126	531	5.5	
	O_2Cr-O	114	477	4.9	
	Cr-Br	78	326	3.4	7
	Cr-Cl	86	360	3.8	
	Cr-F	92	385	4	
	Cr-H	66	276	2.8	
	Cr-I	68	285	2.9	
Cobalt	Co-Co	39	163	1.7	7
Copper	Cu-Cu	45.5	190.4	2.0	1
	Cu-F	70	293	3.0	
	FCu-F	87.6	366	3.8	
	Cu-H	66	276	2.9	
	Cu-O	113	473	4.9	2
		113	473	4.9	3
		113	473	4.9	4
		95	397	4.1	7
	Cu-Br	78	326	3.4	
	Cu-Cl	83	347	3.6	
	Cu-I	46	192	2	
	Cu-S	71	297	3.1	
Curium	no information				

Element	Bond	Dissociation Energy			Ref.
		Kcal/mole	KJ/mole	eV	
Dysprosium	Dy-O	150	628	6.5	5
	Dy-F	148	619	6.4	6
		131	548	5.7	
Einsteinium	no information				
Erbium	Er-O	152	636	6.6	5
		148	619	6.4	6
	Er-F	140	586	6.1	
Europium	Eu-O	134	561	5.8	5
		132	552	5.7	6
	Eu-F	131	548	5.7	
	Eu-S	81	339	3.5	
Fermium	no information				
Fluorine	F-F	37.0	154.8	1.6	1
	F-H	135	565	5.9	
	F-O	37	155	1.6	
	(Also see specific elements)				
Francium	no information				
Gadolinium	Gd-O	171	715	7.4	5
		171	715	7.4	6
		161	674	7.0	7
	Gd-F	155	649	6.7	6
	Gd-S	123	515	5.3	
Gallium	Ga-Br	104	435	4.5	1
	Ga-Cl	114	477	4.9	
	Ga-F	144	602	6.2	
	Ga-Ga	27	113	1.2	
	Ga-H	65	272	2.8	
	Ga-I	84	351	3.6	
	Ga-O	59	247	2.6	
	Ga-OH	103	431	4.5	

Element	Bond	Dissociation Energy			Ref.
		Kcal/mole	KJ/mole	eV	
Germanium	Ge–Br	60	251	2.6	1
	Ge–C	109	456	4.7	
	Ge–Cl	81	339	3.5	
	Ge–F	115	484	5.0	
	Ge–Ge	65	272	2.8	
	Ge–H	76	318	3.3	
	Ge–O	60	669	6.9	
	Ge–S	133	556	5.8	
Gold	Au–Au	51.5	215.5	2.2	1
	Au–Cl	69	289	3.0	
	Au–H	68	285	2.9	
Hafnium	Hf–O	185	774	8.0	1
Holmium	Ho–O	150	628	6.5	5
		153	640	6.6	6
		175	732	7.6	7
	Ho–F	134	561	5.8	6
	Ho–S	95	397	4.1	
Hydrogen	H–H	103.25	432.00	4.5	1

(See also specific elements)

Element	Bond	Dissociation Energy			Ref.
Indium	In–Br	97	406	4.2	1
	In–Cl	104	435	4.5	
	In–F	125	523	5.4	
	In–H	57	238	2.5	
	In–I	81	339	3.5	
	In–In	24	100	1.0	
	In–O	25	105	1.1	
	In–S	35	146	1.5	
	In–OH	86	360	3.7	2
Iodine	I–H	70.4	294.6	3.1	1
	I–I	35.60	148.95	1.5	
	I–O	43	180	1.9	

(See also specific elements)

Element	Bond	Dissociation Energy			Ref.
		Kcal/mole	KJ/mole	eV	
Iridium	no information				
Iron	Fe-O	92	385	4.0	3
		99	414	4.3	4
		99	414	4.3	7
	Fe-Br	58	243	2.5	7
	Fe-Cl	83	347	3.6	
Lanthanum	La-La	57.6	241.0	2.5	1
	La-O	187	782	8.1	
	La-S	137	573	5.9	
Lead	Pb-O	90	377	3.9	1
	Pb-S	76	318	3.3	
	Pb-Pb	18	75	0.8	7
	Pb-Br	58	243	2.5	
	Pb-Cl	71	297	3.1	
	Pb-F	69	289	3	
	Pb-H	41	172	1.8	
	Pb-I	46	192	2	
Lithium	Li-Br	100	418	4.3	1
	Li-Cl	111	464	4.8	
	Li-F	137	573	5.9	
	Li-I	83	347	3.6	
	Li-O	79	335	3.4	
	Li-OH	101	423	4.4	
	Li-Li	26	109	1.1	7
	Li-H	56	234	2.4	
Lutetium	Lu-O	166	695	7.2	5
		164	686	7.1	6
	Lu-S	120	502	5.2	
Magnesium	Mg-F	110	460	4.8	1
	Mg-O	90	377	3.9	
	Mg-OH	56	234	2.4	
	FMg-F	138	577	6.0	6
	Mg-Cl	74	310	3.2	
		81	339	3.5	7
	ClMg-Cl	77	322	3.3	6

Element	Bond	Dissociation Energy			Ref.
		Kcal/mole	KJ/mole	eV	
Magnesium	Mg-Br	74	310	3.2	7
(continued)	Mg-H	46	192	2.0	
	Mg-S	<55	<230	<2.4	
Manganese	Mn-Br	70	293	3.0	1
	Mn-Cl	80	335	3.5	
	Mn-I	66.7	279.1	2.9	
	Mn-Mn	<21	<88	.9	
	Mn-O	96	402	4.2	
	Mn-S	71.4	298.7	3.1	
	Mn-F	120	502	5.2	7
	Mn-H	55	230	2.4	
Mendelevium	no information				
Mercury	Hg-Br	16.4	68.6	.7	1
	Hg-Cl	23	96	1.0	
	Hg-S	64	268	2.8	
	Hg-H	8.7	36	.4	6
	Hg-F	31	130	1.3	
	Hg-Cl	23	96	1.0	
	Hg-I	8.1	34	.4	
Molybdenum	Mo-O	116	485	5.0	1
	OMo-O	160	669	6.9	
	O_2Mo-O	134	561	5.8	
Neodymium	Nd-O	165	690	7.2	1
	Nd-F	130	545	5.6	
	FNd-F	146	611	6.3	6
	Nd-S	112	469	4.9	6
Neptunium	no information				
Nickel	Ni-Br	85	356	3.7	1
	Ni-Cl	83	347	3.6	
	Ni-I	69	289	3.0	
	Ni-Ni	54.5	228.0	2.4	
	Ni-O	97	406	4.2	2
		86	360	3.7	6

Element	Bond	Dissociation Energy			Ref.
		Kcal/mole	KJ/mole	eV	
Nickel	Ni-F	88	368	3.8	7
(continued)	Ni-H	60	251	2.6	
Niobium	Nb-O	180	753	7.8	2
		92	385	4.0	3
		161	674	7	7
		181	757	7.8	6
	ONb-O	162	678	7.0	
Nitrogen	N-H	85	365	3.7	1
	HN-H	89	372	3.9	
	H_2N-H	103	431	4.5	
	N-N	225.07	941.69	9.8	
	NO-N	113.5	474.9	4.9	
	N-O	150	628	6.5	7
		(See also specific elements)			
Nobelium	no information				
Osmium	O_3Os-O	108	452	4.7	1
Oxygen	O-H	101.3	423.8	4.4	1
	HO-H	118.0	493.7	5.1	
	O-O	117.97	493.59	5.1	
		(See also specific elements)			
Palladium	Pd-O	67	280	2.9	6
	Pd-Pd	<33	<138	<1.4	7
Phosphorus	P-O	141.5	592.0	6.1	1
	P-P	115	481	5.0	
	P-F	104	435	4.5	7
	P-H	81	339	3.5	
	P-N	164	686	7.1	
	P-S	120	502	5.2	

Element	Bond	Dissociation Energy			Ref.
		Kcal/mole	KJ/mole	eV	
Platinum	Pt-C	150	628	6.5	7
	Pt-H	83	347	3.6	
Plutonium	no information				
Polonium	Po-Po	44	184	1.9	7
Potassium	K-Br	90.5	378.7	3.9	1
	K-Cl	101	423	4.4	
	K-F	117	490	5.1	
	K-I	78	326	3.4	
	K-OH	81	339	3.5	
	K-H	43	180	1.9	7
Praseodymium	Pr-O	171	715	7.4	2
		182	761	7.9	5
		182	761	7.9	6
		171	715	7.4	7
Promethium	no information				
Protactinium	no information				
Radium	Ra-Cl	81	339	3.5	7
Rhenium	no information				
Rhodium	Rh-O	101	423	4.4	6
	Rh-C	139	582	6	7
Rubidium	Rb-Br	92	385	4.0	1
	Rb-Cl	106	444	4.6	
	Rb-F	117	490	5.1	
	Rb-I	79	331	3.4	
	Rb-O	83	347	3.6	
	Rb-Rb	11	46	.47	7
	Rb-H	39	163	1.7	

Element	Bond	Dissociation Energy			Ref.
		Kcal/mole	KJ/mole	eV	
Ruthenium	Ru-O	104	435	4.5	1
		42	176	1.8	7
Samarium	Sm-F	131	548	5.7	5
	Sm-O	141	590	6.1	
		139	582	6.0	6
Scandium	Sc-Sc	25.9	108.4	1.1	1
	Sc-F	121	506	5.2	
	Sc-O	160	669	6.9	2,7
		139	582	6.0	3
		139	582	6.0	4
		141	590	6.1	5
		162	678	7.0	6
	Sc-S	114	477	4.9	
	Sc-Cl	92	385	4.0	
Selenium	Se-O	81	339	3.5	2
		81	339	3.5	3
		100	418	4.3	6,7
	Se-Se	73	305	3.2	
Silicon	Si-Si	75	314	3.4	7
	Si-H	74	310	3.2	1
	Si-O	192	803	8.3	2
		185	774	8.0	4
		191	799	8.3	6
		187	782	8.1	7
	Si-Br	81	339	3.5	
	Si-C	103	431	4.5	
	Si-Cl	104	435	4.5	
	Si-F	115	481	5.0	
	Si-N	104	435	4.5	
	Si-S	147	615	6.4	
	OSi-O	112	469	4.9	6
	Si-F	125	523	5.4	
	Si-Cl	91	381	3.9	
	Si-Br	85	356	3.7	
	Si-S	147	615	6.4	
	Si-N	104	435	4.5	
	Si-C	102	427	4.4	

Element	Bond	Dissociation Energy			Ref.
		Kcal/mole	KJ/mole	eV	
Silver	Ag–Ag	38	159	1.6	1
	Ag–Br	69	289	3.0	
	Ag–Cl	75	314	3.3	
	Ag–F	84	351	3.6	7
	Ag–H	55	230	2.4	1
	Ag–I	68.7	287.4	3.0	
	Ag–O	57	238	2.5	
		32	134	1.4	
		57	238	2.5	2
		32	134	1.4	3
		42	176	1.8	4
		46	192	2	7
Sodium	Na–Br	87.5	366.1	3.8	1
	Na–Cl	98	410	4.2	
	Na–I	71	297	3.1	
	Na–O	65	272	2.8	
	Na–OH	77	322	3.3	
		90	377	3.9	2
	Na–Na	17	71	0.75	7
	Na–F	114	477	4.9	
	Na–H	47	197	2.0	
Strontium	Sr–Cl	80	335	3.5	1
	ClSr–Cl	100	418	4.3	
	Sr–F	129	540	5.6	
	FSr–F	135	565	5.9	
	Sr–O	110	460	4.8	
	Sr–OH	98	410	4.2	
	(OH)Sr–OH	117	490	5.1	
	Sr–S	75	314	3.2	
	Sr–H	39	163	1.7	6
Sulfur	S–H	83.5	349.4	3.6	1
	HS–H	90	377	3.9	
	S–O	123.6	517.1	5.4	
	OS–O	130.8	547.3	5.7	
	O_2S–O	81.9	342.7	3.6	
	S–S	101.5	424.7	4.4	

(See also specific elements)

Element	Bond	Dissociation Energy			Ref.
		Kcal/mole	KJ/mole	eV	
Tantalum	Ta-O	193.7	810	8.4	1
		194	812	8.4	2
	OTa-O	160	669	6.9	6
Technetium	no information				
Tellurium	Te-Te	52	218	2.3	1
	Te-O	63	264	2.7	2
		62	259	2.7	3
		90	377	3.9	6,7
	Te-S	80	335	3.5	
Terbium	Tb-O	171	715	7.4	5
		171	715	7.4	6
Thallium	Tl-Br	78	326	3.4	1
	Tl-Cl	87	364	3.8	
	Tl-F	105	439	4.6	
	Tl-I	67	280	2.9	
	Tl-Tl	<21	<88	<0.9	7
	Tl-O is probably unstable				
Thorium	Th-O	196	820	8.5	1
	OTh-O	<184	<770	<8.0	
Thulium	Tm-O	139	582	6.0	5
		134	561	5.8	6
Tin	Sn-O	130	544	5.6	1
	Sn-S	111	464	4.8	
	Sn-Sn	46	192	2.0	7
	Sn-Br	46	192	2.0	
	Sn-Cl	74	310	3.2	
	Sn-F	90	376	3.9	
	Sn-H	62	259	2.7	
Titanium	Ti-Ti	<58	<243	<2.5	1
	Ti-O	156	653	6.8	
	OTi-O	123	519	5.4	6

Element	Bond	Dissociation Energy			Ref.
		Kcal/mole	KJ/mole	eV	
Titanium	Ti-S	106	444	4.6	6
(continued)	Ti-C	$\leq$127	<531	$\leq$ 5.8	
Tungsten	W-O	158	661	6.9	1
	OW-O	150	628	6.5	
	O_2W-O	142	594	6.2	
Uranium	U-O	179	749	7.8	1
	OU-O	161	674	7.0	
	O_2U-O	153	640	6.6	
	U-S	134	561	5.8	
Vanadium	V-O	147.6	617.6	6.4	1
	OV-O	147.6	617.6	6.4	
Ytterbium	Yb-O	90	377	3.9	5
		122	510	5.3	7
	Yb-H	37	155	1.6	
Yttrium	Y-Y	37.3	156.1	1.6	1
	Y-O	167	699	7.2	2
		162	678	7	3
		208	870	9	4
		169	707	7.3	5
		168	703	7.3	6,7
	Y-F	86	360	3.7	6
	Y-Cl	81	339	3.5	6,7
	OY-O	58	243	2.5	6
	Y-S	125	523	5.4	6
Zinc	Zn-S	48	201	2.1	1
	Zn-O	92	385	4	2
		92	385	4	3
		65	272	2.8	7
	Zn-Zn	6	25	0.25	
	Zn-Cl	49	205	2.1	
	Zn-H	20	84	0.8	
	Zn-I	32	134	1.4	

Element	Bond	Dissociation Energy			Ref.
		Kcal/mole	KJ/mole	eV	
Zirconium	Zr-O	180	753	7.8	1
		181	757	7.8	2,7
		180	753	7.8	3
		180	753	7.8	4
	OZr-O	150	628	6.5	6
	Zr-Cl	115	481	5.0	
	ClZr-Cl	129	540	5.6	
	Zr-N	124	519	5.4	

1. B. deB. Darwent, "Bond Dissociation Energies in Simple Molecules,"
    NSRDS-NBS 31, U.S. Government Printing Office, Washington, D.C.,
    1970.

2. B.V. L'vov, "Atomic Absorption Spectroscopy," Israel Program
    for Scientific Translations, Jerusalem, 1969.

3. I. Rubeska and B. Moldan, "Atomic Absorption Spectrophotometry."
    Iliffe Books, London, 1969.

4. P.W.J.M. Boumans, "Theory of Spectrochemical Excitation."
    Plenum, New York, 1966.

5. R.N. Kniseley, V.A. Fassel and C.C. Butler in "Analytical Flame
    Spectroscopy." Ed. R. Mavrodineanu, Springer-Verlag, New York,
    1970.

6. Calculated from data in one or more of the following:

    D.D. Wagman, W.H. Evans, V.B. Parker, I. Halow, S.M. Bailey and
    R.H. Schumm, "Selected Values of Chemical Thermodynamic Prop-
    erties." NBS Technical Note 270-3, U.S. Government Printing
    Office, Washington, D.C., 1968.

    Ibid., NBS Technical Note 270-4, U.S. Government Printing Office,
    Washington, D.C., 1969.

    Ibid., NBS Technical Note 270-5, U.S. Government Printing Office,
    Washington, D.C., 1971.

    V.B. Parker, D.D. Wagman and W.H. Evans, "Selected Values of
    Chemical Thermodynamic Properties." NBS Technical Note 270-6,
    U.S. Government Printing Office, Washington, D.C., 1971.

R.H. Schumm, D.D. Wagman, S. Bailey, W.H. Evans and V.B. Parker, "Selected Values of Chemical Thermodynamic Properties." NBS Technical Note 270-7, U.S. Government Printing Office, Washington, D.C., 1973.

7. A.G. Gaydon, "Dissociation Energies." Chapman and Hall, Ltd., London, 1968.

Free Atom Fractions, The β Factor

The free atom fraction, or β factor, is defined by equation 1.

$$\beta = \frac{[M] \ \text{atomic}}{\Sigma [M] \ \text{all forms}} \tag{1}$$

where [M] atomic is the concentration of free atoms and [M] all forms of the metal M, i.e.

[M] all forms = [M] atomic + [MO] + [MOH] + [M$^+$] + etc.

As always, the measurement in all forms of atomic spectroscopy is produced by the atoms in the free state only. The β factor then is an indication of the utility of a particular flame for a particular analyte. A β factor which approaches one is the ideal. It should be pointed out that useful analytical analysis can be performed with β factors as low as about 0.01; however, if another flame can produce , a β factor of 0.1 a 10-fold increase in sensitivity would be expected.

Unfortunately, the methods of experimentally measuring the β factor are generally frought with assumptions, and analytical flames are difficult as sample cells in terms of homogeneity. Further, the accuracy of the transition probability (Chapter III.B.1) often limits the calculation. However, the measured and calculated β factors are an important indicator as to the applicability of a flame for a particular analyte.

Table III.B.6 presents this data for 44 elements in several flames. The flame code is given in Table III.B.1. The temperature given is either the measured temperature for the burner system used on an assumed temperature for the calculation. Information concerning the use of ion suppression and further assumptions are given in the footnotes along with the references.

Table III.B.6

A TABULATION OF REPORTED VALUES OF THE ATOMIZATION FACTOR (BETA) IN FLAMES

Element	Flame	Temperature (°K)	Beta		Reference
Aluminum	Ac/N	2950	0.13		4
Al	Ac/N	3000	0.29		9
	H/N	2900	<0.0001		4
	Ac/A	2450	<0.00005		4
	Ac/A	2450	<0.0001		9
	H/A	2000	<0.00008		4
	Ac/A	2450	<0.0001		5
	Ac/N	2950	0.97	(B)	10
Antimony Sb	Ac/A	2450	0.03		8
Arsenic As	Ac/A	2450	0.0002		8
Barium	Ac/N	2950	0.17		4
Ba	Ac/N	3000	0.074		9
	H/N	2900	0.0046		4
	Ac/A	2450	0.0018		4
	Ac/A	2450	0.0011		5
	Ac/A	2410	0.0009	(A)	1
	Ac/A	2480	0.002		6,7
	Ac/A	2500	0.0026	(A)	2
	Ac/A	2480	0.0031		6
	Ac/A	2450	0.0034		9
	H/A	2000	0.005		4
Beryllium	Ac/N	2950	0.095		4
Be	H/N	2900	0.0004		4
	Ac/A	2450	0.00004		4
	H/A	2000	0.00002		4
	Ac/N	2950	0.98	(C)	10
Bismuth	Ac/N	2950	0.35		4
Bi	H/N	2900	0.26		4
	Ac/A	2450	0.17		4
	H/A	2000	0.63		4
Boron	Ac/N	2950	0.0035		4
B	H/N	2900	<0.001		4

Element	Flame	Temperature ($^{\circ}K$)	Beta	Reference
Boron (cont)	Ac/A	2450	<0.0006	4
	H/A	2000	<0.001	4
Cadmium	Ac/N	2950	0.56	4
Cd	Ac/N	3000	0.60	9
	H/N	2900	0.62	4
	Ac/A	2450	0.38	4
	Ac/A	2450	0.50	5
	Ac/A	2450	0.80	9
	H/A	2000	0.37	4
Calcium	Ac/N	2950	0.52 (B)	4
Ca	Ac/N	3000	0.34 (C)	9
	H/N	2900	0.036(B)	4
	Ac/A	2450	0.07 (B)	4
	Ac/A	2450	0.14 (C)	5
	Ac/A	2410	0.066(C)(A)	1
	Ac/A	2500	0.070(B)(A)	2
	Ac/A	2480	0.067(C)	6
	Ac/A	2480	0.05 (B)	6,7
	Ac/A	2450	0.052(C)	9
	H/A	2000	0.15 (B)	4
Cesium	Ac/A	2450	0.02	8
Cs				
Chromium	Ac/N	2950	0.63	4
Cr	Ac/N	3000	1.02	9
	H/N	2900	0.042	4
	Ac/A	2450	0.071	4
	Ac/A	2450	0.064	5
	Ac/A	2410	0.065 (A)	1
	Ac/A	2480	0.13	6,7
	Ac/A	2450	0.53	9
	H/A	2000	0.31	4
Cobalt	Ac/N	2950	0.25	4
Co	Ac/N	3000	0.11	9
	H/N	2900	0.28	4
	Ac/A	2450	0.28	4
	Ac/A	2450	0.41	4
	Ac/A	2450	0.023	9
	H/A	2000	0.21	4

Element	Flame	Temperature (°K)	Beta	Reference
Copper	Ac/N	2950	0.66	4
Cu	Ac/N	3000	0.49	9
	H/N	2900	0.92	4
	H/O/Ar	2350	0.97(E)	3
	Ac/A	2450	0.88	4
	Ac/A	2450	0.98	5
	Ac/A	2410	1.00(A)	1
	Ac/A	2480	0.82	6,7
	Ac/A	2500	1.00(A)	2
	Ac/A	2450	0.40	9
	H/A	2000	0.96	4
	Ac/N	2950	1.00(B)	10
Gallium	Ac/N	2950	0.73	4
Ga	H/N	2900	0.16	4
	Ac/A	2450	0.16	4
	Ac/A	2450	0.16	5
	H/A	2000	0.45	4
Germanium	Ac/A	2450	0.001	8
Ge				
Gold	Ac/N	2950	0.27	4
Au	Ac/N	3000	0.16	9
	H/N	2900	0.43	4
	Ac/A	2450	<0.001	5
	Ac/A	2450	0.40	4
	Ac/A	2500	0.63(A)	2
	Ac/A	2450	0.21	9
	H/A	2000	0.54	4
Indium	Ac/N	2950	0.93	4
In	Ac/N	3000	0.37	9
	H/N	2900	0.61	4
	Ac/A	2450	0.67	4
	Ac/A	2450	0.67	5
	Ac/A	2450	0.10	9
	H/A	2000	1.07	4
Iridium	Ac/A	2450	0.1	8
Ir				

Element	Flame	Temperature (°K)	Beta	Reference
Iron	Ac/N	2950	0.83	4
Fe	H/N	2900	0.91	4
	Ac/A	2450	0.84	4
	Ac/A	2450	0.66	5
	Ac/A	2410	0.38(A)	1
	H/N	2000	0.82	4
	Ac/N	2950	1.00(C)	10
Lead	Ac/N	2950	0.84	4
Pb	H/N	2900	0.93	4
	Ac/A	2450	0.77	4
	Ac/A	2450	0.44	5
	H/A	2000	1.03	4
Lithium	Ac/N	2950	0.34 (B)	4
Li	H/N	2900	0.094(B)	4
	Ac/A	2450	0.12 (B)	4
	Ac/A	2450	0.20 (C)	5
	Ac/A	2480	0.20 (B)	6,7
	Ac/A	2500	0.26 (B)(A)	2
	Ac/A	2480	0.21 (C)	6
	H/A	2000	0.14 (B)	4
	Ac/N	2950	0.96 (B)	10
Magnesium	Ac/N	2950	0.88	4
Mg	Ac/N	3000	1.07	9
	H/N	2900	0.97	4
	H/O/Ar	2350	0.29(E)	3
	Ac/A	2450	1.06	4
	Ac/A	2450	0.59	5
	Ac/A	2410	0.64(A)	1
	Ac/A	2500	0.84(A)	2
	Ac/A	2450	1.05	9
	H/A	2000	0.87	4
	Ac/N	2950	0.99(C)	10
Manganese	Ac/N	2950	0.77	4
Mn	Ac/N	3000	0.39	9
	H/N	2900	0.54	4
	H/O/Ar	2350	0.26(E)	3
	Ac/A	2450	0.62	4
	Ac/A	2450	0.45	5
	Ac/A	2480	1.0	6,7
	Ac/A	2500	0.70(A)	2
	Ac/A	2410	0.59(A)	1
	Ac/A	2480	0.93	6
	H/A	2000	0.75	4

Element	Flame	Temperature (°K)	Beta	Reference
Mercury Hg	Ac/A	2450	0.04	8
Molybdenum Mo	Ac/A	2450	0.03	8
Nickel Ni	Ac/A	2450	1	8
Palladium Pd	Ac/A	2450	1	8
Platinum Pt	Ac/A	2450	0.4	8
Potassium K	Ac/N	2950	0.12(B)	4
	H/N	2900	0.20(B)	4
	Ac/A	2480	0.45(C)	6
	Ac/A	2450	0.32(B)	4
	Ac/A	2500	0.28(D)(A)	2
	Ac/A	2450	0.25(C)	5
	Ac/A	2480	0.7 (B)	6,7
	H/A	2000	0.40(B)	4
Rhodium Rd	Ac/A	2450	1	8
Rubidium Rb	Ac/A	2500	0.16(D)(A)	2
Ruthenium Ru	Ac/A	2450	0.3	8
Selenium Se	Ac/A	2450	0.0001	8
Silicon Si	Ac/N	2950	0.055	4
	H/N	2900	<0.001	4
	Ac/A	2450	<0.001	4
	H/A	2000	<0.003	4
	Ac/N	2950	0.065-0.12(C)	10

Element	Flame	Temperature ($^\circ$K)	Beta	Reference
Silver	Ac/N	2950	0.57	4
Ag	H/N	2900	0.72	4
	H/O/Ar	2350	1.00(E)	3
	Ac/A	2450	0.70	4
	Ac/A	2450	0.66	5
	H/A	2000	0.85	4
Sodium	Ac/N	2950	0.97(B)	4
Na	Ac/N	3000	0.32(C)	9
	H/N	2900	0.90(B)	4
	Ac/A	2450	1.04(B)	4
	Ac/A	2450	1.00(B)	5
	Ac/A	2500	0.53(B)(A)	2
	Ac/A	2480	1.00(C)	6
	Ac/A	2450	0.63(C)	9
	H/A	2000	1.06(B)	4
	Ac/N	2950	0.99(B)	10
Strontium	Ac/N	2950	0.26	4
Sr	Ac/N	3000	0.57	9
	H/N	2900	0.039	4
	H/O/Ar	2350	0.094(E)	3
	Ac/A	2450	0.063	4
	Ac/A	2450	0.13	5
	Ac/A	2480	0.10	6,7
	Ac/A	2410	0.075(A)	1
	Ac/A	2450	0.068	9
	H/A	2000	0.17	4
Tellurium	Ac/A	2450	0.01	8
Te				
Thallium	Ac/N	2950	0.55	4
Tl	H/N	2900	0.61	4
	Ac/A	2450	0.52	4
	Ac/A	2450	0.36	5
	H/A	2000	0.68	4
Tin	Ac/N	2950	0.82	4
Sn	Ac/N	3000	0.35	9
	H/N	2900	0.059	4
	Ac/A	2450	0.043	4
	Ac/A	2450	<0.0001	5

Element	Flame	Temperature (°K)	Beta	Reference
Tin (cont)	Ac/A	2500	0.078(A)	2
	Ac/A	2450	0.061	9
	H/A	2000	0.38	4
Titanium Ti	Ac/N	2950	0.11	4
	H/N	2900	<0.003	4
	Ac/A	2450	<0.001	4
	H/A	2000	<0.002	4
	Ac/N	2950	0.33-0.49(C)	10
Tungsten W	Ac/A	2450	0.004	8
	Ac/N	2950	0.71 (C)	10
Vanadium V	Ac/N	2950	0.32	4
	H/N	2900	0.0081	4
	Ac/A	2450	0.015	4
	Ac/A	2450	0.0004	9
	H/A	2000	0.018	4
Zinc Zn	Ac/N	3000	0.49	9
	Ac/A	2450	0.45	5
	Ac/A	2500	1.10(A)	2
	Ac/A	2450	0.66	9

Notes:

A. Assuming Beta for copper is 1.00.

B. Ionization totally suppressed.

C. No Ionization suppressant added.

D. Ionization could not be completely suppressed.

E. Assuming Beta for Silver is 1.00.

References:

1. P.J.Th. Zeegers, W.P. Townsend and J.D. Winefordner, Spectro-Chimica Acta, 24B, 243 (1969).

2. J.B. Willis, Spectrochimica Acta, 25B, 487 (1970).

3. D.S. Smyly, W.P. Townsend, P.J.Th. Zeegers and J.D. Winefordner, Spectrochimica Acta, 26B, 531 (1971).

4. L. de Galan and G.F. Samaey, Spectrochimica Acta, 25B, 245 (1970).

5. L. de Galan and J.D. Winefordner, JQSRT, 7, 251 (1967).

6. E. Hinnov and H. Kohn, J. Opt. Soc. Am., 47, 156 (1957).

7. F.W. Hofmann and H. Kohn, J. Opt. Soc. Am., 51, 512 (1961).

8. M.L. Parsons and P.M. McElfresh, Appl. Spectrosc., 26, 472 (1972).

9. S.R. Koirtyohann and E.E. Pickett, "The Efficiency of Atom Formation in Premixed Flames" in "XIII Spectroscopium Internationale Ottawa," Adam Hilger Ltd., London, 1967, p. 270.

10. J. O. Rasmuson, V. A. Fassel, and R. N. Kniseley, Spectrochim. Acta, 28B, 365 (1973).

Spectral Line Width Data

In fundamental studies, particularly, and in many practical studies as well, information concerning the line width of a spectral transition is quite useful. In analytical flame spectroscopy the width at half-intensity for a spectral line is comprised of three types of line broadening:

1. Natural broadening, $\delta\lambda_N$.

Natural broadening is essentially due to the Heisenberg uncertainty principle and is related to the atomic transition probability according to Equation 1.

$$\delta\lambda_N = \frac{\lambda_o^2}{2\pi c} (A_{ji} + A_{ih}) \tag{1}$$

where the transition is from the ith to the jth levels, λ_o is the wavelength of the transition, c is the velocity of light, and A is the transition probability of the appropriate state.

The $\delta\lambda_N$ values are quite small, generally on the order of 10^{-5} to 10^{-4} Å, and under normal flame conditions do not contribute significantly to the overall line widths.(1).

2. Doppler Broadening, $\delta\lambda_D$.

This type of spectral broadening is due to the random motion in the flame and can be described by a gaussian curve. The half-intensity line width is given by Equation 2.

$$\delta\lambda_D = \frac{2(2R\ln 2)^{\frac{1}{2}}}{c} \lambda_o (T/A_r)^{\frac{1}{2}} \tag{2}$$

where R is the gas constant, T is the absolute temperature, and A_r is the atomic weight of the analyte species.

This type of broadening essentially accounts for the total broadening from HCL's and perhaps EDL's, and contributes significantly to the line broadening in flames. Line widths from HCL's should be approximately an order of magnitude smaller than the $\delta\lambda_D$ calculated for the common analytical flames listed in Tables III.B.7-a,b, and c.

3. Collisional Broadening, $\delta\lambda_C$.

This type of broadening is due to 'collisions' or near collisions which cause electronic level perturbations. It is described by a Lorentzian curve. The half-intensity line width is given by Equation 3.

$$\delta\lambda_C = \frac{2\lambda_o^2 \sigma_c^2 P_f}{\pi c k T} [2\pi R T (1/A_{r_a} + 1/A_{r_f})]^{\frac{1}{2}} \tag{3}$$

where σ_c^2 is the collisional cross section in units of area, P_f is the pressure of the foreign species, A_{r_a} is the atomic weight of the analyte species, A_{r_f} is the average atomic (or molecular) weight of the foreign species, and the other terms are defined above.

This equation can be solved by assuming the analyte species to be at infinite dilution (which is a very good assumption in analytical flame work), the flame gases to be at atmospheric pressure, have a knowledge of the major compounds present in the flame gases (See Table I.C.2), and finally know the value of the collisional cross section, σ_c^2. Experimental measurements of the collisional cross section have ranged from about 50 to 120 $\AA^2$ for quite a few elements under a variety of flame conditions with many being in the range of 100 $\AA^2$; therefore, a low ($50\AA^2$) and a high ($120\AA^2$) estimate was used in calculating $\delta\lambda_c$'s for Tables III.B.7-a,b,&c.

   4.   Total line width in Analytical Flames, $\delta\lambda$.

Because of the fact that the broadening effects are described by distinctly different mathematical curves, the total line width is not capable of absolute calculation; however, an empirical equation which gives a good approximation of the Voight profile (an expression which mathematically approximates the true shape of a spectral line, but cannot be absolutely solved is given in Equation 4.(2).

$$\delta\lambda = \frac{\delta\lambda_C}{2} + [(\frac{\delta\lambda_C}{2}) + \delta\lambda_D^2]^{\frac{1}{2}}$$ (4)

where all terms have been defined above. This equation was used with both the low and high estimates of $\delta\lambda_C$ for use in Tables III.B.7-a,b,&c. Constants and values of a fundamental nature were all made to conform with $\delta\lambda$ values being in $\AA$ units. It is sometimes desirable to have $\delta\nu$ information in frequency units. It must be remembered that in effect $\delta\lambda$ is a derivative value, and a simple conversion cannot be made, the correct equation for converting from $\delta\lambda$ in units of $\AA$, nm, etc., to $\delta\nu$ in frequency units (sec^{-1}) is in Equation 5.

$$\delta\lambda_{(sec^{-1})} = \frac{c}{\lambda_o^2} \delta\lambda_{(\AA)}$$ (5)

Of course the velocity of light must have consistent units with λ_o and $\delta\lambda_{(\AA)}$.

   5.   The damping constant, a.

Also of fundamental importance is the ratio of the collisional broadening to the doppler broadening, which is directly related to the damping constant, a, and is expressed by Equation 6.

$$a = \frac{\delta\lambda_C}{\delta\lambda_D} (\ell n2)^{\frac{1}{2}}$$ (6)

This value has importance in the determination of many other parameters and is therefore estimated once again with a high and low
$\delta\lambda_C$ value for Tables III.B.7-a,b,&c.

6. Tables.

The tables presented in this section cover the most common
transitions for the same 53 elements for which the spectral data
is given in Table III.B.1. The wavelengths once again conform
with the N.B.S. Tables (references 1, 2, and 4 in Chapter III.B.1).
All $\delta\lambda$ values are in Å units and flame parameters were assumed to
be as follows:

Flame	Temperature (°K)	A_{r_f} (g/mol)
Nitrous Oxide-acetylene	2950	22.59
Air-acetylene	2450	27.84
Air-hydrogen	2300	16.21

The average molecular weight of the flames, A_{r_f}, were calculated from the same data as the flame composition data in Table I.C.2.

Finally, a figure has been added which gives the relationship
between the line broadening factor, δ, which has theoretical importance in AAS (Equation 8, Chapter III.A) and the damping constant, a.

References

1. M. L. Parsons, W. J. McCarthy, and J. D. Winefordner, Appl.
   Spectrosc., 20, 223 (1966).

2. E. E. Whiting, J. Quant. Spectr. Radiative Transfer, 8, 1379
   (1968).

Table III.B.7-a

Line Width Data for the
Air-Acetylene Flame

Wavelength (Å)	$\delta\lambda_D$	$\delta\lambda_{C_L}$	$\delta\lambda_{C_H}$	$\delta\lambda_{T_L}$	$\delta\lambda_{T_H}$	a_L	a_H
	(Line Widths All In Å)					(No Units)	
ALUMINUM							
3092.71	.021	.009	.022	.026	.035	.37	.88
3092.84	.021	.009	.022	.026	.035	.37	.88
3944.01	.027	.015	.036	.036	.051	.47	1.12
3961.52	.027	.015	.037	.036	.051	.47	1.13
ANTIMONY							
2068.33	.007	.003	.008	.008	.012	.41	.97
2175.81	.007	.004	.009	.009	.012	.43	1.02
2311.47	.007	.004	.010	.010	.014	.45	1.09
2598.05	.008	.005	.012	.011	.016	.51	1.22
ARSENIC							
1936.96	.008	.003	.007	.010	.012	.31	.75
1971.97	.008	.003	.007	.010	.013	.32	.77
2349.84	.010	.004	.011	.012	.016	.38	.92
2492.91	.010	.005	.012	.013	.018	.40	.97
2860.44	.012	.007	.016	.015	.022	.46	1.11
2898.71	.012	.007	.016	.016	.022	.47	1.13
BARIUM							
4554.03	.014	.016	.037	.024	.042	.94	2.25
5535.48	.017	.023	.055	.032	.060	1.14	2.73
BERYLLIUM							
2348.61	.028	.008	.018	.032	.038	.23	.55

Line Width Data for the Air-Acetylene Flame

Wavelength (Å)	$\delta\lambda_D$	$\delta\lambda_{C_L}$	$\delta\lambda_{C_H}$	$\delta\lambda_{T_L}$	$\delta\lambda_{T_H}$	a_L	a_H
		(Line Widths All In Å)				(No Units)	
BISMUTH							
2061.70	.005	.003	.007	.007	.010	.51	1.22
2230.61	.005	.004	.009	.008	.011	.55	1.32
3024.64	.007	.007	.016	.011	.019	.75	1.79
3067.72	.008	.007	.016	.012	.019	.76	1.81
4722.19	.012	.016	.039	.022	.042	1.16	2.79
BORON							
2496.77	.027	.008	.019	.031	.038	.25	.60
2497.72	.027	.008	.019	.031	.038	.25	.60
CADMIUM							
2288.02	.008	.004	.010	.010	.014	.43	1.04
3261.06	.011	.008	.019	.016	.024	.62	1.48
CALCIUM							
3933.66	.022	.014	.033	.030	.044	.52	1.25
4226.73	.024	.016	.038	.033	.049	.56	1.34
CESIUM							
8521.10	.026	.054	.131	.065	.136	1.73	4.15
8943.50	.028	.060	.144	.071	.149	1.82	4.36
CHROMIUM							
3578.69	.018	.011	.026	.024	.035	.51	1.23
3593.49	.018	.011	.026	.024	.035	.51	1.23
4254.35	.021	.015	.037	.030	.046	.61	1.46
5206.04	.026	.023	.055	.039	.065	.74	1.79
COBALT							
2407.25	.011	.005	.012	.014	.018	.36	.86
2424.93	.011	.005	.012	.014	.018	.36	.87
2528.97	.012	.005	.013	.015	.020	.38	.91
3453.50	.016	.010	.024	.022	.032	.52	1.24
3526.85	.016	.010	.025	.022	.033	.53	1.26
4252.31	.020	.015	.036	.028	.045	.63	1.52

Line Width Data for the
Air-Acetylene Flame

Wavelength (Å)	$\delta\lambda_D$	$\delta\lambda_{C_L}$	$\delta\lambda_{C_H}$	$\delta\lambda_{T_L}$	$\delta\lambda_{T_H}$	a_L	a_H
		(Line Widths All In Å)				(No Units)	
COPPER							
3247.54	.014	.009	.021	.019	.028	.50	1.19
3273.96	.015	.009	.021	.020	.028	.50	1.20
GALLIUM							
2874.24	.012	.007	.016	.016	.023	.45	1.09
2943.64	.012	.007	.017	.016	.023	.47	1.12
2944.18	.013	.007	.017	.016	.023	.47	1.12
4172.06	.018	.014	.034	.026	.041	.66	1.58
GERMANIUM							
2068.65	.009	.003	.008	.010	.014	.33	.80
2651.18	.011	.006	.014	.014	.020	.43	1.02
2651.58	.011	.006	.014	.014	.020	.43	1.02
GOLD							
2427.95	.006	.004	.010	.009	.013	.58	1.40
2675.95	.007	.005	.013	.010	.015	.64	1.54
HAFNIUM							
2866.37	.008	.006	.014	.011	.018	.66	1.58
3072.88	.008	.007	.017	.012	.020	.71	1.70
3682.24	.010	.010	.024	.016	.027	.85	2.03
INDIUM							
3039.36	.010	.007	.017	.014	.022	.58	1.40
3256.09	.011	.008	.019	.016	.024	.62	1.49
4101.76	.014	.013	.031	.021	.036	.78	1.88
4511.31	.015	.015	.037	.025	.042	.86	2.07
IRIDIUM							
2088.82	.005	.003	.008	.007	.010	.50	1.19
2543.97	.007	.005	.011	.009	.014	.60	1.45
2639.42	.007	.005	.012	.010	.015	.63	1.50
2639.71	.007	.005	.012	.010	.015	.63	1.51
3800.12	.010	.011	.025	.016	.029	.90	2.17

Line Width Data for the Air-Acetylene Flame

Wavelength (Å)	$\delta\lambda_D$	$\delta\lambda_{C_L}$	$\delta\lambda_{C_H}$	$\delta\lambda_{T_L}$	$\delta\lambda_{T_H}$	a_L	a_H
		(Line Widths All In Å)				(No Units)	
IRON							
2483.27	.012	.005	.012	.015	.019	.36	.87
2489.75	.012	.005	.012	.015	.020	.36	.88
3719.94	.018	.012	.028	.024	.036	.54	1.31
3859.91	.018	.012	.030	.026	.039	.57	1.36
LANTHANUM							
4187.32	.013	.013	.031	.021	.036	.87	2.08
5455.15	.016	.022	.053	.031	.058	1.13	2.71
5501.34	.017	.023	.054	.031	.059	1.14	2.73
5769.34	.017	.025	.060	.034	.064	1.19	2.86
6578.51	.020	.032	.078	.042	.082	1.36	3.27
LEAD							
2169.99	.005	.003	.008	.007	.011	.53	1.28
2833.06	.007	.006	.014	.010	.017	.70	1.67
3683.48	.009	.010	.024	.015	.027	.90	2.17
4057.83	.010	.012	.029	.018	.032	1.00	2.39
LITHIUM							
3232.63	.044	.016	.038	.052	.067	.31	.73
6707.80	.090	.069	.165	.131	.205	.63	1.52
MAGNESIUM							
2852.13	.021	.008	.019	.025	.032	.33	.79
MANGANESE							
2794.82	.013	.007	.016	.017	.023	.41	.98
2798.27	.013	.007	.016	.017	.023	.41	.98
2801.06	.013	.007	.016	.017	.023	.41	.98
4030.76	.019	.014	.033	.027	.042	.59	1.41
4033.07	.019	.014	.033	.027	.042	.59	1.41
4034.49	.019	.014	.033	.027	.042	.59	1.41

Line Width Data for the
Air-Acetylene Flame

Wavelength (Å)	$\delta\lambda_D$	$\delta\lambda_{C_L}$	$\delta\lambda_{C_H}$	$\delta\lambda_{T_L}$	$\delta\lambda_{T_H}$	a_L	a_H
		(Line Widths All In Å)				(No Units)	
MERCURY							
2536.52	.006	.005	.011	.009	.014	.61	1.47
MOLYBDENUM							
3132.59	.011	.008	.018	.016	.024	.56	1.34
3170.35	.011	.008	.019	.016	.024	.56	1.36
3798.25	.014	.011	.027	.020	.033	.68	1.62
3864.11	.014	.012	.028	.021	.034	.69	1.65
3902.96	.014	.012	.028	.021	.034	.70	1.67
NICKEL							
2310.96	.011	.004	.011	.013	.017	.34	.83
2320.03	.011	.004	.011	.013	.017	.35	.83
3414.76	.016	.010	.023	.021	.031	.51	1.22
3524.54	.016	.010	.025	.022	.033	.53	1.26
NIOBIUM							
3349.10	.012	.009	.021	.017	.027	.59	1.41
4058.94	.015	.013	.031	.023	.037	.71	1.71
4079.73	.015	.013	.031	.023	.037	.72	1.72
OSMIUM							
2909.06	.007	.006	.015	.011	.018	.69	1.65
4420.47	.011	.014	.034	.021	.038	1.05	2.51
PALLADIUM							
2447.91	.008	.005	.011	.011	.016	.45	1.09
2476.42	.009	.005	.011	.011	.016	.46	1.10
3404.58	.012	.009	.021	.017	.026	.63	1.52
3609.55	.012	.010	.024	.018	.029	.67	1.61
3634.70	.012	.010	.024	.019	.030	.67	1.62
PLATINUM							
2659.45	.007	.005	.012	.010	.015	.64	1.53
3064.71	.008	.007	.016	.012	.020	.73	1.76

Line Width Data for the Air-Acetylene Flame

Wavelength (Å)	$\delta\lambda_D$	$\delta\lambda_{C_L}$	$\delta\lambda_{C_H}$	$\delta\lambda_{T_L}$	$\delta\lambda_{T_H}$	a_L	a_H
		(Line Widths All In Å)				(No Units)	
POTASSIUM							
7664.91	.043	.052	.126	.077	.139	1.00	2.41
7698.98	.044	.053	.127	.077	.140	1.01	2.42
RHENIUM							
2287.51	.006	.004	.009	.008	.012	.54	1.29
3460.46	.009	.009	.021	.014	.024	.81	1.95
3464.73	.009	.009	.021	.014	.024	.81	1.95
RHODIUM							
3434.89	.012	.009	.022	.017	.027	.63	1.51
3692.36	.013	.010	.025	.019	.031	.68	1.62
4374.80	.015	.015	.035	.024	.041	.80	1.92
RUBIDIUM							
7800.23	.030	.048	.115	.062	.122	1.33	3.19
7947.60	.030	.050	.119	.064	.126	1.35	3.25
RUTHENIUM							
3498.94	.012	.009	.023	.018	.028	.64	1.53
3728.03	.013	.011	.026	.020	.031	.68	1.63
SCANDIUM							
3907.49	.021	.013	.032	.028	.042	.53	1.28
3911.81	.021	.013	.032	.028	.042	.53	1.28
4020.40	.021	.014	.034	.029	.044	.55	1.32
6305.67	.033	.035	.083	.055	.095	.86	2.07
SELENIUM							
1960.26	.008	.003	.007	.009	.012	.32	.78
2039.85	.008	.003	.008	.010	.013	.34	.81
SILICON							
2516.11	.017	.006	.015	.020	.026	.30	.72

Line Width Data for the Air-Acetylene Flame

Wavelength (Å)	$\delta\lambda_D$	$\delta\lambda_{C_L}$	$\delta\lambda_{C_H}$	$\delta\lambda_{T_L}$	$\delta\lambda_{T_H}$	a_L	a_H
			(Line Widths All In Å)			(No Units)	
SILVER							
3280.68	.011	.008	.020	.016	.025	.61	1.47
3382.89	.012	.009	.021	.017	.026	.63	1.51
SODIUM							
5889.95	.044	.035	.084	.065	.103	.67	1.61
5895.92	.044	.035	.085	.065	.103	.67	1.62
STRONTIUM							
4077.71	.015	.013	.031	.023	.038	.70	1.68
4607.33	.017	.017	.040	.028	.046	.79	1.90
TANTALUM							
2714.67	.007	.005	.013	.010	.016	.63	1.51
4740.16	.012	.016	.040	.023	.043	1.10	2.63
4812.75	.013	.017	.041	.024	.044	1.11	2.67
TELLERIUM							
2142.75	.007	.003	.008	.009	.012	.43	1.03
2383.25	.007	.004	.010	.010	.014	.48	1.14
2385.76	.007	.004	.010	.010	.014	.48	1.14
THALLIUM							
3775.72	.009	.010	.025	.016	.028	.92	2.21
5350.46	.013	.021	.050	.027	.053	1.31	3.13
TIN							
2246.05	.007	.004	.009	.009	.013	.44	1.05
2354.84	.008	.004	.010	.010	.014	.46	1.10
2839.99	.009	.006	.015	.013	.019	.55	1.32
2863.33	.009	.006	.015	.013	.019	.56	1.33
3034.12	.010	.007	.017	.014	.021	.59	1.41
3175.05	.010	.008	.018	.015	.023	.62	1.48
3262.34	.011	.008	.019	.015	.024	.63	1.52

Line Width Data for the
Air-Acetylene Flame

Wavelength (Å)	$\delta\lambda_D$	$\delta\lambda_{C_L}$	$\delta\lambda_{C_H}$	$\delta\lambda_{T_H}$	$\delta\lambda_T$	a_L	a_H
		(Line Widths All In Å)				(No Units)	
TITANIUM							
3635.46	.019	.011	.027	.025	.037	.51	1.22
3642.68	.019	.011	.027	.025	.037	.51	1.22
3653.50	.019	.011	.027	.025	.037	.51	1.22
3998.64	.020	.014	.033	.028	.043	.56	1.34
TUNGSTEN							
2551.35	.007	.005	.011	.009	.015	.59	1.43
2681.41	.007	.005	.013	.010	.016	.62	1.50
4008.75	.010	.012	.028	.018	.032	.93	2.24
4074.36	.011	.012	.029	.018	.033	.95	2.28
VANADIUM							
3183.41	.016	.009	.021	.021	.029	.45	1.09
3183.98	.016	.009	.021	.021	.029	.45	1.09
3185.40	.016	.009	.021	.021	.029	.45	1.09
4111.78	.020	.014	.034	.029	.044	.58	1.40
4379.24	.022	.016	.039	.031	.049	.62	1.49
YTTRIUM							
3620.94	.014	.010	.025	.020	.031	.63	1.50
4077.38	.015	.013	.031	.023	.037	.71	1.69
4102.38	.015	.013	.032	.023	.038	.71	1.70
6435.00	.024	.032	.078	.045	.085	1.11	2.67
ZINC							
2138.56	.009	.004	.009	.011	.015	.33	.79
ZIRCONIUM							
3519.60	.013	.010	.023	.019	.029	.62	1.48
3547.68	.013	.010	.024	.019	.029	.62	1.49
3601.19	.013	.010	.024	.019	.030	.63	1.51

Table III.B.7-b

Line Width Data for the Nitrous Oxide-Acetylene Flame

Wavelength (Å)	$\delta\lambda_D$	$\delta\lambda_{C_L}$	$\delta\lambda_{C_H}$	$\delta\lambda_{T_L}$	$\delta\lambda_{T_H}$	a_L	a_H
		(Line Widths All In Å)				(No Units)	
ALUMINUM							
3092.71	.023	.009	.021	.028	.036	.32	.77
3092.84	.023	.009	.021	.028	.036	.32	.77
3944.01	.030	.015	.035	.038	.052	.41	.98
3961.52	.030	.015	.035	.038	.052	.41	.99
ANTIMONY							
2068.33	.007	.003	.008	.009	.012	.37	.88
2175.81	.008	.004	.009	.010	.013	.39	.93
2311.47	.008	.004	.010	.010	.014	.41	.98
2598.05	.009	.005	.012	.012	.017	.46	1.11
ARSENIC							
1936.96	.009	.003	.007	.010	.013	.28	.68
1971.97	.009	.003	.007	.011	.013	.29	.69
2349.84	.011	.004	.010	.013	.017	.34	.82
2492.91	.011	.005	.012	.014	.019	.36	.87
2860.44	.013	.006	.015	.016	.023	.42	1.00
2898.71	.013	.007	.016	.017	.023	.42	1.01
BARIUM							
4554.03	.015	.015	.037	.025	.042	.85	2.04
5535.48	.018	.023	.055	.033	.060	1.03	2.48
BERYLLIUM							
2348.61	.030	.007	.017	.034	.040	.19	.47

Line Width Data for the Nitrous Oxide-Acetylene Flame

Wavelength (Å)	$\delta\lambda_D$	$\delta\lambda_{C_L}$	$\delta\lambda_{C_H}$	$\delta\lambda_{T_L}$	$\delta\lambda_{T_H}$	a_L	a_H
		(Line Widths All In Å)				(No Units)	
BISMUTH							
2061.70	.006	.003	.007	.007	.010	.46	1.11
2230.61	.006	.004	.009	.008	.012	.50	1.20
3024.64	.008	.007	.016	.012	.019	.68	1.63
3067.72	.008	.007	.016	.012	.020	.69	1.65
4722.19	.013	.016	.039	.023	.043	1.06	2.55
BORON							
2496.77	.030	.008	.018	.034	.040	.21	.51
2497.72	.030	.008	.018	.034	.040	.21	.51
CADMIUM							
2288.02	.008	.004	.009	.011	.014	.39	.94
3261.06	.012	.008	.019	.017	.025	.56	1.34
CALCIUM							
3933.66	.024	.013	.032	.032	.045	.46	1.10
4226.73	.026	.015	.037	.035	.050	.49	1.19
CESIUM							
8521.10	.029	.054	.130	.067	.136	1.57	3.77
8943.50	.030	.060	.143	.072	.149	1.65	3.95
CHROMIUM							
3578.69	.019	.011	.025	.025	.036	.46	1.10
3593.49	.019	.011	.026	.025	.036	.46	1.10
4254.35	.023	.015	.036	.032	.047	.54	1.30
5206.04	.028	.022	.054	.041	.066	.66	1.59
COBALT							
2407.25	.012	.005	.011	.015	.019	.32	.77
2424.93	.012	.005	.011	.015	.019	.32	.78
2528.97	.013	.005	.012	.016	.020	.34	.81
3453.50	.018	.010	.023	.023	.033	.46	1.10
3526.85	.018	.010	.024	.024	.034	.47	1.13
4252.31	.022	.015	.035	.030	.045	.57	1.36

Line Width Data for the
Nitrous Oxide-Acetylene Flame

Wavelength (Å)	$\delta\lambda_D$	$\delta\lambda_{C_L}$	$\delta\lambda_{C_H}$	$\delta\lambda_{T_L}$	$\delta\lambda_{T_H}$	a_L	a_H
		(Line Widths All In Å)				(No Units)	
COPPER							
3247.54	.016	.008	.020	.021	.029	.45	1.07
3273.96	.016	.009	.021	.021	.029	.45	1.08
GALLIUM							
2874.24	.013	.007	.016	.017	.023	.41	.98
2943.64	.014	.007	.017	.018	.024	.42	1.00
2944.18	.014	.007	.017	.018	.024	.42	1.00
4172.06	.019	.014	.033	.028	.042	.59	1.42
GERMANIUM							
2068.65	.009	.003	.008	.011	.014	.30	.72
2651.18	.012	.006	.013	.015	.020	.38	.92
2651.58	.012	.006	.013	.015	.020	.38	.92
GOLD							
2427.95	.007	.004	.010	.009	.014	.53	1.27
2675.95	.007	.005	.013	.010	.016	.59	1.41
HAFNIUM							
2866.37	.008	.006	.014	.012	.018	.60	1.44
3072.88	.009	.007	.017	.013	.021	.64	1.54
3682.24	.011	.010	.024	.017	.028	.77	1.85
INDIUM							
3039.36	.011	.007	.017	.015	.022	.53	1.26
3256.09	.012	.008	.019	.016	.025	.56	1.35
4101.76	.015	.013	.030	.023	.037	.71	1.70
4511.31	.016	.015	.037	.026	.043	.78	1.87
IRIDIUM							
2088.82	.006	.003	.008	.008	.011	.45	1.08
2543.97	.007	.005	.011	.010	.015	.55	1.32
2639.42	.007	.005	.012	.010	.016	.57	1.37
2639.71	.007	.005	.012	.010	.016	.57	1.37
3800.12	.011	.011	.025	.017	.029	.82	1.97

Line Width Data for the
Nitrous Oxide-Acetylene Flame

Wavelength (Å)	$\delta\lambda_D$	$\delta\lambda_{C_L}$	$\delta\lambda_{C_H}$	$\delta\lambda_{T_L}$	$\delta\lambda_{T_H}$	a_L	a_H
	(Line Widths All In Å)					(No Units)	
IRON							
2483.27	.013	.005	.012	.016	.020	.32	.78
2489.75	.013	.005	.012	.016	.020	.33	.78
3719.94	.019	.011	.027	.026	.037	.49	1.17
3859.91	.020	.012	.029	.027	.039	.50	1.21
LANTHANUM							
4187.32	.014	.013	.031	.022	.037	.79	1.89
5455.15	.018	.022	.053	.032	.059	1.02	2.46
5501.34	.018	.023	.054	.033	.060	1.03	2.48
5769.34	.019	.025	.059	.035	.065	1.08	2.60
6578.51	.022	.032	.077	.043	.083	1.23	2.96
LEAD							
2169.99	.006	.003	.008	.008	.011	.49	1.17
2833.06	.008	.006	.014	.011	.017	.63	1.52
3683.48	.010	.010	.024	.016	.027	.82	1.98
4057.83	.011	.012	.029	.018	.032	.91	2.18
LITHIUM							
3232.63	.048	.015	.036	.056	.069	.26	.62
6707.80	.099	.064	.154	.136	.202	.54	1.29
MAGNESIUM							
2852.13	.023	.008	.019	.027	.034	.29	.69
MANGANESE							
2794.82	.015	.006	.015	.018	.024	.36	.87
2798.27	.015	.006	.015	.018	.024	.36	.87
2801.06	.015	.006	.015	.018	.024	.36	.87
4030.76	.021	.013	.032	.029	.042	.52	1.26
4033.07	.021	.013	.032	.029	.043	.52	1.26
4034.49	.021	.013	.032	.029	.043	.52	1.26

Line Width Data for the
Nitrous Oxide-Acetylene Flame

Wavelength (Å)	$\delta\lambda_D$	$\delta\lambda_{C_L}$	$\delta\lambda_{C_H}$	$\delta\lambda_{T_L}$	$\delta\lambda_{r_H}$	a_L	a_H
		(Line Widths All In Å)				(No Units)	
MERCURY							
2536.52	.007	.005	.011	.010	.015	.56	1.34
MOLYBDENUM							
3132.59	.012	.008	.018	.017	.024	.50	1.21
3170.35	.013	.008	.018	.017	.025	.51	1.22
3798.25	.015	.011	.027	.022	.033	.61	1.47
3864.11	.015	.011	.027	.022	.034	.62	1.49
3902.96	.016	.012	.028	.022	.035	.63	1.51
NICKEL							
2310.96	.012	.004	.010	.014	.018	.31	.74
2320.03	.012	.004	.010	.014	.018	.31	.74
3414.76	.017	.009	.023	.023	.032	.45	1.09
3524.54	.018	.010	.024	.024	.034	.47	1.13
NIOBIUM							
3349.10	.014	.009	.021	.019	.027	.53	1.28
4058.94	.016	.013	.030	.024	.038	.64	1.55
4079.73	.016	.013	.031	.024	.038	.65	1.55
OSMIUM							
2909.06	.008	.006	.015	.012	.018	.63	1.50
4420.47	.012	.014	.034	.021	.038	.95	2.29
PALLADIUM							
2447.91	.009	.005	.011	.012	.016	.41	.99
2476.42	.009	.005	.011	.012	.016	.42	1.00
3404.58	.013	.009	.021	.018	.027	.57	1.37
3609.55	.014	.010	.024	.019	.030	.61	1.45
3634.70	.014	.010	.024	.020	.030	.61	1.46
PLATINUM							
2659.45	.007	.005	.012	.010	.016	.58	1.39
3064.71	.009	.007	.016	.013	.020	.67	1.60

Line Width Data for the Nitrous Oxide-Acetylene Flame

Wavelength (Å)	$\delta\lambda_D$	$\delta\lambda_{C_L}$	$\delta\lambda_{C_H}$	$\delta\lambda_{T_L}$	$\delta\lambda_{T_H}$	a_L	a_H
		(Line Widths All In Å)				(No Units)	
POTASSIUM							
7664.91	.048	.051	.122	.080	.139	.89	2.13
7698.98	.048	.051	.123	.080	.140	.89	2.14
RHENIUM							
2287.51	.007	.004	.009	.009	.013	.49	1.17
3460.46	.010	.009	.021	.015	.025	.74	1.77
3464.73	.010	.009	.021	.015	.025	.74	1.77
RHODIUM							
3434.89	.013	.009	.022	.018	.028	.57	1.36
3692.36	.014	.010	.025	.020	.031	.61	1.47
4374.80	.017	.015	.035	.026	.042	.72	1.74
RUBIDIUM							
7800.23	.033	.047	.113	.064	.122	1.20	2.87
7947.60	.033	.049	.118	.066	.126	1.22	2.93
RUTHENIUM							
3498.94	.014	.009	.022	.019	.029	.57	1.38
3728.03	.014	.011	.025	.021	.032	.61	1.47
SCANDIUM							
3907.49	.023	.013	.031	.030	.043	.47	1.14
3911.81	.023	.013	.031	.030	.043	.47	1.14
4020.40	.023	.014	.033	.031	.045	.49	1.17
6305.67	.037	.034	.081	.057	.095	.77	1.84
SELENIUM							
1960.26	.009	.003	.007	.010	.013	.29	.70
2039.85	.009	.003	.008	.011	.014	.30	.73
SILICON							
2516.11	.018	.006	.014	.022	.027	.26	.63

Line Width Data for the
Nitrous Oxide-Acetylene Flame

Wavelength (Å)	$\delta\lambda_D$	$\delta\lambda_{C_L}$	$\delta\lambda_{C_H}$	$\delta\lambda_{T_L}$	$\delta\lambda_{T_H}$	a_L	a_H
		(Line Widths All In Å)				(No Units)	
SILVER							
3280.68	.012	.008	.020	.017	.026	.55	1.33
3382.89	.013	.009	.021	.018	.027	.57	1.37
SODIUM							
5889.95	.048	.034	.081	.068	.103	.59	1.41
5895.92	.048	.034	.081	.068	.103	.59	1.41
STRONTIUM							
4077.71	.017	.013	.031	.025	.038	.63	1.52
4607.33	.019	.016	.039	.029	.047	.71	1.71
TANTALUM							
2714.67	.008	.005	.013	.011	.017	.57	1.37
4740.16	.014	.016	.039	.024	.044	1.00	2.40
4812.75	.014	.017	.041	.025	.045	1.01	2.43
TELLERIUM							
2142.75	.007	.003	.008	.009	.013	.39	.93
2383.25	.008	.004	.010	.011	.015	.43	1.04
2385.76	.008	.004	.010	.011	.015	.43	1.04
THALLIUM							
3775.72	.010	.010	.025	.017	.029	.84	2.02
5350.46	.015	.021	.050	.028	.054	1.19	2.86
TIN							
2246.05	.008	.004	.009	.010	.014	.39	.95
2354.84	.008	.004	.010	.011	.015	.41	.99
2839.99	.010	.006	.015	.014	.020	.50	1.20
2863.33	.010	.006	.015	.014	.020	.50	1.21
3034.12	.011	.007	.017	.015	.022	.53	1.28
3175.05	.011	.008	.018	.016	.024	.56	1.34
3262.34	.012	.008	.019	.016	.025	.57	1.37

Line Width Data for the
Nitrous Oxide-Acetylene Flame

Wavelength (Å)	$\delta\lambda_D$	$\delta\lambda_{C_L}$	$\delta\lambda_{C_H}$	$\delta\lambda_{T_L}$	$\delta\lambda_{T_H}$	a_L	a_H
			(Line Widths All In Å)			(No Units)	
TITANIUM							
3635.46	.020	.011	.027	.027	.038	.45	1.08
3642.68	.020	.011	.027	.027	.038	.45	1.08
3653.50	.021	.011	.027	.027	.038	.45	1.09
3998.64	.022	.013	.032	.030	.044	.50	1.19
TUNGSTEN							
2551.35	.007	.005	.011	.010	.015	.54	1.30
2681.41	.008	.005	.013	.011	.016	.57	1.37
4008.75	.012	.012	.028	.019	.032	.85	2.04
4074.36	.012	.012	.029	.019	.033	.86	2.07
VANADIUM							
3183.41	.017	.008	.020	.022	.030	.40	.97
3183.98	.017	.008	.020	.022	.030	.40	.97
3185.40	.017	.008	.020	.022	.030	.40	.97
4111.78	.022	.014	.034	.030	.045	.52	1.25
4379.24	.024	.016	.038	.033	.050	.55	1.33
YTTRIUM							
3620.94	.015	.010	.024	.021	.031	.56	1.35
4077.38	.017	.013	.031	.024	.038	.64	1.53
4102.38	.017	.013	.031	.025	.039	.64	1.54
6435.00	.027	.032	.077	.047	.085	1.00	2.41
ZINC							
2138.56	.010	.004	.009	.012	.016	.30	.71
ZIRCONIUM							
3519.60	.014	.010	.023	.020	.030	.55	1.33
3547.68	.014	.010	.023	.020	.030	.56	1.34
3601.19	.015	.010	.024	.020	.031	.57	1.36

Table III.B.7-c

Line Width Data for the
Air-Hydrogen Flame

Wavelength (Å)	$\delta\lambda_D$	$\delta\lambda_{C_L}$	$\delta\lambda_{C_H}$	$\delta\lambda_{T_L}$	$\delta\lambda_{T_H}$	a_L	a_H
		(Line Widths All In Å)				(No Units)	
ALUMINUM							
3092.71	.020	.011	.027	.027	.038	.45	1.09
3092.84	.020	.011	.027	.027	.038	.45	1.09
3944.01	.026	.018	.044	.037	.056	.58	1.39
3961.52	.026	.018	.044	.037	.056	.58	1.40
ANTIMONY							
2068.33	.006	.004	.010	.009	.013	.54	1.30
2175.81	.007	.005	.011	.009	.014	.57	1.37
2311.47	.007	.005	.013	.010	.016	.61	1.46
2598.05	.008	.007	.016	.012	.019	.68	1.64
ARSENIC							
1936.96	.008	.004	.009	.010	.014	.41	.99
1971.97	.008	.004	.009	.010	.014	.42	1.01
2349.84	.009	.006	.013	.013	.018	.50	1.20
2492.91	.010	.006	.015	.014	.020	.53	1.28
2860.44	.011	.008	.020	.016	.025	.61	1.47
2898.71	.012	.009	.021	.017	.026	.62	1.48
BARIUM							
4554.03	.013	.020	.049	.027	.052	1.26	3.03
5535.48	.016	.030	.072	.037	.075	1.53	3.68
BERYLLIUM							
2348.61	.027	.009	.020	.031	.039	.26	.63

Line Width Data for the Air-Hydrogen Flame

Wavelength (Å)	$\delta\lambda_D$	$\delta\lambda_{C_L}$	$\delta\lambda_{C_H}$	$\delta\lambda_{T_L}$	$\delta\lambda_{T_H}$	a_L	a_H
		(Line Widths All In Å)				(No Units)	
BISMUTH							
2061.70	.005	.004	.010	.007	.012	.69	1.66
2230.61	.005	.005	.011	.008	.014	.75	1.80
3024.64	.007	.009	.021	.013	.023	1.01	2.44
3067.72	.007	.009	.022	.013	.024	1.03	2.47
4722.19	.011	.021	.051	.026	.054	1.58	3.80
BORON							
2496.77	.026	.009	.022	.031	.039	.29	.70
2497.72	.026	.009	.022	.031	.039	.29	.70
CADMIUM							
2288.02	.007	.005	.012	.010	.016	.58	1.39
3261.06	.011	.010	.025	.017	.029	.83	1.98
CALCIUM							
3933.66	.021	.017	.041	.031	.050	.66	1.58
4226.73	.023	.020	.047	.035	.056	.71	1.70
CESIUM							
8521.10	.025	.071	.170	.079	.174	2.33	5.58
8943.50	.027	.078	.188	.086	.191	2.44	5.86
CHROMIUM							
3578.69	.017	.014	.032	.025	.040	.66	1.59
3593.49	.017	.014	.033	.025	.040	.66	1.59
4254.35	.020	.019	.046	.032	.054	.79	1.89
5206.04	.025	.029	.069	.043	.077	.96	2.31
COBALT							
2407.25	.011	.006	.014	.014	.020	.47	1.12
2424.93	.011	.006	.015	.014	.020	.47	1.13
2528.97	.011	.007	.016	.015	.022	.49	1.18
3453.50	.015	.012	.030	.023	.036	.67	1.61
3526.85	.016	.013	.031	.024	.038	.68	1.64
4252.31	.019	.019	.045	.031	.052	.82	1.98

Line Width Data for the
Air-Hydrogen Flame

Wavelength (Å)	$\delta\lambda_D$	$\delta\lambda_{C_L}$	$\delta\lambda_{C_H}$	$\delta\lambda_{T_L}$	$\delta\lambda_{T_H}$	a_L	a_H
		(Line Widths All In Å)				(No Units)	
COPPER							
3247.54	.014	.011	.026	.020	.032	.65	1.56
3273.96	.014	.011	.027	.021	.033	.65	1.57
GALLIUM							
2874.24	.012	.008	.020	.017	.026	.60	1.43
2943.64	.012	.009	.021	.017	.027	.61	1.46
2944.18	.012	.009	.021	.017	.027	.61	1.46
4172.06	.017	.018	.043	.028	.049	.86	2.08
GERMANIUM							
2068.65	.008	.004	.010	.011	.015	.44	1.05
2651.18	.011	.007	.017	.015	.022	.56	1.34
2651.58	.011	.007	.017	.015	.022	.56	1.34
GOLD							
2427.95	.006	.006	.014	.009	.016	.79	1.90
2675.95	.007	.007	.016	.011	.019	.87	2.10
HAFNIUM							
2866.37	.007	.008	.019	.012	.022	.89	2.15
3072.88	.008	.009	.022	.014	.024	.96	2.30
3682.24	.009	.013	.031	.018	.034	1.15	2.76
INDIUM							
3039.36	.010	.009	.022	.015	.026	.78	1.87
3256.09	.010	.010	.025	.017	.029	.83	2.00
4101.76	.013	.017	.040	.024	.044	1.05	2.52
4511.31	.014	.020	.048	.028	.052	1.15	2.77
IRIDIUM							
2088.82	.005	.004	.010	.008	.012	.67	1.62
2543.97	.006	.006	.015	.010	.017	.82	1.97
2639.42	.007	.007	.016	.011	.018	.85	2.04
2639.71	.007	.007	.016	.011	.018	.85	2.04
3800.12	.009	.014	.033	.019	.036	1.23	2.94

Line Width Data for the
Air-Hydrogen Flame

Wavelength (Å)	$\delta\lambda_D$	$\delta\lambda_{C_L}$	$\delta\lambda_{C_H}$	$\delta\lambda_{T_L}$	$\delta\lambda_{T_H}$	a_L	a_H
		(Line Widths All In Å)				(No Units)	
IRON							
2483.27	.011	.006	.016	.015	.022	.47	1.13
2489.75	.011	.006	.016	.015	.022	.47	1.13
3719.94	.017	.014	.035	.026	.042	.71	1.69
3859.91	.018	.016	.037	.027	.045	.73	1.76
LANTHANUM							
4187.32	.012	.017	.041	.023	.044	1.17	2.80
5455.15	.016	.029	.070	.036	.073	1.52	3.65
5501.34	.016	.029	.071	.037	.074	1.53	3.68
5769.34	.017	.032	.078	.040	.081	1.61	3.86
6578.51	.019	.042	.101	.050	.105	1.83	4.40
LEAD							
2169.99	.005	.005	.011	.008	.013	.73	1.74
2833.06	.007	.008	.018	.012	.021	.95	2.27
3683.48	.009	.013	.031	.017	.033	1.23	2.95
4057.83	.010	.016	.038	.020	.040	1.36	3.25
LITHIUM							
3232.63	.042	.018	.042	.052	.068	.35	.83
6707.80	.087	.076	.182	.133	.217	.72	1.73
MAGNESIUM							
2852.13	.020	.010	.023	.025	.035	.41	.97
MANGANESE							
2794.82	.013	.008	.020	.018	.026	.53	1.26
2798.27	.013	.008	.020	.018	.026	.53	1.27
2801.06	.013	.008	.020	.018	.026	.53	1.27
4030.76	.019	.017	.041	.029	.048	.76	1.82
4033.07	.019	.017	.041	.029	.048	.76	1.83
4034.49	.019	.017	.041	.029	.048	.76	1.83

Line Width Data for the
Air-Hydrogen Flame

Wavelength (Å)	$\delta\lambda_D$	$\delta\lambda_{C_L}$	$\delta\lambda_{C_H}$	$\delta\lambda_{T_L}$	$\delta\lambda_{T_H}$	a_L	a_H
		(Line Widths All In Å)				(No Units)	
MERCURY							
2536.52	.006	.006	.015	.010	.017	.83	2.00
MOLYBDENUM							
3132.59	.011	.010	.023	.017	.028	.74	1.78
3170.35	.011	.010	.024	.017	.028	.75	1.80
3798.25	.013	.014	.035	.022	.039	.90	2.16
3864.11	.014	.015	.036	.023	.040	.91	2.20
3902.96	.014	.015	.036	.023	.041	.92	2.22
NICKEL							
2310.96	.010	.006	.013	.014	.019	.45	1.07
2320.03	.010	.006	.013	.014	.019	.45	1.08
3414.76	.015	.012	.029	.023	.036	.66	1.59
3524.54	.016	.013	.031	.024	.038	.68	1.64
NIOBIUM							
3349.10	.012	.011	.027	.019	.031	.78	1.88
4058.94	.014	.016	.040	.025	.044	.95	2.27
4079.73	.015	.017	.040	.025	.045	.95	2.29
OSMIUM							
2909.06	.007	.008	.020	.012	.022	.93	2.24
4420.47	.011	.019	.045	.024	.048	1.42	3.41
PALLADIUM							
2447.91	.008	.006	.014	.012	.018	.61	1.45
2476.42	.008	.006	.015	.012	.018	.61	1.47
3404.58	.011	.011	.028	.018	.032	.84	2.02
3609.55	.012	.013	.031	.020	.035	.89	2.14
3634.70	.012	.013	.031	.020	.036	.90	2.16
PLATINUM							
2659.45	.007	.007	.016	.011	.019	.86	2.07
3064.71	.008	.009	.022	.013	.024	1.00	2.39

Line Width Data for the
Air-Hydrogen Flame

Wavelength (Å)	$\delta\lambda_D$	$\delta\lambda_{C_L}$	$\delta\lambda_{C_H}$	$\delta\lambda_{T_L}$	$\delta\lambda_{T_H}$	a_L	a_H
		(Line Widths All In Å)				(No Units)	
POTASSIUM							
7664.91	.042	.064	.155	.085	.165	1.27	3.06
7698.98	.042	.065	.156	.086	.167	1.28	3.07
RHENIUM							
2287.51	.006	.005	.012	.009	.014	.73	1.75
3460.46	.009	.012	.028	.016	.030	1.10	2.64
3464.73	.009	.012	.028	.016	.030	1.10	2.64
RHODIUM							
3434.89	.012	.012	.028	.019	.032	.84	2.01
3692.36	.013	.014	.032	.021	.037	.90	2.16
4374.80	.015	.019	.046	.027	.050	1.07	2.56
RUBIDIUM							
7800.23	.029	.061	.147	.073	.152	1.76	4.22
7947.60	.030	.064	.153	.075	.158	1.79	4.30
RUTHENIUM							
3498.94	.012	.012	.029	.019	.033	.85	2.03
3728.03	.013	.014	.033	.021	.037	.90	2.17
SCANDIUM							
3907.49	.020	.016	.039	.030	.048	.68	1.64
3911.81	.020	.016	.040	.030	.048	.68	1.64
4020.40	.021	.017	.042	.031	.050	.70	1.69
6305.67	.032	.043	.103	.060	.112	1.10	2.65
SELENIUM							
1960.26	.008	.004	.009	.010	.014	.43	1.03
2039.85	.008	.004	.010	.010	.014	.44	1.07
SILICON							
2516.11	.016	.007	.018	.020	.027	.37	.90

Line Width Data for the Air-Hydrogen Flame

Wavelength (Å)	$\delta\lambda_D$	$\delta\lambda_{C_L}$	$\delta\lambda_{C_H}$	$\delta\lambda_{T_L}$	$\delta\lambda_{T_H}$	a_L	a_H
	(Line Widths All In Å)					(No Units)	
SILVER							
3280.68	.011	.011	.026	.017	.030	.82	1.96
3382.89	.011	.011	.027	.018	.031	.84	2.02
SODIUM							
5889.95	.042	.042	.100	.068	.116	.82	1.98
5895.92	.042	.042	.100	.068	.116	.83	1.98
STRONTIUM							
4077.71	.015	.017	.040	.025	.045	.93	2.23
4607.33	.017	.021	.051	.031	.056	1.05	2.52
TANTALUM							
2714.67	.007	.007	.017	.011	.019	.85	2.05
4740.16	.012	.022	.052	.027	.055	1.49	3.57
4812.75	.012	.022	.054	.028	.056	1.51	3.63
TELLERIUM							
2142.75	.007	.004	.011	.009	.014	.57	1.38
2383.25	.007	.006	.013	.011	.017	.64	1.53
2385.76	.007	.006	.013	.011	.017	.64	1.54
THALLIUM							
3775.72	.009	.014	.033	.018	.035	1.25	3.01
5350.46	.013	.027	.066	.033	.068	1.78	4.26
TIN							
2246.05	.007	.005	.012	.010	.015	.58	1.40
2354.84	.007	.005	.013	.011	.016	.61	1.47
2839.99	.009	.008	.019	.014	.023	.74	1.77
2863.33	.009	.008	.019	.014	.023	.74	1.78
3034.12	.010	.009	.022	.015	.025	.79	1.89
3175.05	.010	.010	.024	.016	.027	.82	1.98
3262.34	.010	.010	.025	.017	.029	.85	2.03

Line Width Data for the Air-Hydrogen Flame

Wavelength (Å)	$\delta\lambda_D$	$\delta\lambda_{C_L}$	$\delta\lambda_{C_H}$	$\delta\lambda_{T_L}$	$\delta\lambda_{T_H}$	a_L	a_H
		(Line Widths All In Å)				(No Units)	
TITANIUM							
3635.46	.018	.014	.034	.026	.042	.65	1.56
3642.68	.018	.014	.034	.026	.042	.65	1.56
3653.50	.018	.014	.034	.027	.042	.65	1.57
3998.64	.020	.017	.041	.030	.049	.72	1.72
TUNGSTEN							
2551.35	.006	.006	.015	.010	.017	.81	1.94
2681.41	.007	.007	.017	.011	.019	.85	2.03
4008.75	.010	.015	.037	.020	.040	1.27	3.04
4074.36	.010	.016	.038	.021	.041	1.29	3.09
VANADIUM							
3183.41	.015	.011	.026	.022	.033	.58	1.40
3183.98	.015	.011	.026	.022	.033	.58	1.40
3185.40	.015	.011	.026	.022	.033	.58	1.40
4111.78	.020	.018	.043	.031	.051	.75	1.81
4379.24	.021	.020	.049	.034	.057	.80	1.93
YTTRIUM							
3620.94	.013	.013	.032	.021	.036	.83	1.99
4077.38	.015	.017	.040	.025	.045	.93	2.24
4102.38	.015	.017	.041	.026	.045	.94	2.26
6435.00	.023	.042	.100	.052	.105	1.47	3.54
ZINC							
2138.56	.009	.005	.011	.012	.016	.43	1.04
ZIRCONIUM							
3519.60	.013	.012	.030	.020	.034	.82	1.96
3547.68	.013	.013	.030	.021	.035	.82	1.97
3601.19	.013	.013	.031	.021	.036	.83	2.00

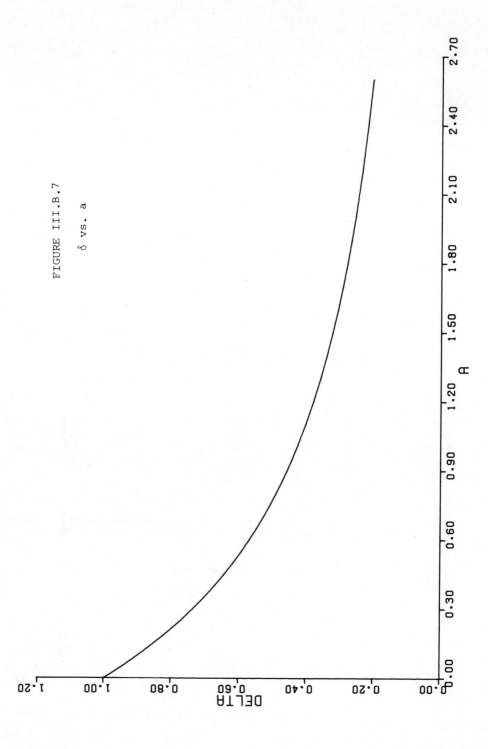

FIGURE III.B.7

δ vs. a

Spectral Interference Data

    In addition to the tables in Chapter I.E., it is felt that
certain additional information would be useful to the analytical
spectroscopist. Any time that two transitions are not resolved
by the monochromator, regardless of their origin, a potential spectral inteference is faced. Therefore, a tabulation of close spectral lines has been included from the data in Table III.B.1 for
the transitions observed in AAS (Table III.B.8-a), AES (Table III.
B.8-b) and AFS (Table III.B.8-c).

    Furthermore, in order to appreciate the potential spectral
inteference from the flame background emission, and to graphically
show the relationship of the commonly used analysis lines to the
flame background, as well as to each other, Figures III.B.8-a
through x have been included. The flames presented are the air-
acetylene, the air-hydrogen, and the nitrous oxide-acetylene. The
spectra were made using a 0.25 meter Jarrell-Ash monochromator
(blazed at 3000Å) with a 1P28 photomultiplier tube, a Jarrell-
Ash d.c. electrometer, and a Heath recorder. The parameters were
readjusted from flame to flame; therefore, their relative intensities are not comparable. In the air-acetylene and nitrous oxide-
acetylene flames the upper spectra was made with a 10 fold increase in sensitivity.

Table III.B.8-a

Close Lines in AA (0.5Å)

Wavelength (Å)	Element	Wavelength (Å)	Element	Difference (Å)
2021.21	BISMUTH	2021.38	GOLD	.17
2068.33	ANTIMONY	2068.65	GERMANIUM	.32
2088.82	IRIDIUM	2088.84	BORON	.02
2165.09	COPPER	2165.17	PLATINUM	.08
2268.91	TIN	2269.10	ALUMINUM	.19
2269.10	ALUMINUM	2269.22	ALUMINUM	.12
2274.38	PLATINUM	2274.49	COBALT	.11
2274.49	COBALT	2274.62	RHENIUM	.13
2337.49	NICKEL	2337.82	NICKEL	.33
2372.77	IRIDIUM	2373.12	ALUMINUM	.35
2373.12	ALUMINUM	2373.35	ALUMINUM	.23
2373.35	ALUMINUM	2373.35	ALUMINUM	.00
2424.56	OSMIUM	2424.93	COBALT	.37
2424.93	COBALT	2424.97	OSMIUM	.04
2427.95	GOLD	2428.10	STRONTIUM	.15
2428.10	STRONTIUM	2428.58	RHENIUM	.48
2476.42	PALLADIUM	2476.84	OSMIUM	.42
2476.84	OSMIUM	2476.87	NICKEL	.03
2481.18	IRIDIUM	2481.44	TUNGSTEN	.26
2483.27	IRON	2483.39	TIN	.12
2489.75	IRON	2490.12	PLATINUM	.37
2497.72	BORON	2497.96	GERMANIUM	.24
2500.70	GALLIUM	2501.13	IRON	.43
2506.90	SILICON	2506.90	VANADIUM	.00
2506.90	VANADIUM	2507.38	VANADIUM	.48
2507.38	VANADIUM	2507.45	TANTALUM	.07
2519.20	SILICON	2519.62	VANADIUM	.42
2524.11	SILICON	2524.29	IRON	.18
2526.22	VANADIUM	2526.35	TANTALUM	.13
2528.51	SILICON	2528.97	COBALT	.46
2575.10	ALUMINUM	2575.40	ALUMINUM	.30
2608.20	TANTALUM	2608.63	TANTALUM	.43
2612.63	OSMIUM	2613.07	TUNGSTEN	.44
2613.65	LEAD	2613.82	TUNGSTEN	.17
2613.82	TUNGSTEN	2614.18	LEAD	.36

Close Lines in AA (0.5Å)

Wavelength (Å)	Element	Wavelength (Å)	Element	Difference (Å)
3018.04	OSMIUM	3018.31	HAFNIUM	.27
3020.49	IRON	3020.53	HAFNIUM	.04
3020.53	HAFNIUM	3020.64	IRON	.11
3020.64	IRON	3021.07	IRON	.43
3039.06	GERMANIUM	3039.36	INDIUM	.30
3058.66	OSMIUM	3059.09	IRON	.43
3067.41	HAFNIUM	3067.72	BISMUTH	.31
3092.71	ALUMINUM	3092.84	ALUMINUM	.13
3101.55	NICKEL	3101.88	NICKEL	.33
3191.57	TUNGSTEN	3191.99	TITANIUM	.42
3232.63	LITHIUM	3232.96	NICKEL	.33
3255.69	SCANDIUM	3256.09	INDIUM	.40
3258.56	INDIUM	3258.60	OSMIUM	.04
3262.29	OSMIUM	3262.75	OSMIUM	.46
3273.63	SCANDIUM	3273.96	COPPER	.33
3280.60	RHODIUM	3280.68	SILVER	.08
3412.34	COBALT	3412.63	COBALT	.29
3440.61	IRON	3440.99	IRON	.38
3461.65	NICKEL	3462.04	RHODIUM	.39
3465.80	COBALT	3465.86	IRON	.06
3502.28	COBALT	3502.52	RHODIUM	.24
3502.52	RHODIUM	3502.62	COBALT	.10
3513.48	COBALT	3513.64	IRIDIUM	.16
3529.43	THALLIUM	3529.81	COBALT	.38
3575.36	COBALT	3575.79	ZIRCONIUM	.43
3634.70	PALLADIUM	3634.93	RUTHENIUM	.23
3704.54	LANTHANUM	3704.70	VANADIUM	.16
3745.56	IRON	3745.90	IRON	.34
3787.06	NIOBIUM	3787.48	NIOBIUM	.42
3798.12	NIOBIUM	3798.25	MOLYBDENUM	.13
3855.37	VANADIUM	3855.84	VANADIUM	.47
3863.87	ZIRCONIUM	3864.11	MOLYBDENUM	.24
3933.38	SCANDIUM	3933.66	CALCIUM	.28
3968.26	ZIRCONIUM	3968.47	CALCIUM	.21
4077.38	YTTRIUM	4077.71	STRONTIUM	.33

Close Lines in AA (0.5Å)

Wavelength (Å)	Element	Wavelength (Å)	Element	Difference (Å)
2636.90	TANTALUM	2637.13	OSMIUM	.23
2639.42	IRIDIUM	2639.71	IRIDIUM	.29
2641.10	TITANIUM	2641.49	GOLD	.39
2644.11	OSMIUM	2644.26	TITANIUM	.15
2646.64	TITANIUM	2646.89	PLATINUM	.25
2651.18	GERMANIUM	2651.58	GERMANIUM	.40
2656.54	TUNGSTEN	2656.61	TANTALUM	.07
2659.45	PLATINUM	2659.87	GALLIUM	.42
2661.24	TIN	2661.34	TANTALUM	.10
2661.89	TANTALUM	2661.98	IRIDIUM	.09
2691.31	TANTALUM	2691.34	GERMANIUM	.03
2706.51	TIN	2706.77	SCANDIUM	.26
2709.23	THALLIUM	2709.63	GERMANIUM	.40
2710.26	INDIUM	2710.67	THALLIUM	.41
2714.64	OSMIUM	2714.67	TANTALUM	.03
2718.90	TUNGSTEN	2719.03	IRON	.13
2719.03	IRON	2719.04	PLATINUM	.01
2719.65	GALLIUM	2720.04	OSMIUM	.39
2720.04	OSMIUM	2720.52	IRON	.48
2720.52	IRON	2720.90	IRON	.38
2838.17	OSMIUM	2838.63	OSMIUM	.46
2850.76	OSMIUM	2850.98	TANTALUM	.22
2879.11	TUNGSTEN	2879.39	TUNGSTEN	.28
2896.01	TUNGSTEN	2896.45	TUNGSTEN	.44
2904.41	HAFNIUM	2904.75	HAFNIUM	.34
2918.32	THALLIUM	2918.58	HAFNIUM	.26
2944.18	GALLIUM	2944.40	TUNGSTEN	.22
2946.98	TUNGSTEN	2947.38	TUNGSTEN	.40
2947.38	TUNGSTEN	2947.88	IRON	.50
2947.88	IRON	2948.24	TITANIUM	.36
2950.68	HAFNIUM	2950.88	NIOBIUM	.20
2973.13	IRON	2973.24	IRON	.11
2980.75	SCANDIUM	2980.81	HAFNIUM	.06
2988.95	SCANDIUM	2988.95	RUTHENIUM	.00
3011.75	ZIRCONIUM	3012.00	NICKEL	.25

Close Lines in AA (0.5Å)

Wavelength (Å)	Element	Wavelength (Å)	Element	Difference (Å)
4137.10	NIOBIUM	4137.59	NIOBIUM	.49
4215.52	STRONTIUM	4215.56	RUBIDIUM	.04
8943.50	CESIUM			

Table III.B.8-b

Close Lines in AE (0.5Å)

Wavelength (Å)	Element	Wavelength (Å)	Element	Difference (Å)
2269.10	ALUMINUM	2269.22	ALUMINUM	.12
2424.56	OSMIUM	2424.97	OSMIUM	.41
2612.63	OSMIUM	2613.07	TUNGSTEN	.44
2613.65	LEAD	2613.82	TUNGSTEN	.17
2613.82	TUNGSTEN	2614.18	LEAD	.36
2651.18	GERMANIUM	2651.58	GERMANIUM	.40
2659.45	PLATINUM	2659.87	GALLIUM	.42
2691.31	TANTALUM	2691.34	GERMANIUM	.03
2706.51	TIN	2706.77	SCANDIUM	.26
2800.87	ZINC	2801.06	MANGANESE	.19
2850.76	OSMIUM	2850.98	TANTALUM	.22
2980.75	SCANDIUM	2980.81	HAFNIUM	.06
2988.95	SCANDIUM	2988.95	RUTHENIUM	.00
3018.04	OSMIUM	3018.31	HAFNIUM	.27
3020.49	IRON	3020.64	IRON	.15
3020.64	IRON	3021.07	IRON	.43
3043.12	VANADIUM	3043.56	VANADIUM	.44
3067.41	HAFNIUM	3067.72	BISMUTH	.31
3092.71	ALUMINUM	3092.84	ALUMINUM	.13
3255.69	SCANDIUM	3256.09	INDIUM	.40
3273.63	SCANDIUM	3273.96	COPPER	.33
3412.34	COBALT	3412.63	COBALT	.29
3460.46	RHENIUM	3460.77	PALLADIUM	.31
3461.65	NICKEL	3462.04	RHODIUM	.39
3472.40	HAFNIUM	3472.54	NICKEL	.14
3502.28	COBALT	3502.52	RHODIUM	.24
3513.48	COBALT	3513.64	IRIDIUM	.16
3519.24	THALLIUM	3519.60	ZIRCONIUM	.36
3552.69	YTTRIUM	3553.08	PALLADIUM	.39
3566.10	ZIRCONIUM	3566.37	NICKEL	.27
3575.36	COBALT	3575.79	ZIRCONIUM	.43
3592.92	YTTRIUM	3593.02	RUTHENIUM	.10
3593.02	RUTHENIUM	3593.49	CHROMIUM	.47
3596.18	RUTHENIUM	3596.19	RHODIUM	.01
3634.70	PALLADIUM	3634.93	RUTHENIUM	.23

Close Lines in AE (0.5Å)

Wavelength (Å)	Element	Wavelength (Å)	Element	Difference (Å)
3657.99	RHODIUM	3658.10	TITANIUM	.11
3704.54	LANTHANUM	3704.70	VANADIUM	.16
3704.70	VANADIUM	3705.04	VANADIUM	.34
3790.15	NIOBIUM	3790.32	VANADIUM	.17
3798.12	NIOBIUM	3798.25	MOLYBDENUM	.13
3799.91	VANADIUM	3800.12	IRIDIUM	.21
3817.84	VANADIUM	3818.24	VANADIUM	.40
3822.89	VANADIUM	3823.21	VANADIUM	.32
3855.37	VANADIUM	3855.84	VANADIUM	.47
3867.60	VANADIUM	3867.98	TUNGSTEN	.38
3875.90	VANADIUM	3876.09	VANADIUM	.19
3929.53	ZIRCONIUM	3929.88	TITANIUM	.35
3929.88	TITANIUM	3930.30	IRON	.42
3933.38	SCANDIUM	3933.66	CALCIUM	.28
3968.26	ZIRCONIUM	3968.47	CALCIUM	.21
4008.75	TUNGSTEN	4008.93	TITANIUM	.18
4032.98	GALLIUM	4033.07	MANGANESE	.09
4047.21	POTASSIUM	4047.63	YTTRIUM	.42
4047.63	YTTRIUM	4047.79	SCANDIUM	.16
4077.38	YTTRIUM	4077.71	STRONTIUM	.33
4101.76	INDIUM	4102.16	VANADIUM	.40
4102.16	VANADIUM	4102.38	YTTRIUM	.22
4104.87	LANTHANUM	4105.17	VANADIUM	.30
4123.57	VANADIUM	4123.81	NIOBIUM	.24
4128.07	VANADIUM	4128.31	YTTRIUM	.24
4137.04	LANTHANUM	4137.10	NIOBIUM	.06
4139.44	NIOBIUM	4139.71	NIOBIUM	.27
4215.52	STRONTIUM	4215.56	RUBIDIUM	.04
4305.92	TITANIUM	4306.21	VANADIUM	.29
4408.20	VANADIUM	4408.51	VANADIUM	.31
6239.41	SCANDIUM	6239.78	SCANDIUM	.37
6242.81	VANADIUM	6243.10	VANADIUM	.29

Table III.B.8-c

Close Lines in AF (0.5Å)

Wavelength (Å)	Element	Wavelength (Å)	Element	Difference (Å)
2269.10	ALUMINUM	2269.22	ALUMINUM	.12
2288.02	CADMIUM	2288.12	ARSENIC	.10
2373.12	ALUMINUM	2373.35	ALUMINUM	.23
2373.35	ALUMINUM	2373.35	ALUMINUM	.00
2424.56	OSMIUM	2424.93	COBALT	.37
2424.93	COBALT	2424.97	OSMIUM	.04
2476.42	PALLADIUM	2476.84	OSMIUM	.42
2506.90	SILICON	2506.90	VANADIUM	.00
2528.51	SILICON	2528.97	COBALT	.46
2575.10	ALUMINUM	2575.40	ALUMINUM	.30
2612.63	OSMIUM	2613.07	TUNGSTEN	.44
2613.65	LEAD	2613.82	TUNGSTEN	.17
2613.82	TUNGSTEN	2614.18	LEAD	.36
2627.91	BISMUTH	2628.03	PLATINUM	.12
2650.86	PLATINUM	2651.18	GERMANIUM	.32
2651.18	GERMANIUM	2651.58	GERMANIUM	.40
2659.45	PLATINUM	2659.87	GALLIUM	.42
2691.31	TANTALUM	2691.34	GERMANIUM	.03
2706.51	TIN	2706.77	SCANDIUM	.26
2719.03	IRON	2719.04	PLATINUM	.01
2780.22	ARSENIC	2780.52	BISMUTH	.30
2850.76	OSMIUM	2850.98	TANTALUM	.22
2933.06	MANGANESE	2933.55	TANTALUM	.49
2980.75	SCANDIUM	2980.81	HAFNIUM	.06
2988.95	SCANDIUM	2988.95	RUTHENIUM	.00
2988.95	RUTHENIUM	2989.03	BISMUTH	.08
3020.49	IRON	3020.64	IRON	.15
3039.06	GERMANIUM	3039.36	INDIUM	.30
3067.41	HAFNIUM	3067.72	BISMUTH	.31
3092.71	ALUMINUM	3092.84	ALUMINUM	.13
3101.55	NICKEL	3101.88	NICKEL	.33
3396.85	RHODIUM	3397.21	BISMUTH	.36
3461.65	NICKEL	3462.04	RHODIUM	.39
3575.36	COBALT	3575.79	ZIRCONIUM	.43
3593.02	RUTHENIUM	3593.49	CHROMIUM	.47

Close Lines in AF (0.5Å)

Wavelength (Å)	Element	Wavelength (Å)	Element	Difference (Å)
3634.70	PALLADIUM	3634.93	RUTHENIUM	.23
3798.05	RUTHENIUM	3798.25	MOLYBDENUM	.20
3798.90	RUTHENIUM	3799.35	RUTHENIUM	.45
3855.37	VANADIUM	3855.84	VANADIUM	.47
4032.98	GALLIUM	4033.07	MANGANESE	.09
4139.44	NIOBIUM	4139.71	NIOBIUM	.27
4408.20	VANADIUM	4408.51	VANADIUM	.31

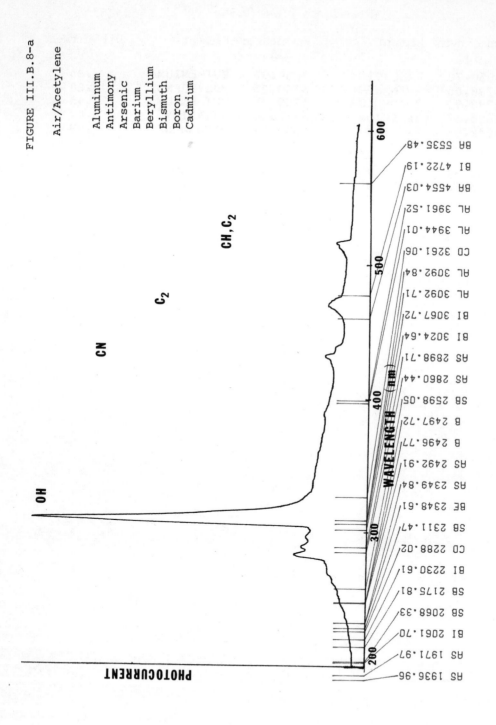

FIGURE III.B.8-a

Air/Acetylene

Aluminum
Antimony
Arsenic
Barium
Beryllium
Bismuth
Boron
Cadmium

FIGURE III.B.8-b

Air/Acetylene

Calcium
Cerium
Chromium
Cobalt
Copper
Dysprosium
Erbium

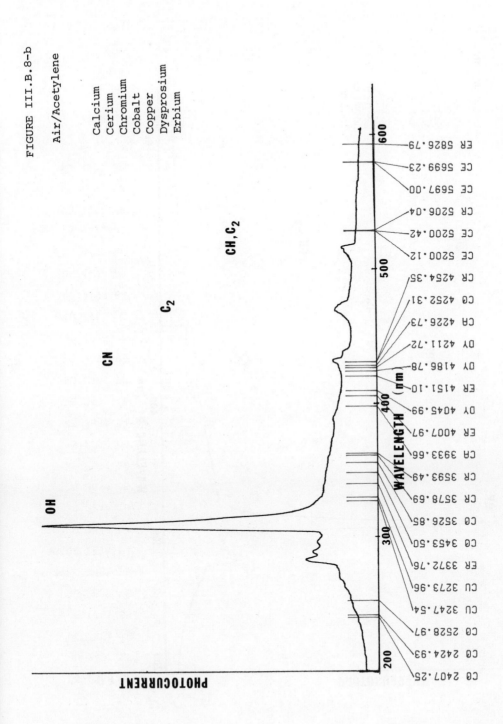

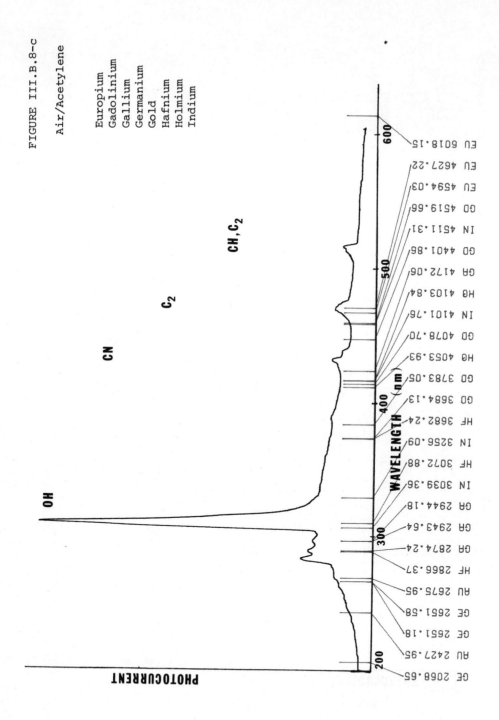

FIGURE III.B.8-c

Air/Acetylene

Europium
Gadolinium
Gallium
Germanium
Gold
Hafnium
Holmium
Indium

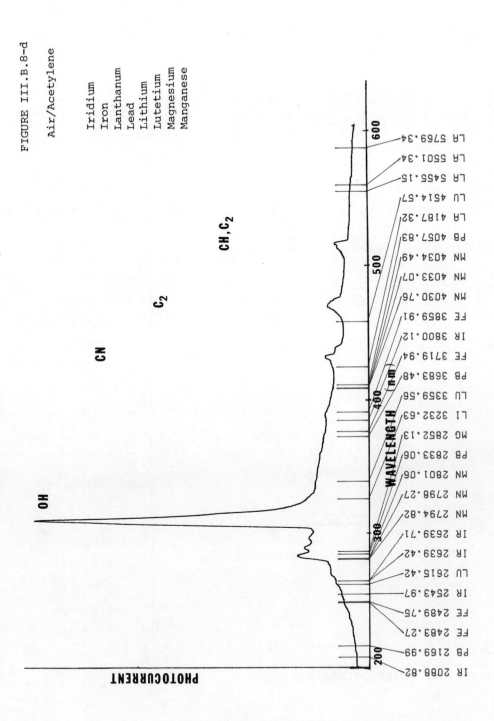

FIGURE III.B.8-d

Air/Acetylene

Iridium
Iron
Lanthanum
Lead
Lithium
Lutetium
Magnesium
Manganese

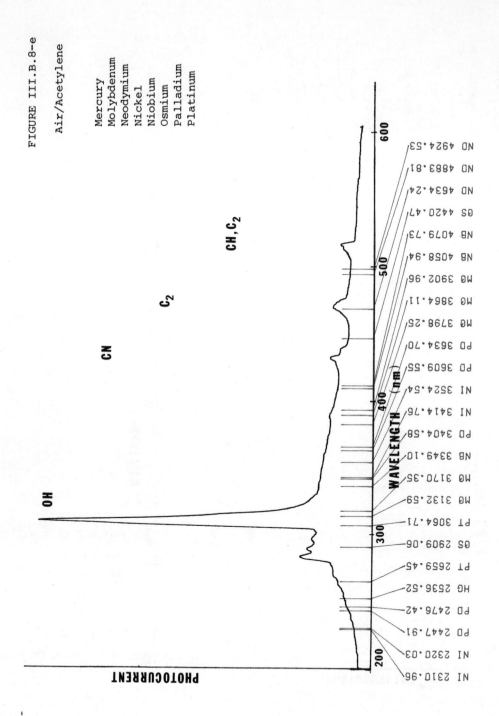

FIGURE III.B.8-e

Air/Acetylene

Mercury
Molybdenum
Neodymium
Nickel
Niobium
Osmium
Palladium
Platinum

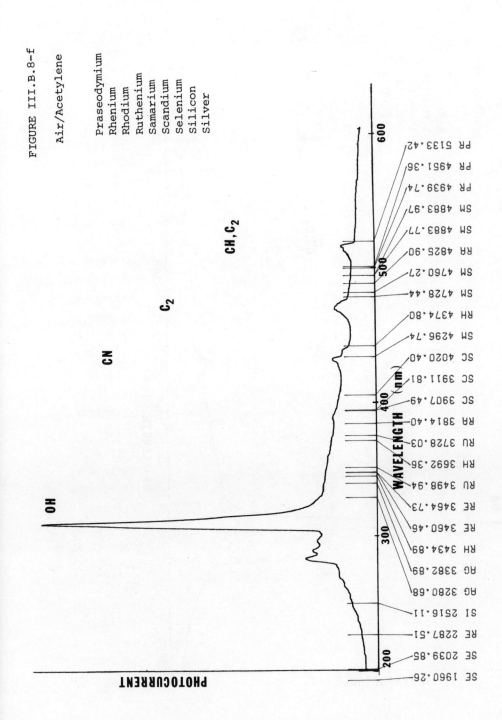

FIGURE III.B.8-f

Air/Acetylene

Praseodymium
Rhenium
Rhodium
Ruthenium
Samarium
Scandium
Selenium
Silicon
Silver

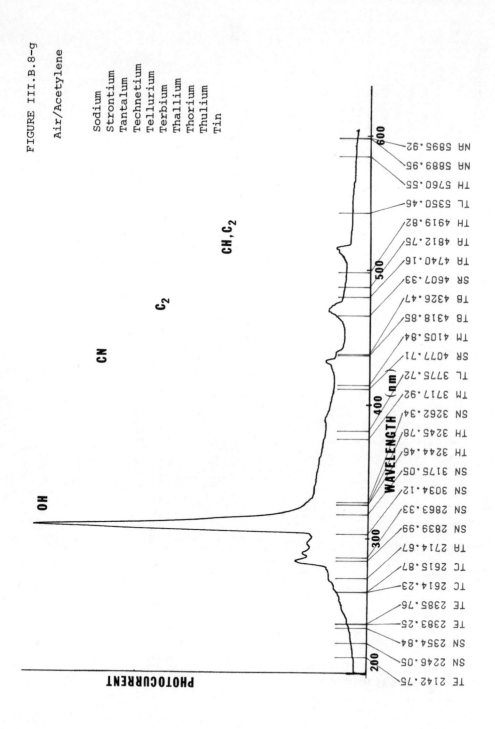

FIGURE III.B.8-g

Air/Acetylene

Sodium
Strontium
Tantalum
Technetium
Tellurium
Terbium
Thallium
Thorium
Thulium
Tin

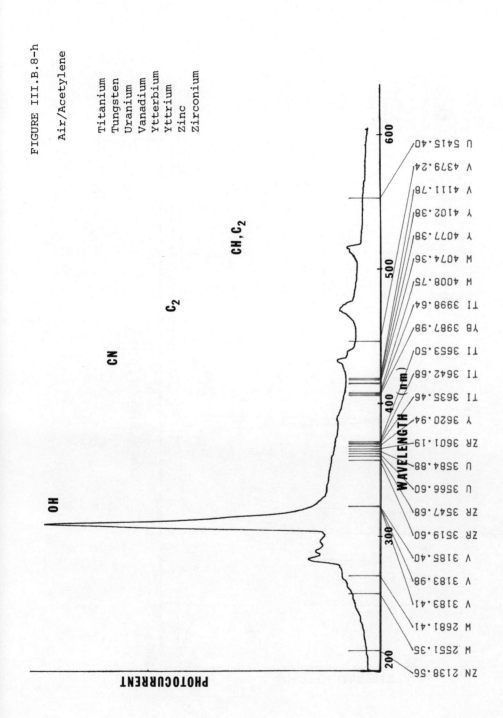

FIGURE III.B.8-h

Air/Acetylene

Titanium
Tungsten
Uranium
Vanadium
Ytterbium
Yttrium
Zinc
Zirconium

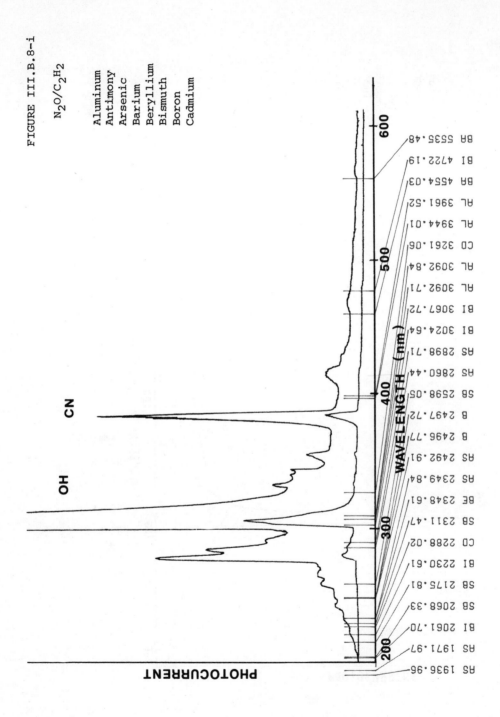

FIGURE III.B.8-i

N_2O/C_2H_2

Aluminum
Antimony
Arsenic
Barium
Beryllium
Bismuth
Boron
Cadmium

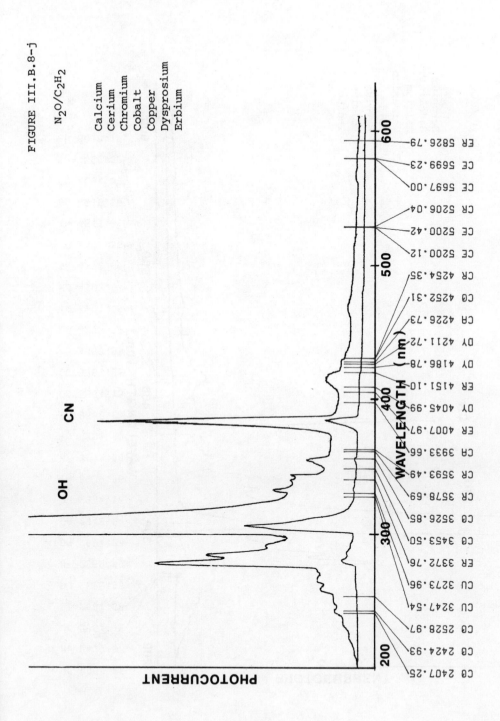

FIGURE III.B.8-j

N_2O/C_2H_2

Calcium
Cerium
Chromium
Cobalt
Copper
Dysprosium
Erbium

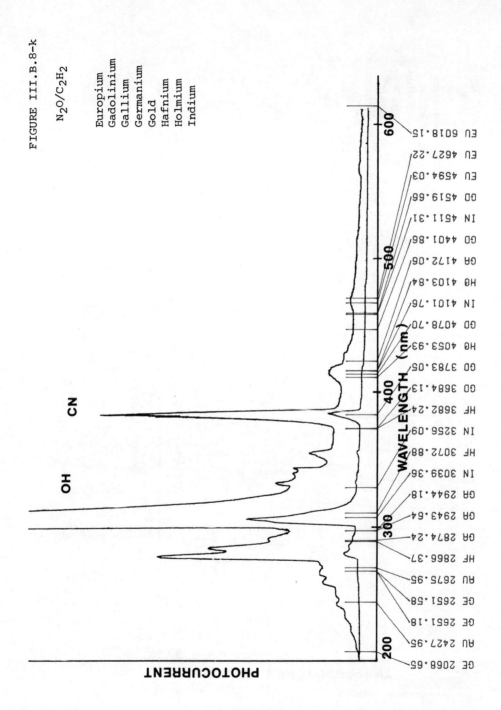

FIGURE III.B.8-k

N_2O/C_2H_2

Europium
Gadolinium
Gallium
Germanium
Gold
Hafnium
Holmium
Indium

EU 6018.15
EU 4627.22
EU 4594.03
GD 4519.66
IN 4511.31
GD 4401.86
GA 4172.06
HO 4103.84
IN 4101.76
GD 4078.70
HO 4053.93
GD 3783.05
GD 3684.13
HF 3682.24
IN 3256.09
HF 3072.88
IN 3039.36
GA 2944.18
GA 2943.64
GA 2874.24
HF 2866.37
AU 2675.95
GE 2651.58
GE 2651.18
AU 2427.95
GE 2068.65

OH

CN

WAVELENGTH (nm)

PHOTOCURRENT

200   300   400   500   600

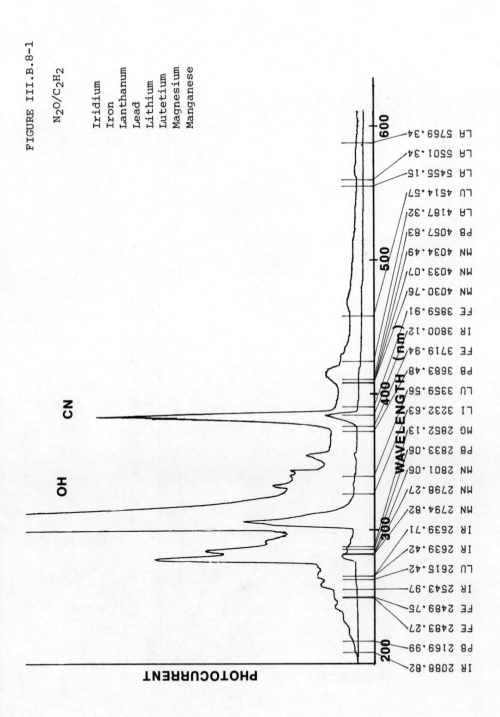

FIGURE III.B.8-1

N_2O/C_2H_2

Iridium
Iron
Lanthanum
Lead
Lithium
Lutetium
Magnesium
Manganese

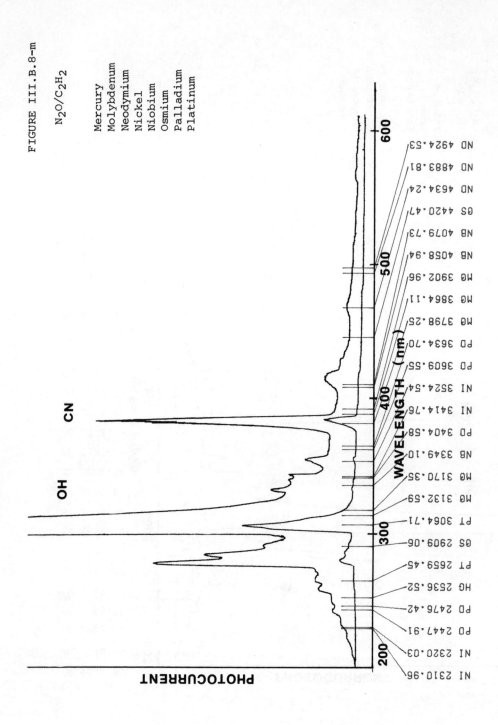

FIGURE III.B.8-m

N_2O/C_2H_2

Mercury
Molybdenum
Neodymium
Nickel
Niobium
Osmium
Palladium
Platinum

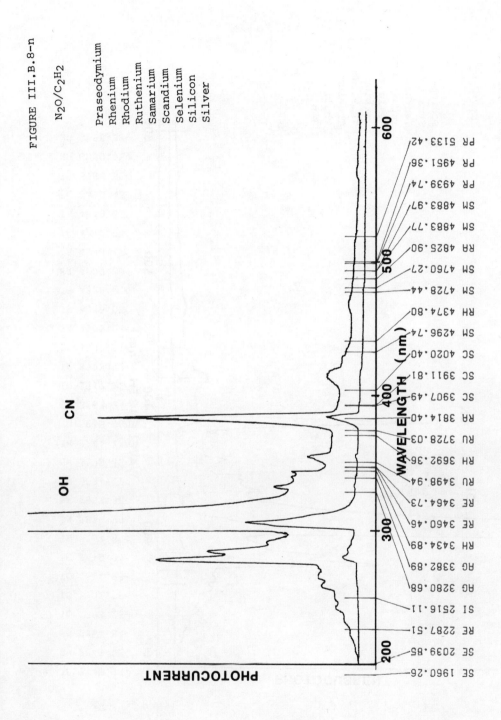

FIGURE III.B.8-n

N_2O/C_2H_2

Praseodymium
Rhenium
Rhodium
Ruthenium
Samarium
Scandium
Selenium
Silicon
Silver

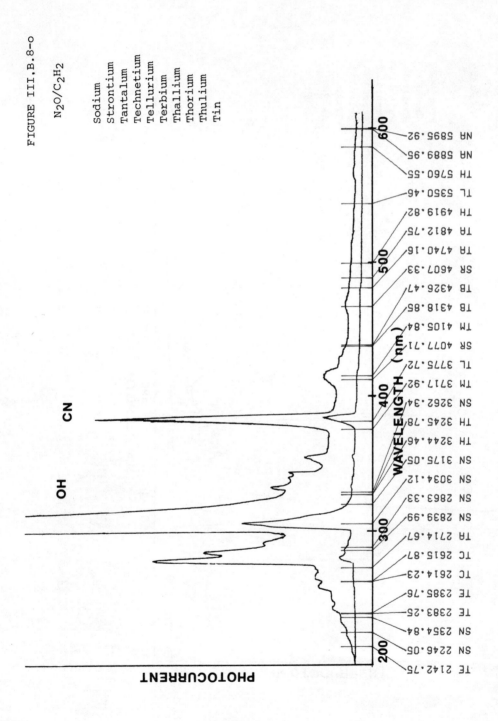

FIGURE III.B.8-o

N_2O/C_2H_2

Sodium
Strontium
Tantalum
Technetium
Tellurium
Terbium
Thallium
Thorium
Thulium
Tin

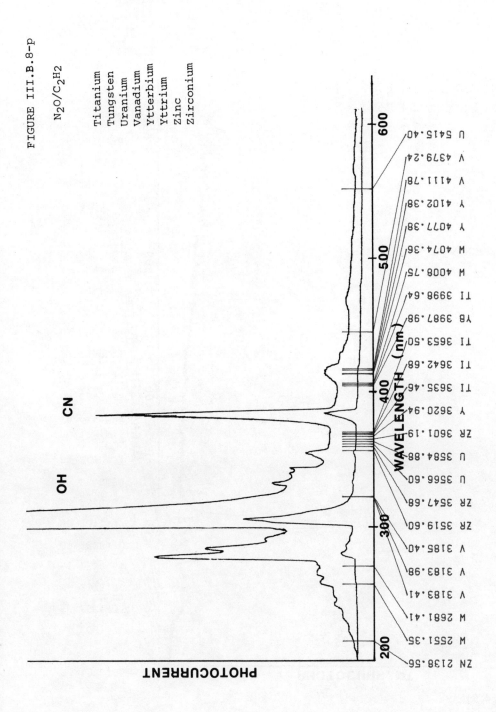

FIGURE III.B.8-p

N_2O/C_2H_2

Titanium
Tungsten
Uranium
Vanadium
Ytterbium
Yttrium
Zinc
Zirconium

FIGURE III.B.8-q

Air/H₂

Aluminum
Antimony
Arsenic
Barium
Beryllium
Bismuth
Boron
Cadmium

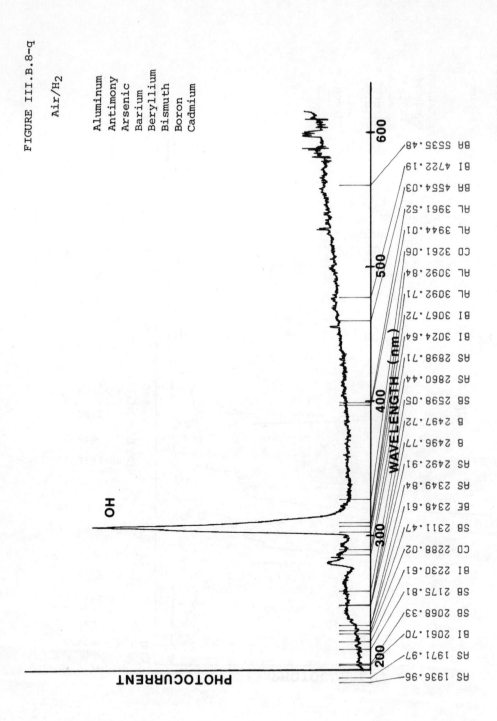

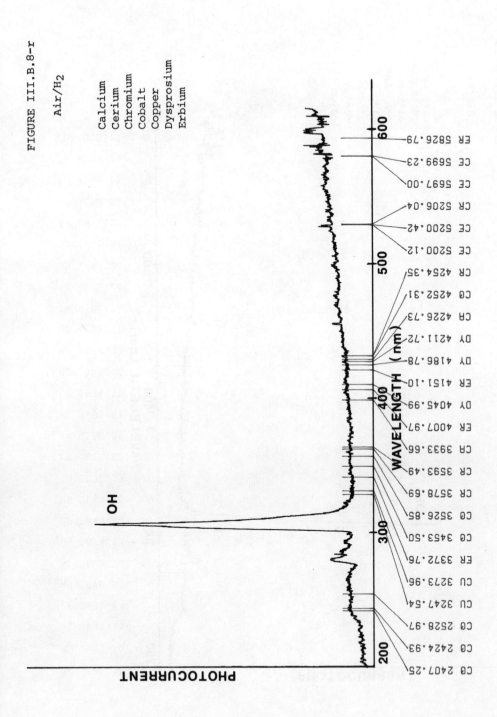

FIGURE III.B.8-r

Air/H₂

Calcium
Cerium
Chromium
Cobalt
Copper
Dysprosium
Erbium

FIGURE III.B.8-s

Air/H₂

Europium
Gadolinium
Gallium
Germanium
Gold
Hafnium
Holmium
Indium

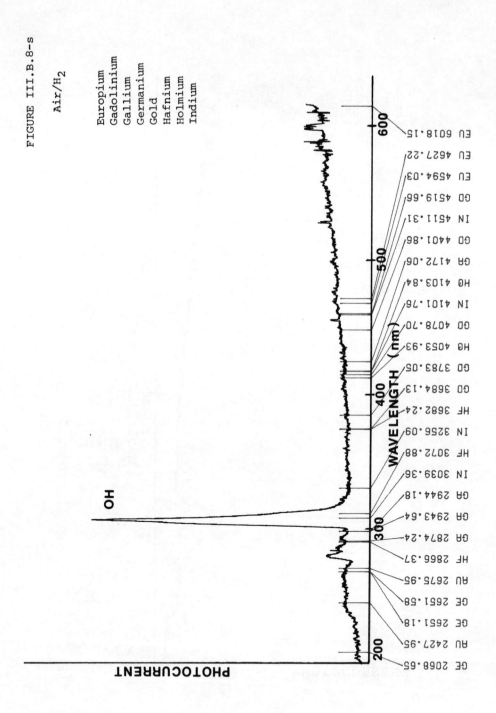

FIGURE III.B.8-t

Air/H₂

Iridium
Iron
Lanthanum
Lead
Lithium
Lutetium
Magnesium
Manganese

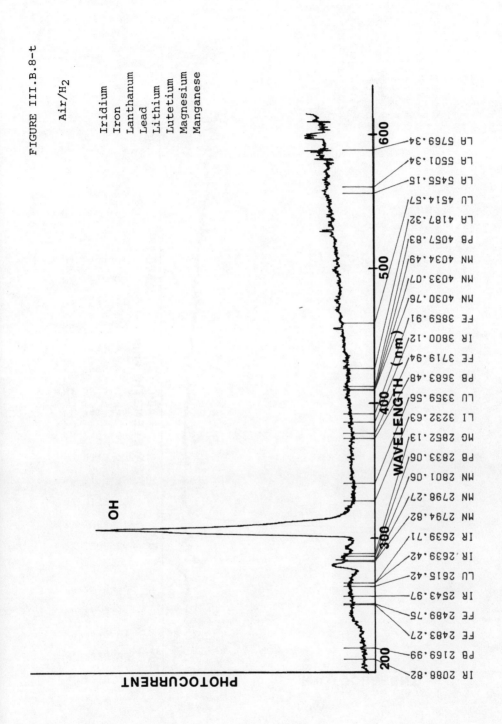

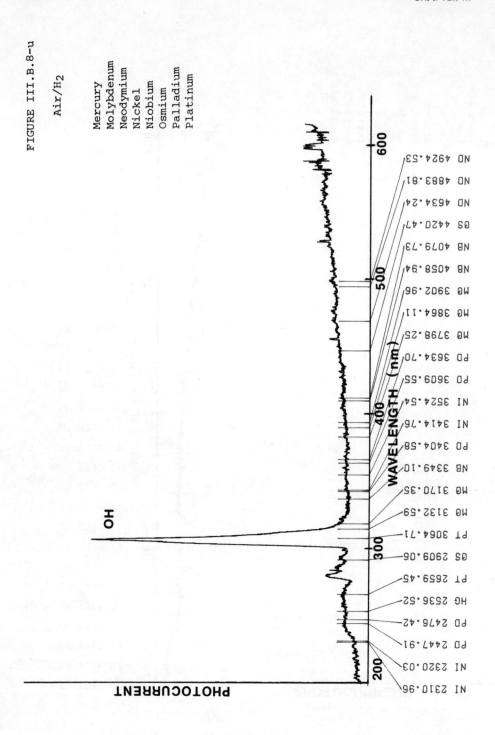

FIGURE III.B.8-u

Air/H$_2$

Mercury
Molybdenum
Neodymium
Nickel
Niobium
Osmium
Palladium
Platinum

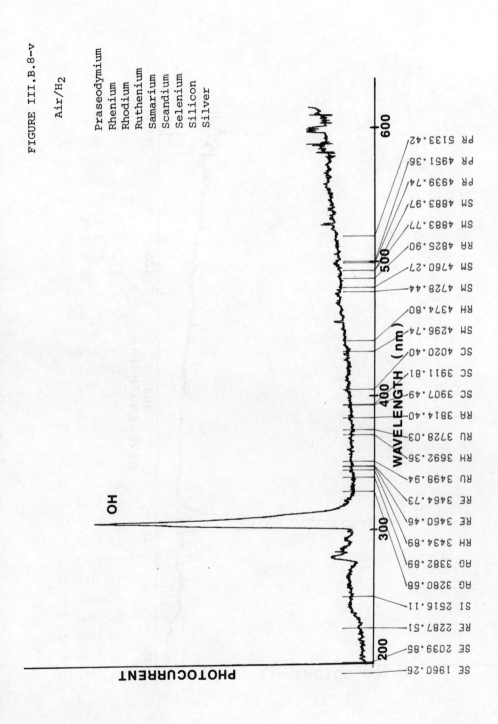

FIGURE III.B.8-v

Air/H₂

Praseodymium
Rhenium
Rhodium
Ruthenium
Samarium
Scandium
Selenium
Silicon
Silver

FIGURE III.B.8-w

Air/H₂

Sodium
Strontium
Tantalum
Technetium
Tellurium
Terbium
Thallium
Thorium
Thulium
Tin

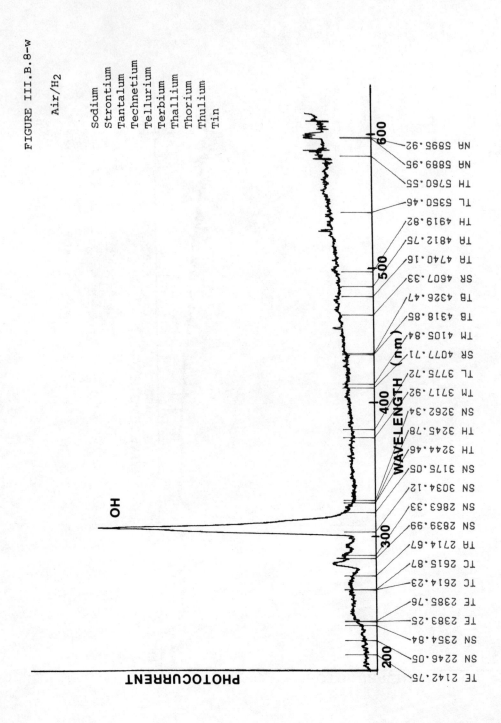

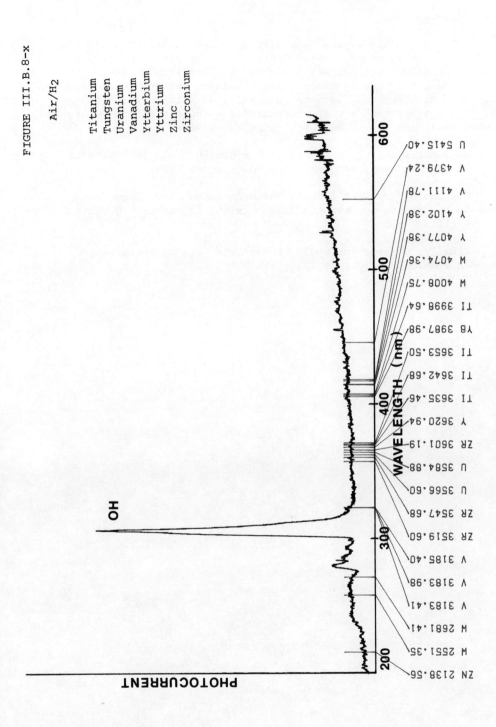

FIGURE III.B.8-x

Air/H2

Titanium
Tungsten
Uranium
Vanadium
Ytterbium
Yttrium
Zinc
Zirconium

Miscellaneous Data

It is felt that several miscellaneous tables are of general use to the practicing analyst. Table III.B.9-a presents the latest values for selected fundamental constants along with units in both the SI and cgs systems. Table III.B.9-b lists selected unit abbreviations, and Tables III.B.9-c & d present conversion factors useful to the analytical spectroscopist; 'c' is concerned with energy conversions, and 'd' with general unit conversion.

Figure III.B.9 gives an energy conversion table for units and equivalents for handy reference and quick comparison.

For the AAS analyst a table has been added to convert from percent absorbence (or percent transmission) to absorbence units (Table III.B.9-e).

Finally, no handbook would be complete without a student's 't' table which is given in Table III.B.9-f.

Table III.B.9-a

SELECTED FUNDAMENTAL CONSTANTS

	Symbol	Value	SI Units	cgs Units	(last sign. figs.)
Speed of light in a vacuum	C	2.997924580	10^8 m·s^{-1}	10^{10} cm·s^{-1}	12
Elementary charge	e	1.6021892	10^{-19} C	10^{-20} emv	46
		4.803242	------	10^{-10} esu	14
Planck constant	h	6.626176	10^{-34} J·s	10^{-27} erg·s	36
Avogadro constant	N_A	6.022045	10^{23} mol^{-1}	10^{23} mol^{-1}	31
Atomic Mass Unit	U	1.6605655	10^{-27} kg	10^{-24} g	86
Electron rest mass	m_e	9.109534	10^{-31} kg	10^{-28} g	47
		5.4858026	10^{-4} U	10^{-4} U	21
Proton rest mass	m_p	1.6726485	10^{-27} kg	10^{-24} g	86
		1.007276470	U	U	11
Neutron rest mass	m_n	1.6749543	10^{-27} kg	10^{-24} g	86
		1.008665012	U	U	37
Rydberg constant	R_∞	1.097373177	10^7 m^{-1}	10^5 cm^{-1}	83
Rohr radium	a_o	5.2917706	10^{-11} m	10^{-9} cm	44
Classical electron radius	r_e	2.8179380	10^{-15} m	10^{-13} cm	70
Compton wavelength of the electron	λ_c	2.4263089	10^{-12} m	10^{-10} cm	40

	Symbol	Value	Units SI	cgs	(last sign. figs.)
Compton wavelength of the proton	$\lambda_{c,p}$	1.3214099	$10^{-15}\,\mathrm{m}$	$10^{-13}\,\mathrm{cm}$	22
Compton wavelength of the neutron	$\lambda_{c,n}$	1.3195909	$10^{-15}\,\mathrm{m}$	$10^{-13}\,\mathrm{cm}$	22
Molar volume of ideal gas at s.t.p.	V_m	22.41383	$10^{-3}\,\mathrm{m}^3 \cdot \mathrm{mol}^{-1}$	$10^{3}\,\mathrm{cm}^3 \cdot \mathrm{mol}^{-1}$	70
Molar gas constant $(T_o \equiv 273.15\mathrm{K}$, $P_o \equiv 1\,\mathrm{atm} \equiv 101325\,P_a)$	R	8.31441	$\mathrm{J} \cdot \mathrm{mol}^{-1}\mathrm{K}^{-1}$	$10^{7}\,\mathrm{erg} \cdot \mathrm{mol}^{-1} \cdot \mathrm{K}^{-1}$	26
	R	8.20568	$10^{-5}\,\mathrm{m}^3 \cdot \mathrm{atm} \cdot \mathrm{mol}^{-1} \cdot \mathrm{K}^{-1}$	$10\,\mathrm{cm}^3 \cdot \mathrm{atm} \cdot \mathrm{mol}^{-1} \cdot \mathrm{K}^{-1}$	26
Boltzmann constant	k	1.380662	$10^{-23}\,\mathrm{J} \cdot \mathrm{K}^{-1}$	$10^{-16}\,\mathrm{erg} \cdot \mathrm{K}^{-1}$	44
Stefan–Boltzmann constant	σ	5.67032	$10^{-8}\,\mathrm{W} \cdot \mathrm{m}^{-2} \cdot \mathrm{K}^{-4}$	$10^{-5}\,\mathrm{erg} \cdot \mathrm{s}^{-1} \cdot \mathrm{cm}^{-2}$	71
First Radiation constant	C_1	3.741832	$10^{-16}\,\mathrm{W} \cdot \mathrm{m}^2$	$10^{-5}\,\mathrm{erg} \cdot \mathrm{cm}^2 \cdot \mathrm{s}^{-1}$	20
Second Radiation constant	C_2	1.438786	$10^{-2}\,\mathrm{m} \cdot \mathrm{K}$	$\mathrm{cm} \cdot \mathrm{K}$	45
Gravitational constant	G	6.6720	$10^{-11}\,\mathrm{m}^3 \cdot \mathrm{s}^{-2} \cdot \mathrm{kg}^{-1}$	$10^{-8}\,\mathrm{cm}^3 \cdot \mathrm{sec}^{-2} \cdot \mathrm{g}^{-1}$	41

Footnote: All constants are based on ^{12}C system.

Reference: E.R. Cohen and B.N. Taylor, J. Phys. Chem. Ref. Data, 2, 663 (1973).

Table III.B.9-b

SELECTED UNIT ABBREVIATIONS[1,2]

Symbol	Unit	Symbol	Unit
u	Atomic Mass Unit	sr	Steradian
C	Coulomb	m	Meter
F	Farad	cm	Centimeter
G	Gauss	kg	Kilogram
H	Henry	g	Gram
H_z	Hertz = cycles/s	Mol	Mole
J	Joule	N	Newton
K	Degrees Kelvin	Å	Angstrom
Pa	Pascal = $N.m^{-2}$	dyn	Dyne
T	Tesla - $10^4 G$	bar	Bar
V	Volt	erg	Erg
Wb	Weber = $T.m^2$	Torr	Millimeter of Hg
W	Watt	S	Second
rad	Radian	eV	Electron Volt

Reference:

1. E. R. Cohen & B. N. Taylor, J. Phys. Chem. Ref. Data,
    2, 663 (1973).

2. C. K. Mann, T. J. Vickers, & W. M. Gulick, "Instrumental
    Analysis," Harper & Roe, 1974, Appendix A.

Table III.B.9-c

SELECTED ENERGY CONVERSION FACTORS
AND EQUIVALENTS

Quantity	Value	Units	1σ (Last Significant Figures)
1 Electron Volt	1.6021892	10^{-19} J	46
		10^{-12} erg	
1eV/h	2.4179696	10^{14} Hz	63
1eV/hc	8.065479	10^5 m^{-1}	21
		10^3 cm^{-1}	
1eV/k	1.160450	10^4 K	36
1 Kilogram	5.609545	10^{29} MeV	16
1 Atomic Mass Unit	931.5016	MeV	26
1 Electron Mass	0.5110034	MeV	14
1 Proton Mass	938.2796	MeV	27
1 Neutron Mass	939.5731	MeV	27

Reference:

E. R. Cohen & B. N. Taylor, J. Phys. Chem. Ref. Data, 2, 663 (1973).

Table III.B.9-d

SELECTED CONVERSION FACTORS

To Convert From	To	Multiply by
Angstroms	Centimeters	1×10^{-8}
	Micrometers	1×10^{-4}
	Nanometers	0.1
Atmospheres	mm of Hg (torr)	760
	dynes.cm^{-2}	1.01325×10^6
	N·m^{-2} (pascal)	1.01325×10^5
Barn	Centimeters^2	10^{-24}
Centimeter	Angstroms	1×10^8
	Meters	0.01
	Micrometers	1×10^4
	Nanometers	1×10^7
Centimeter/Second	Angstroms/Second	1×10^8
	Meters/Second	0.01
Cubic Centimeters· Atmosphere	Joules	0.101325
Cubic Feet/Hour	Liters/Hour	28.31605
	Cubic Centimeters/ Second	7.8657907
Fathoms	Centimeters	182.88
Fortnight	Seconds	1.2096×10^6
Furlongs	Centimeters	2.01168×10^4
Grams/Liter	Parts per Million (density=1 g/ml)	1000
Hours	Days	0.041666
	Minutes	60
	Seconds	3600
Light Year	Kilometers	9.46055×10^{12}
Liters	Cubic Centimeters	1000.028

SELECTED CONVERSION FACTORS

To Convert From	To	Multiply by
Liter Atm.	Joules	101.328
Lumens (at 5550 Å)	Watts	0.0014705882
Meters	Angstrom	1×10^{10}
	Centimeters	100
	Micrometers	10^{6}
	Nanometers	10^{9}
Micrometers (Microns)	Angstroms	10^{4}
	Nanometers	10^{3}
Millimeters of Hg (0°C)	Atmosphere	0.0013157895
	Pascals	1.35951×10^{5}
	Torr	1
Nanometer	Angstroms	10
	Centimeters	10^{-7}
	Micrometers	10^{-3}
Newton	Dynes	10^{5}
Noggins (Brit.)	Cubic Centimeters	142.0652
Ounces/Ton (Short)	Milligrams/Killogram (parts per million)	31.25
Palm	cm	7.62
Parts/Million	Grams/Liter (density=1 g/ml)	10^{-3}
	Milligram/Liter	1
Pounds (avdp.)	Kilograms	0.45359237
Quart (U.S. Liquid)	Liter	0.9463264
Radian	Degrees	57.295779
	Minutes	3437.7468
	Seconds	2.0626481×10^{5}

SELECTED CONVERSION FACTORS

To Convert From	To	Multiply by
Scruple (Apoth.)	Grams	1.2959782
Ton (Short)	Kilograms	1016.0469
Ton (Long)	Kilograms	907.18474
Ton (Metric)	Kilogram	1000
Yard	Meter	0.9144

Reference:

"Handbook of Chemistry and Phys.," 48th Ed., Chemical Rubber Co., 1967.

Figure III.B.9

Å	1000	2000	3000	4000	5000	6000	7000	8000	9000	10000
NM	100	200	300	400	500	600	700	800	900	1000
μM	0.1	0.2	0.3	0.4	0.5	0.6	0.7	0.8	0.9	1.0
KK	100	50	33	25	20	17	14	13	11	10
$\bar{\nu}$ $\times 10^{15}$	3.00	1.50	1.00	0.75	0.60	0.50	0.43	0.37	0.33	0.30
ERG $\times 10^{-11}$	1.99	0.99	0.66	0.50	0.40	0.33	0.28	0.25	0.22	0.20
JOULE $\times 10^{-19}$	19.86	9.93	6.62	4.97	3.97	3.31	2.84	2.48	2.21	1.99
EV	12.40	6.20	4.13	3.10	2.48	2.07	1.77	1.55	1.38	1.24

Table III.B.9-e

VALUES OF ABSORBANCE FOR PERCENT ABSORPTION

To convert percent absorption (%A) to absorbance, find the percent absorption to the nearest whole digit in the left-hand column;
read across to the column located under the tenth of a per cent
desired, and read the value of absorbance. The value of absorbance
corresponding to 26.8% absorption is thus 0.1355. %T is equal to
100 - %A.

%A	.0	.1	.2	.3	.4	.5	.6	.7	.8	.9
0.0	.0000	.0004	.0009	.0013	.0017	.0022	.0026	.0031	.0035	.0039
1.0	.0044	.0048	.0052	.0057	.0061	.0066	.0070	.0074	.0079	.0083
2.0	.0088	.0092	.0097	.0101	.0106	.0110	.0114	.0119	.0123	.0128
3.0	.0132	.0137	.0141	.0146	.0150	.0155	.0159	.0164	.0168	.0173
4.0	.0177	.0182	.0186	.0191	.0195	.0200	.0205	.0209	.0214	.0218
5.0	.0223	.0227	.0232	.0236	.0241	.0246	.0250	.0255	.0259	.0264
6.0	.0269	.0273	.0278	.0283	.0287	.0292	.0297	.0301	.0306	.0311
7.0	.0315	.0320	.0325	.0329	.0334	.0339	.0343	.0348	.0353	.0357
8.0	.0362	.0367	.0372	.0376	.0381	.0386	.0391	.0395	.0400	.0405
9.0	.0410	.0414	.0419	.0424	.0429	.0434	.0438	.0443	.0448	.0453
10.0	.0458	.0462	.0467	.0472	.0477	.0482	.0487	.0491	.0496	.0501
11.0	.0506	.0511	.0516	.0521	.0526	.0531	.0535	.0540	.0545	.0550
12.0	.0555	.0560	.0565	.0570	.0575	.0580	.0585	.0590	.0595	.0600
13.0	.0605	.0610	.0615	.0620	.0625	.0630	.0635	.0640	.0645	.0650
14.0	.0655	.0660	.0665	.0670	.0675	.0680	.0685	.0691	.0696	.0701
15.0	.0706	.0711	.0716	.0721	.0726	.0731	.0737	.0742	.0747	.0752
16.0	.0757	.0762	.0768	.0773	.0778	.0783	.0788	.0794	.0799	.0804
17.0	.0809	.0814	.0820	.0825	.0830	.0835	.0841	.0846	.0851	.0857
18.0	.0862	.0867	.0872	.0878	.0883	.0888	.0894	.0899	.0904	.0910
19.0	.0915	.0921	.0926	.0931	.0937	.0942	.0947	.0953	.0958	.0964
20.0	.0969	.0975	.0980	.0985	.0991	.0996	.1002	.1007	.1013	.1018
21.0	.1024	.1029	.1035	.1040	.1046	.1051	.1057	.1062	.1068	.1073
22.0	.1079	.1085	.1090	.1096	.1101	.1107	.1113	.1118	.1124	.1129
23.0	.1135	.1141	.1146	.1152	.1158	.1163	.1169	.1175	.1180	.1186
24.0	.1192	.1198	.1203	.1209	.1215	.1221	.1226	.1232	.1238	.1244
25.0	.1249	.1255	.1261	.1267	.1273	.1278	.1284	.1290	.1296	.1302
26.0	.1308	.1314	.1319	.1325	.1331	.1337	.1343	.1349	.1355	.1361
27.0	.1367	.1373	.1379	.1385	.1391	.1397	.1403	.1409	.1415	.1421
28.0	.1427	.1433	.1439	.1445	.1451	.1457	.1463	.1469	.1475	.1481
29.0	.1487	.1494	.1500	.1506	.1512	.1518	.1524	.1530	.1537	.1543

%A	.0	.1	.2	.3	.4	.5	.6	.7	.8	.9
30.0	.1549	.1555	.1561	.1568	.1574	.1580	.1586	.1593	.1599	.1605
31.0	.1612	.1618	.1624	.1630	.1637	.1643	.1649	.1656	.1662	.1669
32.0	.1675	.1681	.1688	.1694	.1701	.1707	.1713	.1720	.1726	.1733
33.0	.1739	.1746	.1752	.1759	.1765	.1772	.1778	.1785	.1791	.1798
34.0	.1805	.1811	.1818	.1824	.1831	.1838	.1844	.1851	.1858	.1864
35.0	.1871	.1878	.1884	.1891	.1898	.1904	.1911	.1918	.1925	.1931
36.0	.1938	.1945	.1952	.1959	.1965	.1972	.1979	.1986	.1993	.2000
37.0	.2007	.2013	.2020	.2027	.2034	.2041	.2048	.2055	.2062	.2069
38.0	.2076	.2083	.2090	.2097	.2104	.2111	.2118	.2125	.2132	.2140
39.0	.2147	.2154	.2161	.2168	.2175	.2182	.2190	.2197	.2204	.2211
40.0	.2218	.2226	.2233	.2240	.2248	.2255	.2262	.2269	.2277	.2284
41.0	.2291	.2299	.2306	.2314	.2321	.2328	.2336	.2343	.2351	.2358
42.0	.2366	.2373	.2381	.2388	.2396	.2403	.2411	.2418	.2426	.2434
43.0	.2441	.2449	.2457	.2464	.2472	.2480	.2487	.2495	.2503	.2510
44.0	.2518	.2526	.2534	.2541	.2549	.2557	.2565	.2573	.2581	.2588
45.0	.2596	.2604	.2612	.2620	.2628	.2636	.2644	.2652	.2660	.2668
46.0	.2676	.2684	.2692	.2700	.2708	.2716	.2725	.2733	.2741	.2749
47.0	.2757	.2765	.2774	.2782	.2790	.2798	.2807	.2815	.2823	.2832
48.0	.2840	.2848	.2857	.2865	.2874	.2882	.2890	.2899	.2907	.2916
49.0	.2924	.2933	.2941	.2950	.2958	.2967	.2976	.2984	.2993	.3002
50.0	.3010	.3019	.3028	.3036	.3045	.3054	.3063	.3072	.3080	.3089
51.0	.3098	.3107	.3116	.3125	.3134	.3143	.3152	.3161	.3170	.3179
52.0	.3188	.3197	.3206	.3215	.3224	.3233	.3242	.3251	.3261	.3270
53.0	.3279	.3288	.3298	.3307	.3316	.3325	.3335	.3344	.3354	.3363
54.0	.3372	.3382	.3391	.3401	.3410	.3420	.3429	.3439	.3449	.3458
55.0	.3468	.3478	.3487	.3497	.3507	.3516	.3526	.3536	.3546	.3556
56.0	.3565	.3575	.3585	.3595	.3605	.3615	.3625	.3635	.3645	.3655
57.0	.3665	.3675	.3686	.3696	.3706	.3716	.3726	.3737	.3747	.3757
58.0	.3768	.3778	.3788	.3799	.3809	.3820	.3830	.3840	.3851	.3862
59.0	.3872	.3883	.3893	.3904	.3915	.3925	.3936	.3947	.3958	.3969
60.0	.3979	.3990	.4001	.4012	.4023	.4034	.4045	.4056	.4067	.4078
61.0	.4089	.4101	.4112	.4123	.4134	.4145	.4157	.4168	.4179	.4191
62.0	.4202	.4214	.4225	.4237	.4248	.4260	.4271	.4283	.4295	.4306
63.0	.4318	.4330	.4342	.4353	.4365	.4377	.4389	.4401	.4413	.4425
64.0	.4437	.4449	.4461	.4473	.4485	.4498	.4510	.4522	.4535	.4547
65.0	.4559	.4572	.4584	.4597	.4609	.4622	.4634	.4647	.4660	.4672
66.0	.4685	.4698	.4711	.4724	.4737	.4750	.4763	.4776	.4789	.4802
67.0	.4815	.4828	.4841	.4855	.4868	.4881	.4895	.4908	.4921	.4935
68.0	.4948	.4962	.4976	.4989	.5003	.5017	.5031	.5045	.5058	.5072
69.0	.5086	.5100	.5114	.5129	.5143	.5157	.5171	.5186	.5200	.5214
70.0	.5229	.5243	.5258	.5272	.5287	.5302	.5317	.5331	.5346	.5361
71.0	.5376	.5391	.5406	.5421	.5436	.5452	.5467	.5482	.5498	.5513

%A	.0	.1	.2	.3	.4	.5	.6	.7	.8	.9
72.0	.5528	.5544	.5560	.5575	.5591	.5607	.5622	.5638	.5654	.5670
73.0	.5686	.5702	.5719	.5735	.5751	.5768	.5784	.5800	.5817	.5834
74.0	.5850	.5867	.5884	.5901	.5918	.5935	.5952	.5969	.5986	.6003
75.0	.6021	.6038	.6055	.6073	.6091	.6108	.6126	.6144	.6162	.6180
76.0	.6198	.6216	.6234	.6253	.6271	.6289	.6308	.6326	.6345	.6364
77.0	.6383	.6402	.6421	.6440	.6459	.6478	.6498	.6517	.6536	.6556
78.0	.6576	.6596	.6615	.6635	.6655	.6676	.6696	.6716	.6737	.6757
79.0	.6778	.6799	.6819	.6840	.6861	.6882	.6904	.6925	.6946	.6968
80.0	.6990	.7011	.7033	.7055	.7077	.7100	.7122	.7144	.7167	.7190
81.0	.7212	.7235	.7258	.7282	.7305	.7328	.7352	.7375	.7399	.7423
82.0	.7447	.7471	.7496	.7520	.7545	.7570	.7595	.7620	.7645	.7670
83.0	.7696	.7721	.7747	.7773	.7799	.7825	.7852	.7878	.7905	.7932
84.0	.7959	.7986	.8013	.8041	.8069	.8097	.8125	.8153	.8182	.8210
85.0	.8239	.8268	.8297	.8327	.8356	.8386	.8416	.8447	.8477	.8508
86.0	.8539	.8570	.8601	.8633	.8665	.8697	.8729	.8761	.8794	.8827
87.0	.8861	.8894	.8928	.8962	.8996	.9031	.9066	.9101	.9136	.9172
88.0	.9208	.9245	.9281	.9318	.9355	.9393	.9431	.9469	.9508	.9547
89.0	.9586	.9626	.9666	.9706	.9747	.9788	.9830	.9872	.9914	.9957

TABLE III.B.9-f

STUDENT'S t-DISTRIBUTION

n \ Confidence	90%	95%	99%	99.95%
1	3.078	6.314	31.821	636.619
2	1.886	2.920	6.965	31.598
3	1.638	2.353	4.541	12.924
4	1.533	2.132	3.747	8.610
5	1.476	2.015	3.365	6.869
6	1.440	1.943	3.143	5.959
7	1.415	1.895	2.998	5.408
8	2.397	1.860	2.896	5.041
9	1.383	1.833	2.821	4.781
10	1.372	1.812	2.764	4.587
11	1.363	1.796	2.718	4.437
12	1.356	1.782	2.681	4.318
13	1.350	1.771	2.650	4.221
14	1.345	1.761	2.624	4.140
15	1.341	1.753	2.602	4.073
16	1.337	1.746	2.583	4.015
17	1.333	1.740	2.567	3.965
18	1.330	1.734	2.552	3.922
19	1.328	1.729	2.539	3.883
20	1.325	1.725	2.528	3.850
21	1.323	1.721	2.518	3.819
22	1.321	1.717	2.508	3.792
23	1.319	1.714	2.500	3.767
24	1.318	1.711	2.492	3.745
25	1.316	1.708	2.485	3.725
26	1.315	1.706	2.479	3.707
27	1.314	1.703	2.473	3.690
28	1.313	1.701	2.467	3.674
29	1.311	1.699	2.462	3.659
30	1.310	1.697	2.457	3.646
40	1.303	1.684	2.423	3.551
60	1.296	1.671	2.390	3.460
120	1.289	1.658	2.358	3.373
∞	1.282	1.645	2.326	3.291

n = degrees of freedom

Reference: "CRC Standard Mathematical Tables, 15th Ed., S. M. Selby, ed., Chemical Rubber Co., Cleveland, 1967.